A First Course in Information Theory

A First Course in Information Theory

Raymond W. Yeung
The Chinese University of Hong Kong

Springer

ISBN: 978-0-306-46791-2
ISBN: 0-306-46791-7

springer.com

Yeung: A First Course in Information Theory

First Indian Reprint 2010

ISBN 978-81-8489-703-6

This edition is published by Springer (India) Private Limited, (a part of Springer Science+Business Media), Registered Office: 3rd Floor, Gandharva Mahavidyalaya, 212, Deen Dayal Upadhyaya Marg, New Delhi - 110 002, India.

Printed in India by Rashtriya Printers, Delhi.

To my parents and my family

Foreword

The first course usually is an appetizer. In the case of Raymond Yeung's *A First Course in Information Theory*, however, another delectable dish gets served up in each of the sixteen chapters. Chapters 1 through 7 deal with the basic concepts of entropy and information with applications to lossless source coding. This is the traditional early fare of an information theory text, but Yeung flavors it uniquely. No one since Shannon has had a better appreciation for the mathematical structure of information quantities than Prof. Yeung. In the early chapters this manifests itself in a careful treatment of information measures via both Yeung's analytical theory of I-Measure and his geometrically intuitive information diagrams. (This material, never before presented in a textbook, is rooted in works by G. D. Hu, by H. Dyckman, and by R. Yeung *et al.*) Fundamental interrelations among information measures and Markovianness are developed with precision and unity. New slants are provided on staples like the divergence inequality, the data processing theorem, and Fano's inequality. There is also a clever, Kraft-inequality-free way of proving that the average length of the words in a lossless prefix source code must exceed the source's entropy. An easily digestible treatment of the redundancy of lossless prefix source codes also is served up, an important topic in practice that usually is slighted in textbooks.

The concept of weakly typical sequences is introduced and then used to anchor Yeung's proof of the lossless block source coding theorem. The concept of strongly typical sequences is introduced next. Later extended to joint typicality, this provides a foundation for proving the channel coding theorem in Chapter 8, the lossy source coding (rate-distortion) theorem in Chapter 9, and selected multi-source network coding theorems in Chapter 15. Although the proof of the channel coding theorem follows standard lines, Yeung's tasteful development of the interplay between information quantities and Markovianness readies one's palate for a rigorous proof that feedback around a discrete memoryless channel does not increase its capacity. In most information theory books this basic result of Shannon either does not appear or is relegated to a problem in which the several steps are outlined in order to guide the reader toward the goal. Rate-distortion theory and Shannon's lossy source coding theorem are treated in familiar ways. When proving the latter, one confronts lack of independence of the events $\{(\mathbf{X}, \hat{\mathbf{X}}(i)) \in T^n\}$, where $\mathbf{X}$ is a random source word, $\hat{\mathbf{X}}(i)$ is the ith word in a randomly chosen source code, and T^n is the set of jointly typical vector pairs. In those instances in which this widely unappreciated stumbling block is not overlooked entirely, it usually is addressed via either a non-selfcontained reference or a mammoth problem at the end of the chapter. However, Yeung's thorough earlier development of strong joint typicality concepts allows him to tackle it head-on.

Chapter 10 dishes up a careful treatment of the iterative algorithms for computation of channel capacity and rate-distortion functions pioneered by R. E. Blahut and S. Arimoto, which is generally accepted as today's preferred approach to computational information theory. Moreover, it has the extra advantage that iterative optimization algorithms are finding widespread application to areas as diverse as decoding of turbo and low-density parity-check codes and belief propagation in artificial intelligence and in real and artificial neural nets.

Chapters 11 through 16 are a unique tour de force. In as digestible a fashion as could possibly be expected, Yeung unveils a smorgasbord of topics in modern information theory that heretofore have been available only in research papers generated principally by Yeung and his research collaborators. Chapter 11 is a strong treatment of single-source network coding which develops carefully the relationships between information multicasting and the max-flow min-cut theory. Yeung makes an iron-clad case for how nodes must in general perform coding, not just storing and forwarding. Chapters 12, 13 and 14 on information inequalities of both Shannon and non-Shannon type constitute a definitive presentation of these topics by the master chef himself. Connections with linear programming are exploited, culminating in explication of Information Theory Inequality Prover (ITIP) of R. Yeung and Y.-O. Yan for inequalities of Shannon-type which comes with this book (also WWW-available). This leads, in turn, to the fascinating area of non-Shannon-type information inequalities, pioneered by R. Yeung and Z. Zhang. This material has been found to possess profound implications for the general area of information structures being studied by mathematical logicians and may also contribute to thermodynamics and statistical mechanics wherein the concept of entropy originated and which continue to be heavily concerned with various families of general inequalities. The theory of I-Measure introduced in Chapter 6 provides the essential insight into those of the non-Shannon type inequalities that are discussed here. Multi-source network coding in Chapter 15 is a confounding area in which Yeung and others have made considerable progress but a comprehensive theory remains elusive. Nonetheless, the geometrical framework for information inequalities developed in Chapters 12 and 13 renders a unifying tool for attacking this class of problems. The closing chapter linking entropy to the theory of groups is mouthwateringly provocative, having the potential to become a major contribution of information theory to this renowned branch of mathematics and mathematical physics.

Savor this book; I think you will agree the proof is in the pudding.

Toby Berger
Irwin and Joan Jacobs Professor of Engineering
Cornell University, Ithaca, New York

Preface

Cover and Thomas wrote a book on information theory [52] ten years ago which covers most of the major topics with considerable depth. Their book has since become the standard textbook in the field, and it was no doubt a remarkable success. Instead of writing another comprehensive textbook on the subject, which has become more difficult as new results keep emerging, my goal is to write a book on the fundamentals of the subject in a unified and coherent manner.

During the last ten years, significant progress has been made in understanding the entropy function and information inequalities of discrete random variables. The results along this direction not only are of core interest in information theory, but also have applications in network coding theory, probability theory, group theory, Kolmogorov complexity, and possibly physics. This book is an up-to-date treatment of information theory for discrete random variables, which forms the foundation of the theory at large. There are eight chapters on classical topics (Chapters 1, 2, 3, 4, 5, 8, 9, and 10), five chapters on fundamental tools (Chapters 6, 7, 12, 13, and 14), and three chapters on selected topics (Chapters 11, 15, and 16). The chapters are arranged according to the logical order instead of the chronological order of the results in the literature.

What is in this book

Out of the sixteen chapters in this book, the first thirteen chapters are basic topics, while the last three chapters are advanced topics for the more enthusiastic reader. A brief rundown of the chapters will give a better idea of what is in this book.

Chapter 1 is a very high level introduction to the nature of information theory and the main results in Shannon's original paper in 1948 which founded the field. There are also pointers to Shannon's biographies and his works.

Chapter 2 introduces Shannon's information measures and their basic properties. Useful identities and inequalities in information theory are derived and explained. Extra care is taken in handling joint distributions with zero probability masses. The chapter ends with a section on the entropy rate of a stationary information source.

Chapter 3 is a discussion of zero-error data compression by uniquely decodable codes, with prefix codes as a special case. A proof of the entropy bound for prefix codes which involves neither the Kraft inequality nor the fundamental inequality is given. This proof facilitates the discussion of the redundancy of prefix codes.

Chapter 4 is a thorough treatment of weak typicality. The weak asymptotic equipartition property and the source coding theorem are discussed. An explanation of the fact that a good data compression scheme produces almost i.i.d. bits is given. There is also a brief discussion of the Shannon-McMillan-Breiman theorem.

Chapter 5 introduces a new definition of strong typicality which does not involve the cardinalities of the alphabet sets. The treatment of strong typicality here is more detailed than Berger [21] but less abstract than Csiszár and Körner [55]. A new exponential convergence result is proved in Theorem 5.3.

Chapter 6 is an introduction to the theory of I-Measure which establishes a one-to-one correspondence between Shannon's information measures and set theory. A number of examples are given to show how the use of information diagrams can simplify the proofs of many results in information theory. Most of these examples are previously unpublished. In particular, Example 6.15 is a generalization of Shannon's perfect secrecy theorem.

Chapter 7 explores the structure of the I-Measure for Markov structures. Set-theoretic characterizations of full conditional independence and Markov random field are discussed. The treatment of Markov random field here is perhaps too specialized for the average reader, but the structure of the I-Measure and the simplicity of the information diagram for a Markov chain is best explained as a special case of a Markov random field.

Chapter 8 consists of a new treatment of the channel coding theorem. Specifically, a graphical model approach is employed to explain the conditional independence of random variables. Great care is taken in discussing feedback.

Chapter 9 is an introduction to rate-distortion theory. The version of the rate-distortion theorem here, proved by using strong typicality, is a stronger version of the original theorem obtained by Shannon.

In Chapter 10, the Blahut-Arimoto algorithms for computing channel capacity and the rate-distortion function are discussed, and a simplified proof for convergence is given. Great care is taken in handling distributions with zero probability masses.

Chapter 11 is an introduction to network coding theory. The surprising fact that coding at the intermediate nodes can improve the throughput when an information source is multicast in a point-to-point network is explained. The max-flow bound for network coding with a single information source is explained in detail. Multi-source network coding will be discussed in Chapter 15 after the necessary tools are developed in the next three chapters.

Information inequalities are sometimes called the laws of information theory because they govern the impossibilities in information theory. In Chapter 12, the geometrical meaning of information inequalities and the relation between information inequalities and conditional independence are explained in depth. The framework for information inequalities discussed here is the basis of the next two chapters.

Chapter 13 explains how the problem of proving information inequalities can be formulated as a linear programming problem. This leads to a complete characterization of all information inequalities which can be proved by conventional techniques. These are called Shannon-type inequalities, which can now be proved by the software ITIP which comes with this book. It is also shown how Shannon-type inequalities can be used to tackle the implication problem of conditional independence in probability theory.

All information inequalities we used to know were Shannon-type inequalities. Recently, a few non-Shannon-type inequalities have been discovered. This means that there exist laws in information theory beyond those laid down by Shannon. These inequalities and their applications are explained in depth in Chapter 14.

Network coding theory is further developed in Chapter 15. The situation when more than one information source are multicast in a point-to-point network is discussed. The surprising fact that a multi-source problem is not equivalent to a few single-source problems even when the information sources are mutually independent is clearly explained. Implicit and explicit bounds on the achievable coding rate region are discussed. These characterizations on the achievable coding rate region involve almost all the tools that have been developed earlier in the book, in particular, the framework for information inequalities.

Chapter 16 explains an intriguing relation between information theory and group theory. Specifically, for every information inequality satisfied by any joint distribution, there is a corresponding group inequality satisfied by any finite group and its subgroups, and vice versa. Inequalities of the latter type govern the orders of any finite group and their subgroups. Group-theoretic proofs of Shannon-type information inequalities are given. At the end of this chapter, a group inequality is obtained from a non-Shannon-type inequality discussed in Chapter 14. The meaning and the implication of this inequality are yet to be understood.

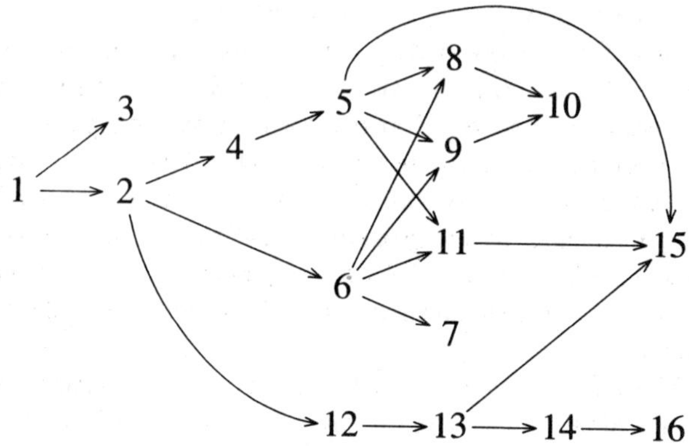

How to use this book

You are recommended to read the chapters according to the above chart. However, you will not have too much difficulty jumping around in the book because there should be sufficient references to the previous relevant sections.

As a relatively slow thinker, I feel uncomfortable whenever I do not reason in the most explicit way. This probably has helped in writing this book, in which all the derivations are from the first principle. In the book, I try to explain all the subtle mathematical details without sacrificing the big picture. Interpretations of the results are usually given before the proofs are presented. The book also contains a large number of examples. Unlike the examples in most books which are supplementary, the examples in this book are essential.

This book can be used as a reference book or a textbook. For a two-semester course on information theory, this would be a suitable textbook for the first semester. This would also be a suitable textbook for a one-semester course if only information theory for discrete random variables is covered. If the instructor also wants to include topics on continuous random variables, this book can be used as a textbook or a reference book in conjunction with another suitable textbook. The instructor will find this book a good source for homework problems because many problems here do not appear in any other textbook. A comprehensive instructor's manual is available upon request. Please contact the author at whyeung@ie.cuhk.edu.hk for information and access.

Just like any other lengthy document, this book for sure contains errors and omissions. To alleviate the problem, an errata will be maintained at the book homepage http://www.ie.cuhk.edu.hk/IT_book/.

RAYMOND W. YEUNG

Acknowledgments

Writing a book is always a major undertaking. This is especially so for a book on information theory because it is extremely easy to make mistakes in the writing if one has not done research work on that particular topic. Besides, the time needed for writing a book cannot be under estimated. Thanks to the generous support of a fellowship from the Croucher Foundation, I have had the luxury of taking a one-year leave in 2000-01 to write this book. I also thank my department for the support which made this arrangement possible.

There are many individuals who have directly or indirectly contributed to this book. First, I am indebted to Toby Berger who taught me information theory and writing. Venkat Anantharam, Dick Blahut, Dave Delchamps, Terry Fine, and Chris Heegard all had much influence on me when I was a graduate student at Cornell.

I am most thankful to Zhen Zhang for his friendship and inspiration. Without the results obtained through our collaboration, this book would not be complete. I would also like to thank Julia Abrahams, Vijay Bhargava, Agnes and Vincent Chan, Tom Cover, Imre Csiszár, Bob Gallager, Bruce Hajek, Te Sun Han, Jim Massey, Alon Orlitsky, Shlomo Shamai, Sergio Verdú, Victor Wei, Frans Willems, and Jack Wolf for their support and encouragement throughout the years. In particular, Imre Csiszár has closely followed my work and has given invaluable feedback. I would also like to thank all the collaborators of my work for their contribution and all the anonymous reviewers for their useful comments.

During the process of writing this book, I received tremendous help from Ning Cai and Fangwei Fu in proofreading the manuscript. Lihua Song and Terence Chan helped compose the figures and gave useful suggestions. Without their help, it would have been quite impossible to finish the book within the set time frame. I also thank Yu He, Andries Hekstra, Siu-Wai Ho, Kingo Kobayashi, Frank Kschischang, Amos Lapidoth, Li Ping, Prakash Narayan, Igal Sason, Wojtek Szpankowski, Emre Telatar, Chunxuan Ye, and Ken Zeger for their valuable inputs. The code for ITIP was written by Ying-On Yan, and Jack Lee helped modify the code and adapt it for the PC. Gordon Yeung helped on numerous Unix and LaTeX problems. The two-dimensional Venn diagram representation for four sets which is repeatedly used in the book was taught to me by Yong Nan Yeh.

On the domestic side, I am most grateful to my wife Rebecca for her love and support. My daughter Shannon has grown from an infant into a toddler during this time. She has added much joy to our family. I also thank my mother-in-law Mrs. Tsang and my sister-in-law Ophelia for coming over from time to time to take care of Shannon. Life would have been a lot more hectic without their generous help.

Contents

A First Course in Information Theory

Chapter 1

THE SCIENCE OF INFORMATION

In a communication system, we try to convey information from one point to another, very often in a noisy environment. Consider the following scenario. A secretary needs to send facsimiles regularly and she wants to convey as much information as possible on each page. She has a choice of the font size, which means that more characters can be squeezed onto a page if a smaller font size is used. In principle, she can squeeze as many characters as desired on a page by using a small enough font size. However, there are two factors in the system which may cause errors. First, the fax machine has a finite resolution. Second, the characters transmitted may be received incorrectly due to noise in the telephone line. Therefore, if the font size is too small, the characters may not be recognizable on the facsimile. On the other hand, although some characters on the facsimile may not be recognizable, the recipient can still figure out the words from the context provided that the number of such characters is not excessive. In other words, it is not necessary to choose a font size such that all the characters on the facsimile are recognizable almost surely. Then we are motivated to ask: What is the maximum amount of meaningful information which can be conveyed on one page of facsimile?

This question may not have a definite answer because it is not very well posed. In particular, we do not have a precise measure of meaningful information. Nevertheless, this question is an illustration of the kind of fundamental questions we can ask about a communication system.

Information, which is not a physical entity but an abstract concept, is hard to quantify in general. This is especially the case if human factors are involved when the information is utilized. For example, when we play Beethoven's violin concerto from a compact disc, we receive the musical information from the loudspeakers. We enjoy this information because it arouses certain kinds of emotion within ourselves. While we receive the same information every

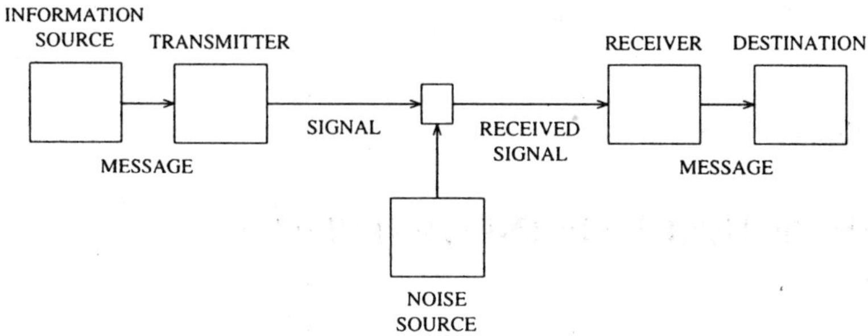

Figure 1.1. Schematic diagram for a general point-to-point communication system.

time we play the same piece of music, the kinds of emotions aroused may be different from time to time because they depend on our mood at that particular moment. In other words, we can derive utility from the same information every time in a different way. For this reason, it is extremely difficult to devise a measure which can quantify the amount of information contained in a piece of music.

In 1948, Bell Telephone Laboratories scientist Claude E. Shannon (1916-2001) published a paper entitled "The Mathematical Theory of Communication" [184] which laid the foundation of an important field now known as information theory. In his paper, the model of a point-to-point communication system depicted in Figure 1.1 is considered. In this model, a message is generated by the information source. The message is converted by the transmitter into a signal which is suitable for transmission. In the course of transmission, the signal may be contaminated by a noise source, so that the received signal may be different from the transmitted signal. Based on the received signal, the receiver then makes an estimate of the message and deliver it to the destination.

In this abstract model of a point-to-point communication system, one is only concerned about whether the message generated by the source can be delivered correctly to the receiver without worrying about how the message is actually used by the receiver. In a way, Shannon's model does not cover all the aspects of a communication system. However, in order to develop a precise and useful theory of information, the scope of the theory has to be restricted.

In [184], Shannon introduced two fundamental concepts about 'information' from the communication point of view. First, information is *uncertainty*. More specifically, if a piece of information we are interested in is deterministic, then it has no value at all because it is already known with no uncertainty. From this point of view, for example, the continuous transmission of a still picture on a television broadcast channel is superfluous. Consequently, an information source is naturally modeled as a random variable or a random process, and probability is employed to develop the theory of information. Second, information to be transmitted is *digital*. This means that the information source should first be converted into a stream of 0's and 1's called bits, and the remaining task is to deliver these bits to the receiver correctly with no reference to their actual meaning. This is the foundation of all modern digital communication systems. In fact, this work of Shannon appears to contain the first published use of the term *bit*, which stands for binary digit.

In the same work, Shannon also proved two important theorems. The first theorem, called the *source coding theorem*, introduces *entropy* as the fundamental measure of information which characterizes the minimum rate of a source code representing an information source essentially free of error. The source coding theorem is the theoretical basis for *lossless* data compression[1]. The second theorem, called the *channel coding theorem*, concerns communication through a noisy channel. It was shown that associated with every noisy channel is a parameter, called the *capacity*, which is strictly positive except for very special channels, such that information can be communicated reliably through the channel as long as the information rate is less than the capacity. These two theorems, which give fundamental limits in point-to-point communication, are the two most important results in information theory.

In science, we study the laws of Nature which must be obeyed by any physical systems. These laws are used by engineers to design systems to achieve specific goals. Therefore, science is the foundation of engineering. Without science, engineering can only be done by trial and error.

In information theory, we study the fundamental limits in communication regardless of the technologies involved in the actual implementation of the communication systems. These fundamental limits are not only used as guidelines by communication engineers, but they also give insights into what optimal coding schemes are like. Information theory is therefore the science of information.

Since Shannon published his original paper in 1948, information theory has been developed into a major research field in both communication theory and applied probability. After more than half a century's research, it is quite impossible for a book on the subject to cover all the major topics with considerable depth. This book is a modern treatment of information theory for discrete random variables, which is the foundation of the theory at large. The book consists of two parts. The first part, namely Chapter 1 to Chapter 13, is a thorough discussion of the basic topics in information theory, including fundamental results, tools, and algorithms. The second part, namely Chapter 14 to Chapter 16, is a selection of advanced topics which demonstrate the use of the tools

[1]A data compression scheme is lossless if the data can be recovered with an arbitrarily small probability of error.

developed in the first part of the book. The topics discussed in this part of the book also represent new research directions in the field.

An undergraduate level course on probability is the only prerequisite for this book. For a non-technical introduction to information theory, we refer the reader to *Encyclopedia Britannica* [65]. In fact, we strongly recommend the reader to first read this excellent introduction before starting this book. For biographies of Claude Shannon, a legend of the 20th Century who had made fundamental contribution to the Information Age, we refer the readers to [38] and [197]. The latter is also a complete collection of Shannon's papers.

Unlike most branches of applied mathematics in which physical systems are studied, abstract systems of communication are studied in information theory. In reading this book, it is not unusual for a beginner to be able to understand all the steps in a proof but has no idea what the proof is leading to. The best way to learn information theory is to study the materials first and come back at a later time. Many results in information theory are rather subtle, to the extent that an expert in the subject may from time to time realize that his/her understanding of certain basic results has been inadequate or even incorrect. While a novice should expect to raise his/her level of understanding of the subject by reading this book, he/she should not be discouraged to find after finishing the book that there are actually more things yet to be understood. In fact, this is exactly the challenge and the beauty of information theory.

Chapter 2

INFORMATION MEASURES

Shannon's information measures refer to entropy, conditional entropy, mutual information, and conditional mutual information. They are the most important measures of information in information theory. In this chapter, we introduce these measures and establish some basic properties they possess. The physical meanings of these measures will be discussed in depth in subsequent chapters. We then introduce informational divergence which measures the "distance" between two probability distributions and prove some useful inequalities in information theory. The chapter ends with a section on the entropy rate of a stationary information source.

2.1 INDEPENDENCE AND MARKOV CHAINS

We begin our discussion in this chapter by reviewing two basic notions in probability: independence of random variables and Markov chain. All the random variables in this book are discrete.

Let X be a random variable taking values in an alphabet $\mathcal{X}$. The probability distribution for X is denoted as $\{p_X(x), x \in \mathcal{X}\}$, with $p_X(x) = \Pr\{X = x\}$. When there is no ambiguity, $p_X(x)$ will be abbreviated as $p(x)$, and $\{p(x)\}$ will be abbreviated as $p(x)$. The *support* of X, denoted by $\mathcal{S}_X$, is the set of all $x \in \mathcal{X}$ such that $p(x) > 0$. If $\mathcal{S}_X = \mathcal{X}$, we say that p is *strictly positive*. Otherwise, we say that p is not strictly positive, or p contains zero probability masses. All the above notations naturally extend to two or more random variables. As we will see, probability distributions with zero probability masses are very delicate in general, and they need to be handled with great care.

DEFINITION 2.1 *Two random variables X and Y are independent, denoted by $X \perp Y$, if*

$$p(x, y) = p(x)p(y) \tag{2.1}$$

for all x and y (i.e., for all $(x, y) \in \mathcal{X} \times \mathcal{Y}$).

For more than two random variables, we distinguish between two types of independence.

DEFINITION 2.2 (MUTUAL INDEPENDENCE) *For $n \geq 3$, random variables $X_1, X_2, \cdots, X_n$ are mutually independent if*

$$p(x_1, x_2, \cdots, x_n) = p(x_1)p(x_2)\cdots p(x_n) \tag{2.2}$$

for all $x_1, x_2, \cdots, x_n$.

DEFINITION 2.3 (PAIRWISE INDEPENDENCE) *For $n \geq 3$, random variables $X_1, X_2, \cdots, X_n$ are pairwise independent if X_i and X_j are independent for all $1 \leq i < j \leq n$.*

Note that mutual independence implies pairwise independence. We leave it as an exercise for the reader to show that the converse is not true.

DEFINITION 2.4 (CONDITIONAL INDEPENDENCE) *For random variables X, Y, and Z, X is independent of Z conditioning on Y, denoted by $X \perp Z|Y$, if*

$$p(x, y, z)p(y) = p(x, y)p(y, z) \tag{2.3}$$

for all x, y, and z, or equivalently,

$$p(x, y, z) = \begin{cases} \frac{p(x,y)p(y,z)}{p(y)} = p(x, y)p(z|y) & \text{if } p(y) > 0 \\ 0 & \text{otherwise.} \end{cases} \tag{2.4}$$

The first definition of conditional independence above is sometimes more convenient to use because it is not necessary to distinguish between the cases $p(y) > 0$ and $p(y) = 0$. However, the physical meaning of conditional independence is more explicit in the second definition.

PROPOSITION 2.5 *For random variables X, Y, and Z, $X \perp Z|Y$ if and only if*

$$p(x, y, z) = a(x, y)b(y, z) \tag{2.5}$$

for all x, y, and z such that $p(y) > 0$.

Proof The 'only if' part follows immediately from the definition of conditional independence in (2.4), so we will only prove the 'if' part. Assume

$$p(x, y, z) = a(x, y)b(y, z) \tag{2.6}$$

for all x, y, and z such that $p(y) > 0$. Then

$$p(x,y) = \sum_z p(x,y,z) = \sum_z a(x,y)b(y,z) = a(x,y)\sum_z b(y,z), \quad (2.7)$$

$$p(y,z) = \sum_x p(x,y,z) = \sum_x a(x,y)b(y,z) = b(y,z)\sum_x a(x,y), \quad (2.8)$$

and

$$p(y) = \sum_z p(y,z) = \left(\sum_x a(x,y)\right)\left(\sum_z b(y,z)\right). \quad (2.9)$$

Therefore,

$$\begin{aligned} \frac{p(x,y)p(y,z)}{p(y)} &= \frac{\left(a(x,y)\sum_z b(y,z)\right)\left(b(y,z)\sum_x a(x,y)\right)}{\left(\sum_x a(x,y)\right)\left(\sum_z b(y,z)\right)} &(2.10)\\ &= a(x,y)b(y,z) &(2.11)\\ &= p(x,y,z). &(2.12)\end{aligned}$$

Hence, $X \perp Z|Y$. The proof is accomplished. □

DEFINITION 2.6 (MARKOV CHAIN) *For random variables $X_1, X_2, \cdots, X_n$, where $n \geq 3$, $X_1 \to X_2 \to \cdots \to X_n$ forms a Markov chain if*

$$\begin{aligned} &p(x_1, x_2, \cdots, x_n)p(x_2)p(x_3)\cdots p(x_{n-1}) \\ &= p(x_1,x_2)p(x_2,x_3)\cdots p(x_{n-1},x_n) \quad (2.13)\end{aligned}$$

for all $x_1, x_2, \cdots, x_n$, or equivalently,

$$p(x_1, x_2, \cdots, x_n) = \begin{cases} p(x_1,x_2)p(x_3|x_2)\cdots p(x_n|x_{n-1}) & \text{if } p(x_2), p(x_3), \cdots, p(x_{n-1}) > 0 \\ 0 & \text{otherwise.} \end{cases} \quad (2.14)$$

We note that $X \perp Z|Y$ is equivalent to the Markov chain $X \to Y \to Z$.

PROPOSITION 2.7 *$X_1 \to X_2 \to \cdots \to X_n$ forms a Markov chain if and only if $X_n \to X_{n-1} \to \cdots \to X_1$ forms a Markov chain.*

Proof This follows directly from the symmetry in the definition of a Markov chain in (2.13). □

In the following, we state two basic properties of a Markov chain. The proofs are left as an exercise.

PROPOSITION 2.8 *$X_1 \to X_2 \to \cdots \to X_n$ forms a Markov chain if and only if*

$$\begin{aligned} &X_1 \to X_2 \to X_3 \\ &(X_1, X_2) \to X_3 \to X_4 \\ &\qquad\vdots \\ &(X_1, X_2, \cdots, X_{n-2}) \to X_{n-1} \to X_n \end{aligned} \tag{2.15}$$

form Markov chains.

PROPOSITION 2.9 *$X_1 \to X_2 \to \cdots \to X_n$ forms a Markov chain if and only if*

$$p(x_1, x_2, \cdots, x_n) = f_1(x_1, x_2) f_2(x_2, x_3) \cdots f_{n-1}(x_{n-1}, x_n) \tag{2.16}$$

for all $x_1, x_2, \cdots, x_n$ such that $p(x_2), p(x_3), \cdots, p(x_{n-1}) > 0$.

Note that Proposition 2.9 is a generalization of Proposition 2.5. From Proposition 2.9, one can prove the following important property of a Markov chain. Again, the details are left as an exercise.

PROPOSITION 2.10 (MARKOV SUBCHAINS) *Let $\mathcal{N}_n = \{1, 2, \cdots, n\}$ and let $X_1 \to X_2 \to \cdots \to X_n$ form a Markov chain. For any subset α of $\mathcal{N}_n$, denote $(X_i, i \in \alpha)$ by X_α. Then for any disjoint subsets $\alpha_1, \alpha_2, \cdots, \alpha_m$ of $\mathcal{N}_n$ such that*

$$k_1 < k_2 < \cdots < k_m \tag{2.17}$$

for all $k_j \in \alpha_j$, $j = 1, 2, \cdots, m$,

$$X_{\alpha_1} \to X_{\alpha_2} \to \cdots \to X_{\alpha_m} \tag{2.18}$$

forms a Markov chain. That is, a subchain of $X_1 \to X_2 \to \cdots \to X_n$ is also a Markov chain.

EXAMPLE 2.11 *Let $X_1 \to X_2 \to \cdots \to X_{10}$ form a Markov chain and $\alpha_1 = \{1, 2\}$, $\alpha_2 = \{4\}$, $\alpha_3 = \{6, 8\}$, and $\alpha_4 = \{10\}$ be subsets of $\mathcal{N}_{10}$. Then Proposition 2.10 says that*

$$(X_1, X_2) \to X_4 \to (X_6, X_8) \to X_{10} \tag{2.19}$$

also forms a Markov chain.

We have been very careful in handling probability distributions with zero probability masses. In the rest of the section, we show that such distributions are very delicate in general. We first prove the following property of a strictly positive probability distribution involving four random variables[1].

PROPOSITION 2.12 *Let X_1, X_2, X_3, and X_4 be random variables such that $p(x_1, x_2, x_3, x_4)$ is strictly positive. Then*

$$\left.\begin{array}{l} X_1 \perp X_4|(X_2, X_3) \\ X_1 \perp X_3|(X_2, X_4) \end{array}\right\} \Rightarrow X_1 \perp (X_3, X_4)|X_2. \tag{2.20}$$

Proof If $X_1 \perp X_4|(X_2, X_3)$, we have

$$p(x_1, x_2, x_3, x_4) = \frac{p(x_1, x_2, x_3)p(x_2, x_3, x_4)}{p(x_2, x_3)}. \tag{2.21}$$

On the other hand, if $X_1 \perp X_3|(X_2, X_4)$, we have

$$p(x_1, x_2, x_3, x_4) = \frac{p(x_1, x_2, x_4)p(x_2, x_3, x_4)}{p(x_2, x_4)}. \tag{2.22}$$

Equating (2.21) and (2.22), we have

$$p(x_1, x_2, x_3) = \frac{p(x_2, x_3)p(x_1, x_2, x_4)}{p(x_2, x_4)}. \tag{2.23}$$

Then

$$\begin{aligned} p(x_1, x_2) &= \sum_{x_3} p(x_1, x_2, x_3) & (2.24) \\ &= \sum_{x_3} \frac{p(x_2, x_3)p(x_1, x_2, x_4)}{p(x_2, x_4)} & (2.25) \\ &= \frac{p(x_2)p(x_1, x_2, x_4)}{p(x_2, x_4)}, & (2.26) \end{aligned}$$

or

$$\frac{p(x_1, x_2, x_4)}{p(x_2, x_4)} = \frac{p(x_1, x_2)}{p(x_2)}. \tag{2.27}$$

Hence from (2.22),

$$p(x_1, x_2, x_3, x_4) = \frac{p(x_1, x_2, x_4)p(x_2, x_3, x_4)}{p(x_2, x_4)} = \frac{p(x_1, x_2)p(x_2, x_3, x_4)}{p(x_2)} \tag{2.28}$$

[1] Proposition 2.12 is called the *intersection* axiom in Bayesian networks. See [160].

for all x_1, x_2, x_3, and x_4, i.e., $X_1 \perp (X_3, X_4)|X_2$. $\square$

If $p(x_1, x_2, x_3, x_4) = 0$ for some x_1, x_2, x_3, and x_4, i.e., p is not strictly positive, the arguments in the above proof are not valid. In fact, the proposition may not hold in this case. For instance, let $X_1 = Y$, $X_2 = Z$, and $X_3 = X_4 = (Y, Z)$, where Y and Z are independent random variables. Then $X_1 \perp X_4|(X_2, X_3)$, $X_1 \perp X_3|(X_2, X_4)$, but $X_1 \not\perp (X_3, X_4)|X_2$. Note that for this construction, p is not strictly positive because $p(x_1, x_2, x_3, x_4) = 0$ if $x_3 \neq (x_1, x_2)$ or $x_4 \neq (x_1, x_2)$.

The above example is somewhat counter-intuitive because it appears that Proposition 2.12 should hold for all probability distributions via a continuity argument. Specifically, such an argument goes like this. For any distribution p, let $\{p_k\}$ be a sequence of strictly positive distributions such that $p_k \to p$ and p_k satisfies (2.21) and (2.22) for all k, i.e.,

$$p_k(x_1, x_2, x_3, x_4)p_k(x_2, x_3) = p_k(x_1, x_2, x_3)p_k(x_2, x_3, x_4) \tag{2.29}$$

and

$$p_k(x_1, x_2, x_3, x_4)p_k(x_2, x_4) = p_k(x_1, x_2, x_4)p_k(x_2, x_3, x_4). \tag{2.30}$$

Then by the proposition, p_k also satisfies (2.28), i.e.,

$$p_k(x_1, x_2, x_3, x_4)p_k(x_2) = p_k(x_1, x_2)p_k(x_2, x_3, x_4). \tag{2.31}$$

Letting $k \to \infty$, we have

$$p(x_1, x_2, x_3, x_4)p(x_2) = p(x_1, x_2)p(x_2, x_3, x_4) \tag{2.32}$$

for all x_1, x_2, x_3, and x_4, i.e., $X_1 \perp (X_3, X_4)|X_2$. Such an argument would be valid if there always exists a sequence $\{p_k\}$ as prescribed. However, the existence of the distribution $p(x_1, x_2, x_3, x_4)$ constructed immediately after Proposition 2.12 simply says that it is not always possible to find such a sequence $\{p_k\}$.

Therefore, probability distributions which are not strictly positive can be very delicate. For strictly positive distributions, we see from Proposition 2.5 that their conditional independence structures are closely related to the factorization problem of such distributions, which has been investigated by Chan and Yeung [43].

2.2 SHANNON'S INFORMATION MEASURES

We begin this section by introducing the *entropy* of a random variable. As we will see shortly, all Shannon's information measures can be expressed as linear combinations of entropies.

DEFINITION 2.13 *The entropy $H(X)$ of a random variable X is defined by*

$$H(X) = -\sum_x p(x) \log p(x). \qquad (2.33)$$

In all definitions of information measures, we adopt the convention that summation is taken over the corresponding support. Such a convention is necessary because $p(x) \log p(x)$ in (2.33) is undefined if $p(x) = 0$.

The base of the logarithm in (2.33) can be chosen to be any convenient real number great than 1. We write $H(X)$ as $H_\alpha(X)$ when the base of the logarithm is α. When the base of the logarithm is 2, the unit for entropy is called a *bit*. When the base of the logarithm is an integer $D \geq 2$, the unit for entropy is D-*it* (D-ary digit). In the context of source coding, the base is usually taken to be the size of the code alphabet. This will be discussed in Chapter 3.

In computer science a bit means an entity which can take the value 0 or 1. In information theory the entropy of a random variable is measured in bits. The reader should distinguish these two meanings of a bit from each other carefully.

Let $g(X)$ be any function of a random variable X. We will denote the expectation of $g(X)$ by $Eg(X)$, i.e.,

$$Eg(X) = \sum_x p(x)g(x), \qquad (2.34)$$

where the summation is over $\mathcal{S}_X$. Then the definition of the entropy of a random variable X can be written as

$$H(X) = -E \log p(X). \qquad (2.35)$$

Expressions of Shannon's information measures in terms of expectations will be useful in subsequent discussions.

The entropy $H(X)$ of a random variable X is a functional of the probability distribution $p(x)$ which measures the average amount of information contained in X, or equivalently, the amount of *uncertainty* removed upon revealing the outcome of X. Note that $H(X)$ depends only on $p(x)$, not on the actual values in $\mathcal{X}$.

For $0 \leq p \leq 1$, define

$$h_b(p) = -p \log p - (1-p) \log(1-p) \qquad (2.36)$$

with the convention $0 \log 0 = 0$, so that $h_b(0) = h_b(1) = 0$. With this convention, $h_b(p)$ is continuous at $p = 0$ and $p = 1$. h_b is called the binary entropy function. For a binary random variable X with distribution $\{p, 1-p\}$,

$$H(X) = h_b(p). \qquad (2.37)$$

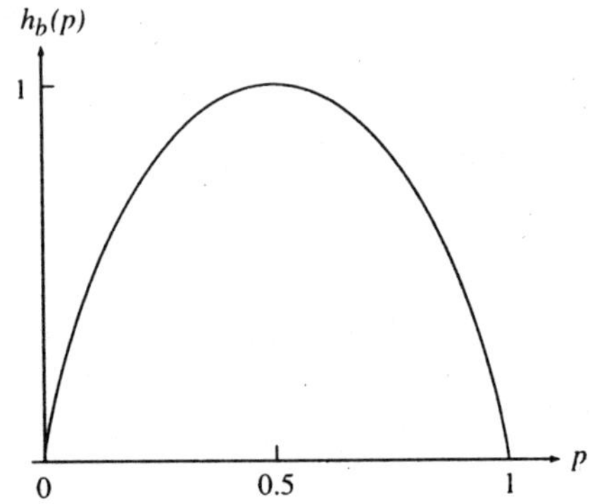

Figure 2.1. $h_b(p)$ versus p in the base 2.

Figure 2.1 shows the graph $h_b(p)$ versus p in the base 2. Note that $h_b(p)$ achieves the maximum value 1 when $p = \frac{1}{2}$.

The definition of the *joint entropy* of two random variables is similar to the definition of the entropy of a single random variable. Extension of this definition to more than two random variables is straightforward.

DEFINITION 2.14 *The joint entropy $H(X,Y)$ of a pair of random variables X and Y is defined by*

$$H(X,Y) = -\sum_{x,y} p(x,y) \log p(x,y) = -E \log p(X,Y). \tag{2.38}$$

For two random variables, we define in the following the *conditional entropy* of one random variable when the other random variable is given.

DEFINITION 2.15 *For random variables X and Y, the conditional entropy of Y given X is defined by*

$$H(Y|X) = -\sum_{x,y} p(x,y) \log p(y|x) = -E \log p(Y|X). \tag{2.39}$$

From (2.39), we can write

$$H(Y|X) = \sum_{x} p(x) \left[-\sum_{y} p(y|x) \log p(y|x) \right]. \tag{2.40}$$

The inner sum is the entropy of Y conditioning on a fixed $x \in \mathcal{S}_X$. Thus we are motivated to express $H(Y|X)$ as

$$H(Y|X) = \sum_{x} p(x) H(Y|X = x), \tag{2.41}$$

where

$$H(Y|X = x) = -\sum_y p(y|x) \log p(y|x). \tag{2.42}$$

Observe that the right hand sides of (2.33) and (2.42) have exactly the same form. Similarly, for $H(Y|X, Z)$, we write

$$H(Y|X, Z) = \sum_z p(z) H(Y|X, Z = z), \tag{2.43}$$

where

$$H(Y|X, Z = z) = -\sum_{x,y} p(x, y|z) \log p(y|x, z). \tag{2.44}$$

PROPOSITION 2.16

$$H(X, Y) = H(X) + H(Y|X) \tag{2.45}$$

and

$$H(X, Y) = H(Y) + H(X|Y). \tag{2.46}$$

Proof Consider

$$\begin{aligned} H(X, Y) &= -E \log p(X, Y) && (2.47) \\ &= -E \log[p(X)p(Y|X)] && (2.48) \\ &= -E \log p(X) - E \log p(Y|X) && (2.49) \\ &= H(X) + H(Y|X). && (2.50) \end{aligned}$$

Note that (2.48) is justified because the summation of the expectation is over $\mathcal{S}_{XY}$, and we have used the linearity of expectation[2] to obtain (2.49). This proves (2.45), and (2.46) follows by symmetry. □

This proposition has the following interpretation. Consider revealing the outcome of a pair of random variables X and Y in two steps: first the outcome of X and then the outcome of Y. Then the proposition says that the total amount of uncertainty removed upon revealing both X and Y is equal to the sum of the uncertainty removed upon revealing X (uncertainty removed in the first step) and the uncertainty removed upon revealing Y once X is known (uncertainty removed in the second step).

[2] See Problem 5 at the end of the chapter.

DEFINITION 2.17 *For random variables X and Y, the mutual information between X and Y is defined by*

$$I(X;Y) = \sum_{x,y} p(x,y) \log \frac{p(x,y)}{p(x)p(y)} = E \log \frac{p(X,Y)}{p(X)p(Y)}. \quad (2.51)$$

Remark $I(X;Y)$ is symmetrical in X and Y.

PROPOSITION 2.18 *The mutual information between a random variable X and itself is equal to the entropy of X, i.e., $I(X;X) = H(X)$.*

Proof This can be seen by considering

$$\begin{aligned} I(X;X) &= E \log \frac{p(X)}{p(X)^2} && (2.52) \\ &= -E \log p(X) && (2.53) \\ &= H(X). && (2.54) \end{aligned}$$

The proposition is proved. □

Remark The entropy of X is sometimes called the self-information of X.

PROPOSITION 2.19

$$\begin{aligned} I(X;Y) &= H(X) - H(X|Y), && (2.55) \\ I(X;Y) &= H(Y) - H(Y|X), && (2.56) \end{aligned}$$

and

$$I(X;Y) = H(X) + H(Y) - H(X,Y). \quad (2.57)$$

The proof of this proposition is left as an exercise.

From (2.55), we can interpret $I(X;Y)$ as the reduction in uncertainty about X when Y is given, or equivalently, the amount of information about X provided by Y. Since $I(X;Y)$ is symmetrical in X and Y, from (2.56), we can as well interpret $I(X;Y)$ as the amount of information about Y provided by X.

The relations between the (joint) entropies, conditional entropies, and mutual information for two random variables X and Y are given in Propositions 2.16 and 2.19. These relations can be summarized by the diagram in Figure 2.2 which is a variation of the Venn diagram[3]. One can check that all the relations

[3]The rectangle representing the universal set in a usual Venn diagram is missing in Figure 2.2.

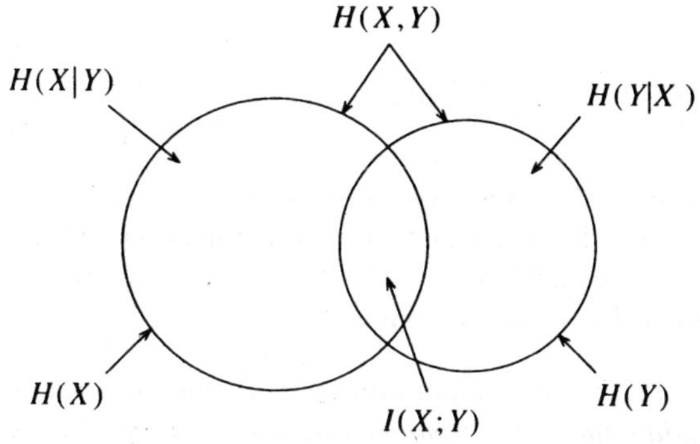

Figure 2.2. Relationship between entropies and mutual information for two random variables.

between Shannon's information measures for X and Y which are shown in Figure 2.2 are consistent with the relations given in Propositions 2.16 and 2.19. This one-to-one correspondence between Shannon's information measures and set theory is not just a coincidence for two random variables. We will discuss this in depth when we introduce the I-Measure in Chapter 6.

Analogous to entropy, there is a conditional version of mutual information called conditional mutual information.

DEFINITION 2.20 *For random variables X, Y and Z, the mutual information between X and Y conditioning on Z is defined by*

$$I(X;Y|Z) = \sum_{x,y,z} p(x,y,z) \log \frac{p(x,y|z)}{p(x|z)p(y|z)} = E \log \frac{p(X,Y|Z)}{p(X|Z)p(Y|Z)}. \tag{2.58}$$

Remark $I(X;Y|Z)$ is symmetrical in X and Y.

Analogous to conditional entropy, we write

$$I(X;Y|Z) = \sum_{z} p(z) I(X;Y|Z=z), \tag{2.59}$$

where

$$I(X;Y|Z=z) = \sum_{x,y} p(x,y|z) \log \frac{p(x,y|z)}{p(x|z)p(y|z)}. \tag{2.60}$$

Similarly, when conditioning on two random variables, we write

$$I(X;Y|Z,T) = \sum_{t} p(t) I(X;Y|Z,T=t) \tag{2.61}$$

where

$$I(X;Y|Z,T=t) = \sum_{x,y,z} p(x,y,z|t) \log \frac{p(x,y|z,t)}{p(x|z,t)p(y|z,t)}. \quad (2.62)$$

Conditional mutual information satisfies the same set of relations given in Propositions 2.18 and 2.19 for mutual information except that all the terms are now conditioned on a random variable Z. We state these relations in the next two propositions. The proofs are omitted.

PROPOSITION 2.21 *The mutual information between a random variable X and itself conditioning on a random variable Z is equal to the conditional entropy of X given Z, i.e., $I(X;X|Z) = H(X|Z)$.*

PROPOSITION 2.22

$$I(X;Y|Z) = H(X|Z) - H(X|Y,Z), \quad (2.63)$$

$$I(X;Y|Z) = H(Y|Z) - H(Y|X,Z), \quad (2.64)$$

and

$$I(X;Y|Z) = H(X|Z) + H(Y|Z) - H(X,Y|Z). \quad (2.65)$$

2.3 CONTINUITY OF SHANNON'S INFORMATION MEASURES

This section is devoted to a discussion of the continuity of Shannon's information measures. To begin with, we first show that all Shannon's information measures are special cases of conditional mutual information. Let Φ be a degenerate random variable, i.e., Φ takes a constant value with probability 1. Consider the mutual information $I(X;Y|Z)$. When $X = Y$ and $Z = \Phi$, $I(X;Y|Z)$ becomes the entropy $H(X)$. When $X = Y$, $I(X;Y|Z)$ becomes the conditional entropy $H(X|Z)$. When $Z = \Phi$, $I(X;Y|Z)$ becomes the mutual information $I(X;Y)$. Thus all Shannon's information measures are special cases of conditional mutual information. Therefore, we only need to discuss the continuity of conditional mutual information.

Recall the definition of conditional mutual information:

$$I_p(X;Y|Z) = \sum_{(x,y,z)\in\mathcal{S}_{XYZ}(p)} p(x,y,z) \log \frac{p(x,y|z)}{p(x|z)p(y|z)}, \quad (2.66)$$

where we have written $I(X;Y|Z)$ and $\mathcal{S}_{XYZ}$ as $I_p(X;Y|Z)$ and $\mathcal{S}_{XYZ}(p)$, respectively to emphasize their dependence on p. Since $\log a$ is continuous in a for $a > 0$, $I_p(X;Y|Z)$ varies continuously with p as long as the support $\mathcal{S}_{XYZ}(p)$ does not change. The problem arises when some positive probability masses become zero or some zero probability masses become positive. Since continuity in the former case implies continuity in the latter case and vice versa, we only need to consider the former case. As $p(x,y,z) \to 0$ and eventually become(s) zero for some x, y, and z, the support $\mathcal{S}_{XYZ}(p)$ is reduced, and there is a discrete change in the number of terms in (2.66). Now for x, y, and z such that $p(x,y,z) > 0$, consider

$$
\begin{aligned}
p(x,y,z)\log\frac{p(x,y|z)}{p(x|z)p(y|z)} &= p(x,y,z)\log\frac{\frac{p(x,y,z)}{p(z)}}{\frac{p(x,z)}{p(z)}\frac{p(y,z)}{p(z)}} && (2.67)\\
&= p(x,y,z)\log\frac{p(x,y,z)p(z)}{p(x,z)p(y,z)} && (2.68)\\
&= p(x,y,z)[\log p(x,y,z) + \log p(z) \\
&\quad - \log p(x,z) - \log p(y,z)]. && (2.69)
\end{aligned}
$$

It can readily be verified by L'Hospital's rule that $p(x,y,z)\log p(x,y,z) \to 0$ as $p(x,y,z) \to 0$. For $p(x,y,z)\log p(z)$, since $p(x,y,z) \le p(z) \le 1$, we have

$$0 \ge p(x,y,z)\log p(z) \ge p(x,y,z)\log p(x,y,z) \to 0. \tag{2.70}$$

Thus $p(x,y,z)\log p(z) \to 0$ as $p(x,y,z) \to 0$. Similarly, we can show that both $p(x,y,z)\log p(x,z)$ and $p(x,y,z)\log p(y,z) \to 0$ as $p(x,y,z) \to 0$. Therefore,

$$p(x,y,z)\log\frac{p(x,y|z)}{p(x|z)p(y|z)} \to 0 \tag{2.71}$$

as $p(x,y,z) \to 0$. Hence, $I(X;Y|Z)$ varies continuously with p even when $p(x,y,z) \to 0$ for some x, y, and z.

To conclude, for any $p(x,y,z)$, if $\{p_k\}$ is a sequence of joint distributions such that $p_k \to p$ as $k \to \infty$, then

$$\lim_{k\to\infty} I_{p_k}(X;Y|Z) = I_p(X;Y|Z). \tag{2.72}$$

2.4 CHAIN RULES

In this section, we present a collection of information identities known as the chain rules which are often used in information theory.

PROPOSITION 2.23 (CHAIN RULE FOR ENTROPY)

$$H(X_1, X_2, \cdots, X_n) = \sum_{i=1}^{n} H(X_i|X_1, \cdots, X_{i-1}). \tag{2.73}$$

Proof The chain rule for $n = 2$ has been proved in Proposition 2.16. We prove the chain rule by induction on n. Assume (2.73) is true for $n = m$, where $m \geq 2$. Then

$$\begin{aligned} & H(X_1, \cdots, X_m, X_{m+1}) \\ &= H(X_1, \cdots, X_m) + H(X_{m+1}|X_1, \cdots, X_m) \qquad (2.74) \\ &= \sum_{i=1}^{m} H(X_i|X_1, \cdots, X_{i-1}) + H(X_{m+1}|X_1, \cdots, X_m) \qquad (2.75) \\ &= \sum_{i=1}^{m+1} H(X_i|X_1, \cdots, X_{i-1}), \qquad (2.76) \end{aligned}$$

where in (2.74) we have used (2.45) by letting $X = (X_1, \cdots, X_m)$ and $Y = X_{m+1}$, and in (2.75) we have used (2.73) for $n = m$. This proves the chain rule for entropy. □

The chain rule for entropy has the following conditional version.

PROPOSITION 2.24 (CHAIN RULE FOR CONDITIONAL ENTROPY)

$$H(X_1, X_2, \cdots, X_n|Y) = \sum_{i=1}^{n} H(X_i|X_1, \cdots, X_{i-1}, Y). \qquad (2.77)$$

Proof This can be proved by considering

$$\begin{aligned} & H(X_1, X_2, \cdots, X_n|Y) \\ &= \sum_{y} p(y) H(X_1, X_2, \cdots, X_n|Y = y) \qquad (2.78) \\ &= \sum_{y} p(y) \sum_{i=1}^{n} H(X_i|X_1, \cdots, X_{i-1}, Y = y) \qquad (2.79) \\ &= \sum_{i=1}^{n} \sum_{y} p(y) H(X_i|X_1, \cdots, X_{i-1}, Y = y) \qquad (2.80) \\ &= \sum_{i=1}^{n} H(X_i|X_1, \cdots, X_{i-1}, Y), \qquad (2.81) \end{aligned}$$

where we have used (2.41) in (2.78) and (2.43) in (2.81), and (2.79) follows from Proposition 2.23. □

PROPOSITION 2.25 (CHAIN RULE FOR MUTUAL INFORMATION)

$$I(X_1, X_2, \cdots, X_n; Y) = \sum_{i=1}^{n} I(X_i; Y|X_1, \cdots, X_{i-1}). \qquad (2.82)$$

Proof Consider

$$
\begin{aligned}
& I(X_1, X_2, \cdots, X_n; Y) \\
&= H(X_1, X_2, \cdots, X_n) - H(X_1, X_2, \cdots, X_n|Y) && (2.83) \\
&= \sum_{i=1}^{n} [H(X_i|X_1, \cdots, X_{i-1}) - H(X_i|X_1, \cdots, X_{i-1}, Y)] && (2.84) \\
&= \sum_{i=1}^{n} I(X_i; Y|X_1, \cdots, X_{i-1}), && (2.85)
\end{aligned}
$$

where in (2.84), we have invoked both Propositions 2.23 and 2.24. The chain rule for mutual information is proved. □

PROPOSITION 2.26 (CHAIN RULE FOR CONDITIONAL MUTUAL INFORMATION) *For random variables* $X_1, X_2, \cdots, X_n$, Y, *and* Z,

$$I(X_1, X_2, \cdots, X_n; Y|Z) = \sum_{i=1}^{n} I(X_i; Y|X_1, \cdots, X_{i-1}, Z). \qquad (2.86)$$

Proof This is the conditional version of the chain rule for mutual information. The proof is similar to that for Proposition 2.24. The details are omitted. □

2.5 INFORMATIONAL DIVERGENCE

Let p and q be two probability distributions on a common alphabet $\mathcal{X}$. We very often want to measure how much p is different from q, and vice versa. In order to be useful, this measure must satisfy the requirements that it is always nonnegative and it takes the zero value if and only if $p = q$. We denote the support of p and q by $\mathcal{S}_p$ and $\mathcal{S}_q$, respectively. The *informational divergence* defined below serves this purpose.

DEFINITION 2.27 *The informational divergence between two probability distributions* p *and* q *on a common alphabet* $\mathcal{X}$ *is defined as*

$$D(p\|q) = \sum_{x} p(x) \log \frac{p(x)}{q(x)} = E_p \log \frac{p(X)}{q(X)}, \qquad (2.87)$$

where E_p *denotes expectation with respect to* p.

In the above definition, in addition to the convention that the summation is taken over $\mathcal{S}_p$, we further adopt the convention $p(x) \log \frac{p(x)}{q(x)} = \infty$ if $q(x) = 0$.

In the literature, informational divergence is also referred to as *relative entropy* or the *Kullback-Leibler distance*. We note that $D(p\|q)$ is not symmetrical

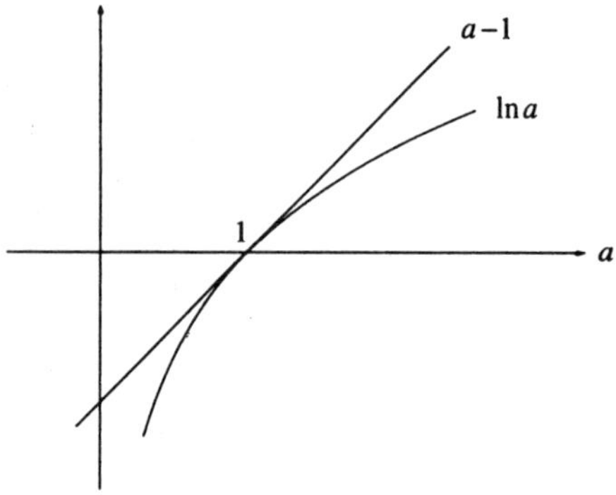

Figure 2.3. The fundamental inequality $\ln a \leq a - 1$.

in p and q, so it is not a true *metric* or "distance." Moreover, $D(\cdot\|\cdot)$ does not satisfy the triangular inequality (see Problem 10).

In the rest of the book, informational divergence will be referred to as *divergence* for brevity. Before we prove that divergence is always nonnegative, we first establish the following simple but important inequality called the *fundamental inequality* in information theory.

LEMMA 2.28 (FUNDAMENTAL INEQUALITY) *For any $a > 0$,*

$$\ln a \leq a - 1 \tag{2.88}$$

with equality if and only if $a = 1$.

Proof Let $f(a) = \ln a - a + 1$. Then $f'(a) = 1/a - 1$ and $f''(a) = -1/a^2$. Since $f(1) = 0$, $f'(1) = 0$, and $f''(1) = -1 < 0$, we see that $f(a)$ attains its maximum value 0 when $a = 1$. This proves (2.88). It is also clear that equality holds in (2.88) if and only if $a = 1$. Figure 2.3 is an illustration of the fundamental inequality. □

COROLLARY 2.29 *For any $a > 0$,*

$$\ln a \geq 1 - \frac{1}{a} \tag{2.89}$$

with equality if and only if $a = 1$.

Proof This can be proved by replacing a by $1/a$ in (2.88). □

We can see from Figure 2.3 that the fundamental inequality results from the convexity of the logarithmic function. In fact, many important results in information theory are also direct or indirect consequences of the convexity of the logarithmic function!

THEOREM 2.30 (DIVERGENCE INEQUALITY) *For any two probability distributions p and q on a common alphabet $\mathcal{X}$,*

$$D(p\|q) \geq 0 \tag{2.90}$$

with equality if and only if $p = q$.

Proof If $q(x) = 0$ for some $x \in \mathcal{S}_p$, then $D(p\|q) = \infty$ and the theorem is trivially true. Therefore, we assume that $q(x) > 0$ for all $x \in \mathcal{S}_p$. Consider

$$\begin{aligned} D(p\|q) &= (\log e) \sum_{x \in \mathcal{S}_p} p(x) \ln \frac{p(x)}{q(x)} && (2.91) \\ &\geq (\log e) \sum_{x \in \mathcal{S}_p} p(x) \left(1 - \frac{q(x)}{p(x)}\right) && (2.92) \\ &= (\log e) \left[\sum_{x \in \mathcal{S}_p} p(x) - \sum_{x \in \mathcal{S}_p} q(x) \right] && (2.93) \\ &\geq (\log e)(1 - 1) && (2.94) \\ &= 0, && (2.95) \end{aligned}$$

where (2.92) results from an application of (2.89), and (2.94) follows from

$$\sum_{x \in \mathcal{S}_p} q(x) \leq 1. \tag{2.96}$$

This proves (2.90). Now for equality to hold in (2.90), equality must hold in (2.92) for all $x \in \mathcal{S}_p$, and equality must hold in (2.94). For the former, we see from Lemma 2.28 that this is the case if and only if $p(x) = q(x)$ for all $x \in \mathcal{S}_p$. This implies

$$\sum_{x \in \mathcal{S}_p} q(x) = \sum_{x \in \mathcal{S}_p} p(x) = 1, \tag{2.97}$$

i.e., (2.94) holds with equality. Finally, since $\sum_x q(x) = 1$ and $q(x) \geq 0$ for all x, $p(x) = q(x)$ for all $x \in \mathcal{S}_p$ implies $q(x) = 0$ for all $x \notin \mathcal{S}_p$, and therefore $p = q$. The theorem is proved. □

We now prove a very useful consequence of the divergence inequality called the *log-sum inequality*.

THEOREM 2.31 (LOG-SUM INEQUALITY) *For positive numbers $a_1, a_2, \cdots$ and nonnegative numbers $b_1, b_2, \cdots$ such that $\sum_i a_i < \infty$ and $0 < \sum_i b_i < \infty$,*

$$\sum_i a_i \log \frac{a_i}{b_i} \geq \left(\sum_i a_i\right) \log \frac{\sum_i a_i}{\sum_i b_i} \tag{2.98}$$

with the convention that $\log \frac{a_i}{0} = \infty$. *Moreover, equality holds if and only if* $\frac{a_i}{b_i} = constant$ *for all* i.

The log-sum inequality can easily be understood by writing it out for the case when there are two terms in each of the summations:

$$a_1 \log \frac{a_1}{b_1} + a_2 \log \frac{a_2}{b_2} \geq (a_1 + a_2) \log \frac{a_1 + a_2}{b_1 + b_2}. \tag{2.99}$$

Proof of Theorem 2.31 Let $a_i' = a_i / \sum_j a_j$ and $b_i' = b_i / \sum_j b_j$. Then $\{a_i'\}$ and $\{b_i'\}$ are probability distributions. Using the divergence inequality, we have

$$\begin{aligned} 0 &\leq \sum_i a_i' \log \frac{a_i'}{b_i'} && (2.100) \\ &= \sum_i \frac{a_i}{\sum_j a_j} \log \frac{a_i / \sum_j a_j}{b_i / \sum_j b_j} && (2.101) \\ &= \frac{1}{\sum_j a_j} \left[\sum_i a_i \log \frac{a_i}{b_i} - \left(\sum_i a_i \right) \log \frac{\sum_j a_j}{\sum_j b_j} \right], && (2.102) \end{aligned}$$

which implies (2.98). Equality holds if and only if $a_i' = b_i'$ for all i, or $\frac{a_i}{b_i} = constant$ for all i. The theorem is proved. □

One can also prove the divergence inequality by using the log-sum inequality (see Problem 15), so the two inequalities are in fact equivalent. The log-sum inequality also finds application in proving the next theorem which gives a lower bound on the divergence between two probability distributions on a common alphabet in terms of the *variational distance* between them.

DEFINITION 2.32 *Let p and q be two probability distributions on a common alphabet* $\mathcal{X}$. *The variational distance between p and q is defined by*

$$d(p, q) = \sum_{x \in \mathcal{X}} |p(x) - q(x)|. \tag{2.103}$$

THEOREM 2.33 (PINSKER'S INEQUALITY)

$$D(p\|q) \geq \frac{1}{2 \ln 2} d^2(p, q). \tag{2.104}$$

The proof of this theorem is left as an exercise (see Problem 18). We will see further applications of the log-sum inequality when we discuss the convergence of some iterative algorithms in Chapter 10.

2.6 THE BASIC INEQUALITIES

In this section, we prove that all Shannon's information measures, namely entropy, conditional entropy, mutual information, and conditional mutual information are always nonnegative. By this, we mean that these quantities are nonnegative for all joint distributions for the random variables involved.

THEOREM 2.34 *For random variables X, Y, and Z,*

$$I(X;Y|Z) \geq 0, \tag{2.105}$$

with equality if and only if X and Y are independent when conditioning on Z.

Proof Observe that

$$\begin{aligned} & I(X;Y|Z) \\ &= \sum_{x,y,z} p(x,y,z) \log \frac{p(x,y|z)}{p(x|z)p(y|z)} && (2.106) \\ &= \sum_{z} p(z) \sum_{x,y} p(x,y|z) \log \frac{p(x,y|z)}{p(x|z)p(y|z)} && (2.107) \\ &= \sum_{z} p(z) D(p_{XY|z} \| p_{X|z} p_{Y|z}), && (2.108) \end{aligned}$$

where we have used $p_{XY|z}$ to denote $\{p(x,y|z), (x,y) \in \mathcal{X} \times \mathcal{Y}\}$, etc. Since for a fixed z, both $p_{XY|z}$ and $p_{X|z}p_{Y|z}$ are joint probability distributions on $\mathcal{X} \times \mathcal{Y}$, we have

$$D(p_{XY|z} \| p_{X|z} p_{Y|z}) \geq 0. \tag{2.109}$$

Therefore, we conclude that $I(X;Y|Z) \geq 0$. Finally, we see from Theorem 2.30 that $I(X;Y|Z) = 0$ if and only if for all $z \in \mathcal{S}_z$,

$$p(x,y|z) = p(x|z)p(y|z), \tag{2.110}$$

or

$$p(x,y,z) = p(x,z)p(y|z) \tag{2.111}$$

for all x and y. Therefore, X and Y are independent conditioning on Z. The proof is accomplished. □

As we have seen in Section 2.3 that all Shannon's information measures are special cases of conditional mutual information, we already have proved that all Shannon's information measures are always nonnegative. The nonnegativity of all Shannon's information measures are called the *basic inequalities*.

For entropy and conditional entropy, we offer the following more direct proof for their nonnegativity. Consider the entropy $H(X)$ of a random variable

X. For all $x \in \mathcal{S}_X$, since $0 < p(x) \leq 1$, $\log p(x) \leq 0$. It then follows from the definition in (2.33) that $H(X) \geq 0$. For the conditional entropy $H(Y|X)$ of random variable Y given random variable X, since $H(Y|X = x) \geq 0$ for each $x \in \mathcal{S}_X$, we see from (2.41) that $H(Y|X) \geq 0$.

PROPOSITION 2.35 *$H(X) = 0$ if and only if X is deterministic.*

Proof If X is deterministic, i.e., there exists $x^* \in \mathcal{X}$ such that $p(x^*) = 1$ and $p(x) = 0$ for all $x \neq x^*$, then $H(X) = -p(x^*)\log p(x^*) = 0$. On the other hand, if X is not deterministic, i.e., there exists $x^* \in \mathcal{X}$ such that $0 < p(x^*) < 1$, then $H(X) \geq -p(x^*)\log p(x^*) > 0$. Therefore, we conclude that $H(X) = 0$ if and only if X is deterministic. □

PROPOSITION 2.36 *$H(Y|X) = 0$ if and only if Y is a function of X.*

Proof From (2.41), we see that $H(Y|X) = 0$ if and only if $H(Y|X = x) = 0$ for each $x \in \mathcal{S}_X$. Then from the last proposition, this happens if and only if Y is deterministic for each given x. In other words, Y is a function of X. □

PROPOSITION 2.37 *$I(X;Y) = 0$ if and only if X and Y are independent.*

Proof This is a special case of Theorem 2.34 with Z being a degenerate random variable. □

One can regard (conditional) mutual information as a measure of (conditional) dependency between two random variables. When the (conditional) mutual information is exactly equal to 0, the two random variables are (conditionally) independent.

We refer to inequalities involving Shannon's information measures only (possibly with constant terms) as *information inequalities*. The basic inequalities are important examples of information inequalities. Likewise, we refer to identities involving Shannon's information measures only as *information identities*. From the information identities (2.45), (2.55), and (2.63), we see that all Shannon's information measures can be expressed as linear combinations of entropies. Specifically,

$$H(Y|X) = H(X,Y) - H(X), \tag{2.112}$$
$$I(X;Y) = H(X) + H(Y) - H(X,Y), \tag{2.113}$$

and

$$I(X;Y|Z) = H(X,Z) + H(Y,Z) - H(X,Y,Z) - H(Z). \tag{2.114}$$

Therefore, an information inequality is an inequality which involves only entropies.

As we will see later in the book, information inequalities form the most important set of tools for proving converse coding theorems in information theory. Except for a few so-called non-Shannon-type inequalities, all known information inequalities are implied by the basic inequalities. Information inequalities will be studied systematically in Chapters 12, 13, and 14. In the next section, we will prove some consequences of the basic inequalities which are often used in information theory.

2.7 SOME USEFUL INFORMATION INEQUALITIES

In this section, we prove some useful consequences of the basic inequalities introduced in the last section. Note that the conditional versions of these inequalities can be proved by techniques similar to those used in the proof of Proposition 2.24.

THEOREM 2.38 (CONDITIONING DOES NOT INCREASE ENTROPY)

$$H(Y|X) \leq H(Y) \tag{2.115}$$

with equality if and only if X and Y are independent.

Proof This can be proved by considering

$$H(Y|X) = H(Y) - I(X;Y) \leq H(Y), \tag{2.116}$$

where the inequality follows because $I(X;Y)$ is always nonnegative. The inequality is tight if and only if $I(X;Y) = 0$, which is equivalent by Proposition 2.37 to X and Y being independent. □

Similarly, it can be shown that $H(Y|X,Z) \leq H(Y|Z)$, which is the conditional version of the above proposition. These results have the following interpretation. Suppose Y is a random variable we are interested in, and X and Z are side-information about Y. Then our uncertainty about Y cannot be increased on the average upon receiving side-information X. Once we know X, our uncertainty about Y again cannot be increased on the average upon further receiving side-information Z.

Remark Unlike entropy, the mutual information between two random variables can be increased by conditioning on a third random variable. We refer the reader to Section 6.4 for a discussion.

THEOREM 2.39 (INDEPENDENCE BOUND FOR ENTROPY)

$$H(X_1, X_2, \cdots, X_n) \leq \sum_{i=1}^{n} H(X_i) \tag{2.117}$$

with equality if and only if X_i, $i = 1, 2, \cdots, n$, *are mutually independent.*

Proof By the chain rule for entropy,

$$H(X_1, X_2, \cdots, X_n) = \sum_{i=1}^{n} H(X_i|X_1, \cdots, X_{i-1}) \tag{2.118}$$

$$\leq \sum_{i=1}^{n} H(X_i), \tag{2.119}$$

where the inequality follows because we have proved in the last theorem that conditioning does not increase entropy. The inequality is tight if and only if it is tight for each i, i.e.,

$$H(X_i|X_1, \cdots, X_{i-1}) = H(X_i) \tag{2.120}$$

for $1 \leq i \leq n$. From the last theorem, this is equivalent to X_i being independent of $X_1, X_2, \cdots, X_{i-1}$ for each i. Then

$$p(x_1, x_2, \cdots, x_n)$$

$$= p(x_1, x_2, \cdots, x_{n-1})p(x_n) \tag{2.121}$$

$$= p(p(x_1, x_2, \cdots, x_{n-2})p(x_{n-1})p(x_n) \tag{2.122}$$

$$\vdots$$

$$= p(x_1)p(x_2) \cdots p(x_n) \tag{2.123}$$

for all $x_1, x_2, \cdots, x_n$, i.e., $X_1, X_2, \cdots, X_n$ are mutually independent.

Alternatively, we can prove the theorem by considering

$$\sum_{i=1}^{n} H(X_i) - H(X_1, X_2, \cdots, X_n)$$

$$= -\sum_{i=1}^{n} E \log p(X_i) + E \log p(X_1, X_2, \cdots, X_n) \tag{2.124}$$

$$= -E \log[p(X_1)p(X_2) \cdots p(X_n)] + E \log p(X_1, X_2, \cdots, X_n) \tag{2.125}$$

$$= E \log \frac{p(X_1, X_2, \cdots, X_n)}{p(X_1)p(X_2) \cdots p(X_n)} \tag{2.126}$$

$$= D(p_{X_1 X_2 \cdots X_n} \| p_{X_1} p_{X_2} \cdots p_{X_n}) \tag{2.127}$$

$$\geq 0, \tag{2.128}$$

where equality holds if and only if

$$p(x_1, x_2, \cdots, x_n) = p(x_1)p(x_2) \cdots p(x_n) \tag{2.129}$$

for all $x_1, x_2, \cdots, x_n$, i.e., $X_1, X_2, \cdots, X_n$ are mutually independent. □

THEOREM 2.40

$$I(X;Y,Z) \geq I(X;Y), \tag{2.130}$$

with equality if and only if $X \to Y \to Z$ forms a Markov chain.

Proof By the chain rule for mutual information, we have

$$I(X;Y,Z) = I(X;Y) + I(X;Z|Y) \geq I(X;Y). \tag{2.131}$$

The above inequality is tight if and only if $I(X;Z|Y) = 0$, or $X \to Y \to Z$ forms a Markov chain. The theorem is proved. □

LEMMA 2.41 *If $X \to Y \to Z$ forms a Markov chain, then*

$$I(X;Z) \leq I(X;Y) \tag{2.132}$$

and

$$I(X;Z) \leq I(Y;Z). \tag{2.133}$$

Before proving this inequality we discuss its meaning. Suppose X is a random variable we are interested in, and Y is an observation of X. If we infer X via Y, our uncertainty about X on the average is $H(X|Y)$. Now suppose we process Y (either deterministically or probabilistically) to obtain a random variable Z. If we infer X via Z, our uncertainty about X on the average is $H(X|Z)$. Since $X \to Y \to Z$ forms a Markov chain, from (2.132), we have

$$\begin{aligned} H(X|Z) &= H(X) - I(X;Z) && (2.134)\\ &\geq H(X) - I(X;Y) && (2.135)\\ &= H(X|Y), && (2.136) \end{aligned}$$

i.e., further processing of Y can only increase our uncertainty about X on the average.

Proof of Lemma 2.41 Assume $X \to Y \to Z$, i.e., $X \perp Z|Y$. By Theorem 2.34, we have

$$I(X;Z|Y) = 0. \tag{2.137}$$

Then

$$\begin{aligned} I(X;Z) &= I(X;Y,Z) - I(X;Y|Z) && (2.138)\\ &\leq I(X;Y,Z) && (2.139)\\ &= I(X;Y) + I(X;Z|Y) && (2.140)\\ &= I(X;Y). && (2.141) \end{aligned}$$

In (2.138) and (2.140), we have used the chain rule for mutual information. The inequality in (2.139) follows because $I(X;Y|Z)$ is always nonnegative, and (2.141) follows from (2.137). This proves (2.132).

Since $X \to Y \to Z$ is equivalent to $Z \to Y \to X$, we also have proved (2.133). This completes the proof of the lemma. □

From Lemma 2.41, we can prove the more general data processing theorem.

THEOREM 2.42 (DATA PROCESSING THEOREM) *If $U \to X \to Y \to V$ forms a Markov chain, then*

$$I(U;V) \leq I(X;Y). \tag{2.142}$$

Proof Assume $U \to X \to Y \to V$. Then by Proposition 2.10, we have $U \to X \to Y$ and $U \to Y \to V$. From the first Markov chain and Lemma 2.41, we have

$$I(U;Y) \leq I(X;Y). \tag{2.143}$$

From the second Markov chain and Lemma 2.41, we have

$$I(U;V) \leq I(U;Y). \tag{2.144}$$

Combining (2.143) and (2.144), we obtain (2.142), proving the theorem. □

2.8 FANO'S INEQUALITY

In the last section, we have proved a few information inequalities involving only Shannon's information measures. In this section, we first prove an upper bound on the entropy of a random variable in terms of the size of the alphabet. This inequality is then used in the proof of Fano's inequality, which is extremely useful in proving converse coding theorems in information theory.

THEOREM 2.43 *For any random variable X,*

$$H(X) \leq \log|\mathcal{X}|, \tag{2.145}$$

where $|\mathcal{X}|$ denotes the size of the alphabet $\mathcal{X}$. This upper bound is tight if and only if X distributes uniformly on $\mathcal{X}$.

Proof Let u be the uniform distribution on $\mathcal{X}$, i.e., $u(x) = |\mathcal{X}|^{-1}$ for all $x \in \mathcal{X}$. Then

$$\begin{aligned}
&\log|\mathcal{X}| - H(X) \\
&= -\sum_{x\in\mathcal{S}_X} p(x)\log|\mathcal{X}|^{-1} + \sum_{x\in\mathcal{S}_X} p(x)\log p(x)
\end{aligned} \tag{2.146}$$

$$= -\sum_{x \in \mathcal{S}_X} p(x) \log u(x) + \sum_{x \in \mathcal{S}_X} p(x) \log p(x) \tag{2.147}$$

$$= \sum_{x \in \mathcal{S}_X} p(x) \log \frac{p(x)}{u(x)} \tag{2.148}$$

$$= D(p\|u) \tag{2.149}$$

$$\geq 0, \tag{2.150}$$

proving (2.145). This upper bound is tight if and if only $D(p\|u) = 0$, which from Theorem 2.30 is equivalent to $p(x) = u(x)$ for all $x \in \mathcal{X}$, completing the proof. □

COROLLARY 2.44 *The entropy of a random variable may take any nonnegative real value.*

Proof Consider a random variable X and let $|\mathcal{X}|$ be fixed. We see from the last theorem that $H(X) = \log |\mathcal{X}|$ is achieved when X distributes uniformly on $\mathcal{X}$. On the other hand, $H(X) = 0$ is achieved when X is deterministic. For any value $0 < a < \log |\mathcal{X}|$, by the intermediate value theorem of continuous functions, there exists a distribution for X such that $H(X) = a$. Then we see that $H(X)$ can take any positive value by letting $|\mathcal{X}|$ be sufficiently large. This accomplishes the proof. □

Remark Let $|\mathcal{X}| = D$, or the random variable X is a D-ary symbol. When the base of the logarithm is D, (2.145) becomes

$$H_D(X) \leq 1. \tag{2.151}$$

Recall that the unit of entropy is the D-it when the logarithm is in the base D. This inequality says that a D-ary symbol can carry at most 1 D-it of information. This maximum is achieved when X has a uniform distribution. We already have seen the binary case when we discuss the binary entropy function $h_b(p)$ in Section 2.2.

We see from Theorem 2.43 that the entropy of a random variable is finite as long as it has a finite alphabet. However, if a random variable has an infinite alphabet, its entropy may or may not be finite. This will be shown in the next two examples.

EXAMPLE 2.45 *Let X be a random variable such that*

$$\Pr\{X - i\} - 2^{-i}, \tag{2.152}$$

$i = 1, 2, \cdots$. *Then*

$$H_2(X) = \sum_{i=1}^{\infty} i 2^{-i} = 2, \tag{2.153}$$

which is finite.

EXAMPLE 2.46 *Let Y be a random variable which takes values in the subset of pairs of integers*

$$\left\{(i,j) : 1 \leq i < \infty \text{ and } 1 \leq j \leq \frac{2^{2^i}}{2^i}\right\} \tag{2.154}$$

such that

$$\Pr\{Y = (i,j)\} = 2^{-2^i} \tag{2.155}$$

for all i and j. First, we check that

$$\sum_{i=1}^{\infty} \sum_{j=1}^{2^{2^i}/2^i} \Pr\{Y = (i,j)\} = \sum_{i=1}^{\infty} 2^{-2^i} \left(\frac{2^{2^i}}{2^i}\right) = 1. \tag{2.156}$$

Then

$$H_2(Y) = -\sum_{i=1}^{\infty} \sum_{j=1}^{2^{2^i}/2^i} 2^{-2^i} \log_2 2^{-2^i} = \sum_{i=1}^{\infty} 1, \tag{2.157}$$

which does not converge.

Let X be a random variable and $\hat{X}$ be an estimate of X which takes value in the same alphabet $\mathcal{X}$. Let the probability of error P_e be

$$P_e = \Pr\{X \neq \hat{X}\}. \tag{2.158}$$

If $P_e = 0$, i.e., $X = \hat{X}$ with probability 1, then $H(X|\hat{X}) = 0$ by Proposition 2.36. Intuitively, if P_e is small, i.e., $X = \hat{X}$ with probability close to 1, then $H(X|\hat{X})$ should be close to 0. Fano's inequality makes this intuition precise.

THEOREM 2.47 (FANO'S INEQUALITY) *Let X and $\hat{X}$ be random variables taking values in the same alphabet $\mathcal{X}$. Then*

$$H(X|\hat{X}) \leq h_b(P_e) + P_e \log(|\mathcal{X}| - 1), \tag{2.159}$$

where h_b is the binary entropy function.

Proof Define a random variable

$$Y = \begin{cases} 0 & \text{if } X = \hat{X} \\ 1 & \text{if } X \neq \hat{X}. \end{cases} \tag{2.160}$$

The random variable Y is an indicator of the error event $\{X \neq \hat{X}\}$, with $\Pr\{Y = 1\} = P_e$ and $H(Y) = h_b(P_e)$. Since Y is a function X and $\hat{X}$,

$$H(Y|X, \hat{X}) = 0. \tag{2.161}$$

Then

$$\begin{aligned}
&H(X|\hat{X}) \\
&= H(X|\hat{X}) + H(Y|X, \hat{X}) && (2.162)\\
&= H(X, Y|\hat{X}) && (2.163)\\
&= H(Y|\hat{X}) + H(X|\hat{X}, Y) && (2.164)\\
&\leq H(Y) + H(X|\hat{X}, Y) && (2.165)\\
&= H(Y) + \sum_{\hat{x} \in \mathcal{X}} \Big[\Pr\{\hat{X} = \hat{x}, Y = 0\} H(X|\hat{X} = \hat{x}, Y = 0) \\
&\quad + \Pr\{\hat{X} = \hat{x}, Y = 1\} H(X|\hat{X} = \hat{x}, Y = 1)\Big]. && (2.166)
\end{aligned}$$

In the above, (2.162) follows from (2.161), (2.165) follows because conditioning does not increase entropy, and (2.166) follows from an application of (2.41). Now X must take the value $\hat{x}$ if $\hat{X} = \hat{x}$ and $Y = 0$. In other words, X is conditionally deterministic given $\hat{X} = \hat{x}$ and $Y = 0$. Therefore, by Proposition 2.35,

$$H(X|\hat{X} = \hat{x}, Y = 0) = 0. \tag{2.167}$$

If $\hat{X} = \hat{x}$ and $Y = 1$, then X must take a value in the set $\{x \in \mathcal{X} : x \neq \hat{x}\}$ which contains $|\mathcal{X}| - 1$ elements. From the last theorem, we have

$$H(X|\hat{X} = \hat{x}, Y = 1) \leq \log(|\mathcal{X}| - 1), \tag{2.168}$$

where this upper bound does not depend on $\hat{x}$. Hence,

$$\begin{aligned}
&H(X|\hat{X}) \\
&\leq h_b(P_e) + \left(\sum_{\hat{x} \in \mathcal{X}} \Pr\{\hat{X} = \hat{x}, Y = 1\}\right) \log(|\mathcal{X}| - 1) && (2.169)\\
&= h_b(P_e) + \Pr\{Y = 1\} \log(|\mathcal{X}| - 1) && (2.170)\\
&= h_b(P_e) + P_e \log(|\mathcal{X}| - 1), && (2.171)
\end{aligned}$$

which completes the proof. □

Very often, we only need the following simplified version when we apply Fano's inequality. The proof is omitted.

COROLLARY 2.48 $H(X|\hat{X}) < 1 + P_e \log |\mathcal{X}|$.

Fano's inequality has the following implication. If the alphabet $\mathcal{X}$ is finite, as $P_e \to 0$, the upper bound in (2.159) tends to 0, which implies $H(X|\hat{X})$ also tends to 0. However, this is not necessarily the case if $\mathcal{X}$ is infinite, which is shown in the next example.

EXAMPLE 2.49 *Let $\hat{X}$ take the value 0 with probability 1. Let Z be an independent binary random variable taking values in $\{0, 1\}$. Define the random variable X by*

$$X = \begin{cases} 0 & \text{if } Z = 0 \\ Y & \text{if } Z = 1, \end{cases} \tag{2.172}$$

where Y is the random variable in Example 2.46 whose entropy is infinity. Let

$$P_e = \Pr\{X \neq \hat{X}\} = \Pr\{Z = 1\}. \tag{2.173}$$

Then

$$\begin{aligned} & H(X|\hat{X}) & (2.174) \\ = \; & H(X) & (2.175) \\ \geq \; & H(X|Z) & (2.176) \\ = \; & \Pr\{Z = 0\}H(X|Z = 0) + \Pr\{Z = 1\}H(X|Z = 1) & (2.177) \\ = \; & (1 - P_e) \cdot 0 + P_e \cdot H(Y) & (2.178) \\ = \; & \infty & (2.179) \end{aligned}$$

for any $P_e > 0$. Therefore, $H(X|\hat{X})$ does not tend to 0 as $P_e \to 0$.

2.9 ENTROPY RATE OF STATIONARY SOURCE

In the previous sections, we have discussed various properties of the entropy of a finite collection of random variables. In this section, we discuss the entropy rate *entropy rate* of a discrete-time information source.

A discrete-time information source $\{X_k, k \geq 1\}$ is an infinite collection of random variables indexed by the set of positive integers. Since the index set is ordered, it is natural to regard the indices as time indices. We will refer to the random variables X_k as *letters*.

We assume that $H(X_k) < \infty$ for all k. Then for any finite subset A of the index set $\{k : k \geq 1\}$, we have

$$H(X_k, k \in A) \leq \sum_{k \in A} H(X_k) < \infty. \tag{2.180}$$

However, it is not meaningful to discuss $H(X_k, k \geq 1)$ because the joint entropy of an infinite collection of letters is infinite except in special cases. On the other hand, since the indices are ordered, we can naturally define the *entropy rate* of an information source, which gives the average entropy per letter of the source.

DEFINITION 2.50 *The entropy rate of an information source* $\{X_k\}$ *is defined by*

$$H_X = \lim_{n\to\infty} \frac{1}{n} H(X_1, X_2, \cdots, X_n) \tag{2.181}$$

when the limit exists.

We show in the next two examples that the entropy rate of a source may or may not exist.

EXAMPLE 2.51 *Let* $\{X_k\}$ *be an i.i.d. source with generic random variable* X. *Then*

$$\begin{aligned} \lim_{n\to\infty} \frac{1}{n} H(X_1, X_2, \cdots, X_n) &= \lim_{n\to\infty} \frac{nH(X)}{n} & (2.182)\\ &= \lim_{n\to\infty} H(X) & (2.183)\\ &= H(X), & (2.184) \end{aligned}$$

i.e., the entropy rate of an i.i.d. source is the entropy of any of its single letters.

EXAMPLE 2.52 *Let* $\{X_k\}$ *be a source such that* X_k *are mutually independent and* $H(X_k) = k$ *for* $k \geq 1$. *Then*

$$\begin{aligned} \frac{1}{n} H(X_1, X_2, \cdots, X_n) &= \frac{1}{n} \sum_{k=1}^{n} k & (2.185)\\ &= \frac{1}{n} \frac{n(n+1)}{2} & (2.186)\\ &= \frac{1}{2}(n+1), & (2.187) \end{aligned}$$

which does not converge as $n \to \infty$ *although* $H(X_k) < \infty$ *for all* k. *Therefore, the entropy rate of* $\{X_k\}$ *does not exist.*

Toward characterizing the asymptotic behavior of $\{X_k\}$, it is natural to consider the limit

$$H'_X = \lim_{n\to\infty} H(X_n | X_1, X_2, \cdots, X_{n-1}) \tag{2.188}$$

if it exists. The quantity $H(X_n | X_1, X_2, \cdots, X_{n-1})$ is interpreted as the conditional entropy of the next letter given that we know all the past history of the source, and H'_X is the limit of this quantity after the source has been run for an indefinite amount of time.

DEFINITION 2.53 *An information source $\{X_k\}$ is stationary if*

$$X_1, X_2, \cdots, X_m \tag{2.189}$$

and

$$X_{1+l}, X_{2+l}, \cdots, X_{m+l} \tag{2.190}$$

have the same joint distribution for any $m, l \geq 1$.

In the rest of the section, we will show that stationarity is a sufficient condition for the existence of the entropy rate of an information source.

LEMMA 2.54 *Let $\{X_k\}$ be a stationary source. Then H'_X exists.*

Proof Since $H(X_n|X_1, X_2, \cdots, X_{n-1})$ is lower bounded by zero for all n, it suffices to prove that $H(X_n|X_1, X_2, \cdots, X_{n-1})$ is non-increasing in n to conclude that the limit H'_X exists. Toward this end, for $n \geq 2$, consider

$$\begin{aligned} & H(X_n|X_1, X_2, \cdots, X_{n-1}) \\ \leq \; & H(X_n|X_2, X_3, \cdots, X_{n-1}) && (2.191) \\ = \; & H(X_{n-1}|X_1, X_2, \cdots, X_{n-2}), && (2.192) \end{aligned}$$

where the last step is justified by the stationarity of $\{X_k\}$. The lemma is proved. □

LEMMA 2.55 (CESÁRO MEAN) *Let a_k and b_k be real numbers. If $a_n \to a$ as $n \to \infty$ and $b_n = \frac{1}{n}\sum_{k=1}^{n} a_k$, then $b_n \to a$ as $n \to \infty$.*

Proof The idea of the lemma is the following. If $a_n \to a$ as $n \to \infty$, then the average of the first n terms in $\{a_k\}$, namely b_n, also tends to a as $n \to \infty$.

The lemma is formally proved as follows. Since $a_n \to a$ as $n \to \infty$, for every $\epsilon > 0$, there exists $N(\epsilon)$ such that $|a_n - a| < \epsilon$ for all $n > N(\epsilon)$. For $n > N(\epsilon)$, consider

$$\begin{aligned} |b_n - a| &= \left| \frac{1}{n} \sum_{i=1}^{n} a_i - a \right| && (2.193) \\ &= \left| \frac{1}{n} \sum_{i=1}^{n} (a_i - a) \right| && (2.194) \\ &\leq \frac{1}{n} \sum_{i=1}^{n} |a_i - a| && (2.195) \\ &= \frac{1}{n} \left(\sum_{i=1}^{N(\epsilon)} |a_i - a| + \sum_{i=N(\epsilon)+1}^{n} |a_i - a| \right) && (2.196) \end{aligned}$$

$$< \frac{1}{n} \sum_{i=1}^{N(\epsilon)} |a_i - a| + \frac{(n - N(\epsilon))\epsilon}{n} \tag{2.197}$$

$$< \frac{1}{n} \sum_{i=1}^{N(\epsilon)} |a_i - a| + \epsilon. \tag{2.198}$$

The first term tends to 0 as $n \to \infty$. Therefore, for any $\epsilon > 0$, by taking n to be sufficiently large, we can make $|b_n - a| < 2\epsilon$. Hence $b_n \to a$ as $n \to \infty$, proving the lemma. □

We now prove that H'_X is an alternative definition/interpretation of the entropy rate of $\{X_k\}$ when $\{X_k\}$ is stationary.

THEOREM 2.56 *For a stationary source $\{X_k\}$, the entropy rate H_X exists, and it is equal to H'_X.*

Proof Since we have proved in Lemma 2.54 that H'_X always exists for a stationary source $\{X_k\}$, in order to prove the theorem, we only have to prove that $H_X = H'_X$. By the chain rule for entropy,

$$\frac{1}{n} H(X_1, X_2, \cdots, X_n) = \frac{1}{n} \sum_{k=1}^{n} H(X_k | X_1, X_2, \cdots, X_{k-1}). \tag{2.199}$$

Since

$$\lim_{k \to \infty} H(X_k | X_1, X_2, \cdots, X_{k-1}) = H'_X \tag{2.200}$$

from (2.188), it follows from Lemma 2.55 that

$$H_X = \lim_{n \to \infty} \frac{1}{n} H(X_1, X_2, \cdots, X_n) = H'_X. \tag{2.201}$$

The theorem is proved. □

In this theorem, we have proved that the entropy rate of a random source $\{X_k\}$ exists under the fairly general assumption that $\{X_k\}$ is stationary. However, the entropy rate of a stationary source $\{X_k\}$ may not carry any physical meaning unless $\{X_k\}$ is also ergodic. This will be explained when we discuss the Shannon-McMillan-Breiman Theorem in Section 4.4.

PROBLEMS

1. Let X and Y be random variables with alphabets $\mathcal{X} = \mathcal{Y} = \{1,2,3,4,5\}$ and joint distribution $p(x,y)$ given by

$$\frac{1}{25}\begin{bmatrix} 1 & 1 & 1 & 1 & 1 \\ 2 & 1 & 2 & 0 & 0 \\ 2 & 0 & 1 & 1 & 1 \\ 0 & 3 & 0 & 2 & 0 \\ 0 & 0 & 1 & 1 & 3 \end{bmatrix}.$$

   Determine $H(X), H(Y), H(X|Y), H(Y|X)$, and $I(X;Y)$.

2. Prove Propositions 2.8, 2.9, 2.10, 2.19, 2.21, and 2.22.

3. Give an example which shows that pairwise independence does not imply mutual independence.

4. Verify that $p(x,y,z)$ as defined in Definition 2.4 is a probability distribution. You should exclude all the zero probability masses from the summation carefully.

5. *Linearity of expectation* It is well-known that expectation is linear, i.e., $E[f(X)+g(Y)] = Ef(X) + Eg(Y)$, where the summation in an expectation is taken over the corresponding alphabet. However, we adopt in information theory the convention that the summation in an expectation is taken over the corresponding support. Justify carefully the linearity of expectation under this convention.

6. Let $C_\alpha = \sum_{n=2}^{\infty} \frac{1}{n(\log n)^\alpha}$.

   a) Prove that

$$C_\alpha \begin{cases} < \infty & \text{if } \alpha > 1 \\ = \infty & \text{if } 0 \le \alpha \le 1. \end{cases}$$

   Then

$$p_\alpha(n) = [C_\alpha n (\log n)^\alpha]^{-1}, \quad n = 2, 3, \cdots$$

   is a probability distribution for $\alpha > 1$.

   b) Prove that

$$H(p_\alpha) \begin{cases} < \infty & \text{if } \alpha > 2 \\ = \infty & \text{if } 1 < \alpha \le 2. \end{cases}$$

7. Prove that $H(p)$ is concave in p, i.e., for $0 \le \lambda \le 1$ and $\bar{\lambda} = 1 - \lambda$,

$$\lambda H(p_1) + \bar{\lambda} H(p_2) \le H(\lambda p_1 + \bar{\lambda} p_2).$$

8. Let $(X,Y) \sim p(x,y) = p(x)p(y|x)$.

a) Prove that for fixed $p(x)$, $I(X;Y)$ is a convex functional of $p(y|x)$.

b) Prove that for fixed $p(y|x)$, $I(X;Y)$ is a concave functional of $p(x)$.

9. Do $I(X;Y) = 0$ and $I(X;Y|Z) = 0$ imply each other? If so, give a proof. If not, give a counterexample.

10. Give an example for which $D(\cdot\|\cdot)$ does not satisfy the triangular inequality.

11. Let X be a function of Y. Prove that $H(X) \leq H(Y)$. Interpret this result.

12. Prove that for any $n \geq 2$,

$$H(X_1, X_2, \cdots, X_n) \geq \sum_{i=1}^{n} H(X_i|X_j, j \neq i).$$

13. Prove that

$$\frac{1}{2}[H(X_1, X_2) + H(X_2, X_3) + H(X_1, X_3)] \geq H(X_1, X_2, X_3).$$

Hint: Sum the identities

$$H(X_1, X_2, X_3) = H(X_j, j \neq i) + H(X_i|X_j, j \neq i)$$

for $i = 1, 2, 3$ and apply the result in Problem 12.

14. Let $\mathcal{N}_n = \{1, 2, \cdots, n\}$ and denote $H(X_i, i \in \alpha)$ by $H(X_\alpha)$ for any subset α of $\mathcal{N}_n$. For $1 \leq k \leq n$, let

$$H_k = \frac{1}{\binom{n-1}{k-1}} \sum_{\alpha:|\alpha|=k} H(X_\alpha).$$

Prove that

$$H_1 \geq H_2 \geq \cdots \geq H_n.$$

This sequence of inequalities, due to Han [93], is a generalization of the independence bound for entropy (Theorem 2.39). See Problem 4 in Chapter 15 for an application of these inequalities.

15. Prove the divergence inequality by using the log-sum inequality.

16. Prove that $D(p\|q)$ is convex in the pair (p, q), i.e., if (p_1, q_1) and (p_2, q_2) are two pairs of probability distributions on a common alphabet, then

$$D(\lambda p_1 + \overline{\lambda} p_2 \| \lambda q_1 + \overline{\lambda} q_2) \leq \lambda D(p_1\|q_1) + \overline{\lambda} D(p_2\|q_2)$$

for all $0 \leq \lambda \leq 1$, where $\overline{\lambda} = 1 - \lambda$.

17. Let p_{XY} and q_{XY} be two probability distributions on $\mathcal{X} \times \mathcal{Y}$. Prove that $D(p_{XY}\|q_{XY}) \geq D(p_X\|q_X)$.

18. *Pinsker's inequality* Let $d(p, q)$ denotes the variational distance between two probability distributions p and q on a common alphabet $\mathcal{X}$. We will determine the largest c which satisfies

$$D(p\|q) \geq cd^2(p, q).$$

   a) Let $A = \{x : p(x) \geq q(x)\}$, $\hat{p} = \{p(A), 1 - p(A)\}$, and $\hat{q} = \{q(A), 1 - q(A)\}$. Show that $D(p\|q) \geq D(\hat{p}\|\hat{q})$ and $d(p, q) = d(\hat{p}, \hat{q})$.

   b) Show that toward determining the largest value of c, we only have to consider the case when $\mathcal{X}$ is binary.

   c) By virtue of b), it suffices to determine the largest c such that

$$p \log \frac{p}{q} + (1 - p) \log \frac{1 - p}{1 - q} - 4c(p - q)^2 \geq 0$$

   for all $0 \leq p, q \leq 1$, with the convention that $0 \log \frac{0}{b} = 0$ for $b \geq 0$ and $a \log \frac{a}{0} = \infty$ for $a > 0$. By observing that equality in the above holds if $p = q$ and considering the derivative of the left hand side with respect to q, show that the largest value of c is equal to $(2 \ln 2)^{-1}$.

19. Find a necessary and sufficient condition for Fano's inequality to be tight.

20. Show that for a stationary source $\{X_k\}$, $\frac{1}{n}H(X_1, X_2, \cdots, X_n)$ is non-increasing in n.

21. Let $\mathbf{p} = \{p_1, p_2, \cdots, p_n\}$ and $\mathbf{q} = \{q_1, q_2, \cdots, q_n\}$ be two probability distributions such that $p_i \geq p_{i'}$ and $q_i \geq q_{i'}$ for all $i < i'$, and $\sum_{i=1}^{m} p_i \leq \sum_{j=1}^{m} q_j$ for all $m = 1, 2, \cdots, n$. Prove that $H(\mathbf{p}) \geq H(\mathbf{q})$. Hint:

   a) Show that for $\mathbf{p} \neq \mathbf{q}$, there exist $1 \leq j < k \leq n$ which satisfy the following:

      i) j is the largest index i such that $p_i < q_i$

      ii) k is the smallest index i such that $i > j$ and $p_i > q_i$

      iii) $p_i = q_i$ for all $j < i < k$.

   b) Consider the distribution $\mathbf{q}^* = \{q_1^*, q_2^*, \cdots, q_n^*\}$ defined by $q_i^* = q_i$ for $i \neq j, k$ and

$$(q_j^*, q_k^*) = \begin{cases} (p_j, q_k + (q_j - p_j)) & \text{if } p_k - q_k \geq q_j - p_j \\ (q_j - (p_k - q_k), p_k) & \text{if } p_k - q_k < q_j - p_j. \end{cases}$$

   Note that either $q_j^* = p_j$ or $q_k^* = p_k$. Show that

i) $q_i^* \geq q_{i'}^*$ for all $i \leq i'$

ii) $\sum_{i=1}^m p_i \leq \sum_{i=1}^m q_i^*$ for all $m = 1, 2, \cdots, n$

iii) $H(\mathbf{q}^*) \geq H(\mathbf{q})$.

c) Prove the result by induction on the Hamming distance between $\mathbf{p}$ and $\mathbf{q}$, i.e., the number of places where $\mathbf{p}$ and $\mathbf{q}$ differ.

(Hardy, Littlewood, and Pólya [97].)

HISTORICAL NOTES

The concept of entropy has its root in thermodynamics. Shannon [184] was the first to use entropy as a measure of information. Informational divergence was introduced by Kullback and Leibler [128], and it has been studied extensively by Csiszár [53] and Amari [11].

The material in this chapter can be found in most textbooks in information theory. The main concepts and results are due to Shannon [184]. Pinsker's inequality is due to Pinsker [164]. Fano's inequality has its origin in the converse proof of the channel coding theorem (to be discussed in Chapter 8) by Fano [66].

Chapter 3

ZERO-ERROR DATA COMPRESSION

In a random experiment, a coin is tossed n times. Let X_i be the outcome of the ith toss, with

$$\Pr\{X_i = \text{HEAD}\} = p \text{ and } \Pr\{X_i = \text{TAIL}\} = 1 - p, \tag{3.1}$$

where $0 \leq p \leq 1$. It is assumed that X_i are i.i.d., and the value of p is known. We are asked to describe the outcome of the random experiment without error (with zero error) by using binary symbols. One way to do this is to encode a HEAD by a '0' and a TAIL by a '1.' Then the outcome of the random experiment is encoded into a binary codeword of length n. When the coin is fair, i.e., $p = 0.5$, this is the best we can do because the probability of every outcome of the experiment is equal to 2^{-n}. In other words, all the outcomes are equally likely.

However, if the coin is biased, i.e., $p \neq 0.5$, the probability of an outcome of the experiment depends on the number of HEADs and the number of TAILs in the outcome. In other words, the probabilities of the outcomes are no longer uniform. It turns out that we can take advantage of this by encoding more likely outcomes into shorter codewords and less likely outcomes into longer codewords. By doing so, it is possible to use fewer than n bits *on the average* to describe the outcome of the random experiment. In particular, in the extreme case in which $p = 0$ or 1, we actually do not need to describe the outcome of the experiment because it is deterministic.

At the beginning of Chapter 2, we mentioned that the entropy $H(X)$ measures the amount of information contained in a random variable X. In this chapter, we substantiate this claim by exploring the role of entropy in the context of zero-error data compression.

3.1 THE ENTROPY BOUND

In this section, we establish that $H(X)$ is a fundamental lower bound on the expected length of the number of symbols needed to describe the outcome of a random variable X with zero error. This is called the *entropy bound.*

DEFINITION 3.1 *A D-ary source code $\mathcal{C}$ for a source random variable X is a mapping from $\mathcal{X}$ to $\mathcal{D}^*$, the set of all finite length sequences of symbols taken from a D-ary code alphabet.*

Consider an information source $\{X_k, k \geq 1\}$, where X_k are discrete random variables which take values in the same alphabet. We apply a source code $\mathcal{C}$ to each X_k and concatenate the codewords. Once the codewords are concatenated, the boundaries of the codewords are no longer explicit. In other words, when the code $\mathcal{C}$ is applied to a source sequence, a sequence of code symbols are produced, and the codewords may no longer be distinguishable. We are particularly interested in *uniquely decodable codes* which are defined as follows.

DEFINITION 3.2 *A code $\mathcal{C}$ is uniquely decodable if for any finite source sequence, the sequence of code symbols corresponding to this source sequence is different from the sequence of code symbols corresponding to any other (finite) source sequence.*

Suppose we use a code $\mathcal{C}$ to encode a source file into a coded file. If $\mathcal{C}$ is uniquely decodable, then we can always recover the source file from the coded file. An important class of uniquely decodable codes, called *prefix codes,* are discussed in the next section. But we first look at an example of a code which is not uniquely decodable.

EXAMPLE 3.3 *Let $\mathcal{X} = \{A, B, C, D\}$. Consider the code $\mathcal{C}$ defined by*

x	$\mathcal{C}(x)$
A	0
B	1
C	01
D	10

Then all the three source sequences AAD, ACA, and AABA produce the code sequence 0010. Thus from the code sequence 0010, we cannot tell which of the three source sequences it comes from. Therefore, $\mathcal{C}$ is not uniquely decodable.

In the next theorem, we prove that for any uniquely decodable code, the lengths of the codewords have to satisfy an inequality called the *Kraft inequality.*

THEOREM 3.4 (KRAFT INEQUALITY) *Let C be a D-ary source code, and let $l_1, l_2, \cdots, l_m$ be the lengths of the codewords. If C is uniquely decodable, then*

$$\sum_{k=1}^{m} D^{-l_k} \leq 1. \tag{3.2}$$

Proof Let N be an arbitrary positive integer, and consider the identity

$$\left(\sum_{k=1}^{m} D^{-l_k}\right)^N = \sum_{k_1=1}^{m} \sum_{k_2=1}^{m} \cdots \sum_{k_N=1}^{m} D^{-(l_{k_1}+l_{k_2}+\cdots+l_{k_N})}. \tag{3.3}$$

By collecting terms on the right-hand side, we write

$$\left(\sum_{k=1}^{m} D^{-l_k}\right)^N = \sum_{i=1}^{Nl_{\max}} A_i D^{-i} \tag{3.4}$$

where

$$l_{\max} = \max_{1\leq k\leq m} l_k \tag{3.5}$$

and A_i is the coefficient of D^{-i} in $\left(\sum_{k=1}^{m} D^{-l_k}\right)^N$. Now observe that A_i gives the total number of sequences of N codewords with a total length of i code symbols. Since the code is uniquely decodable, these code sequences must be distinct, and therefore $A_i \leq D^i$ because there are D^i distinct sequences of i code symbols. Substituting this inequality into (3.4), we have

$$\left(\sum_{k=1}^{m} D^{-l_k}\right)^N \leq \sum_{i=1}^{Nl_{\max}} 1 = Nl_{\max}, \tag{3.6}$$

or

$$\sum_{k=1}^{m} D^{-l_k} \leq (Nl_{\max})^{1/N}. \tag{3.7}$$

Since this inequality holds for any N, upon letting $N \to \infty$, we obtain (3.2), completing the proof. □

Let X be a source random variable with probability distribution

$$\{p_1, p_2, \cdots, p_m\}, \tag{3.8}$$

where $m \geq 2$. When we use a uniquely decodable code C to encode the outcome of X, we are naturally interested in the expected length of a codeword, which is given by

$$L = \sum_i p_i l_i. \tag{3.9}$$

We will also refer to L as the expected length of the code $\mathcal{C}$. The quantity L gives the average number of symbols we need to describe the outcome of X when the code $\mathcal{C}$ is used, and it is a measure of the efficiency of the code $\mathcal{C}$. Specifically, the smaller the expected length L is, the better the code $\mathcal{C}$ is.

In the next theorem, we will prove a fundamental lower bound on the expected length of any uniquely decodable D-ary code. We first explain why this is the lower bound we should expect. In a uniquely decodable code, we use L D-ary symbols on the average to describe the outcome of X. Recall from the remark following Theorem 2.43 that a D-ary symbol can carry at most one D-it of information. Then the maximum amount of information which can be carried by the codeword on the average is $L \cdot 1 = L$ D-its. Since the code is uniquely decodable, the amount of entropy carried by the codeword on the average is $H(X)$. Therefore, we have

$$H_D(X) \leq L. \tag{3.10}$$

In other words, the expected length of a uniquely decodable code is at least the entropy of the source. This argument is rigorized in the proof of the next theorem.

THEOREM 3.5 (ENTROPY BOUND) *Let $\mathcal{C}$ be a D-ary uniquely decodable code for a source random variable X with entropy $H_D(X)$. Then the expected length of $\mathcal{C}$ is lower bounded by $H_D(X)$, i.e.,*

$$L \geq H_D(X). \tag{3.11}$$

This lower bound is tight if and only if $l_i = -\log_D p_i$ for all i.

Proof Since $\mathcal{C}$ is uniquely decodable, the lengths of its codewords satisfy the Kraft inequality. Write

$$L = \sum_i p_i \log_D D^{l_i} \tag{3.12}$$

and recall from Definition 2.33 that

$$H_D(X) = -\sum_i p_i \log_D p_i. \tag{3.13}$$

Then

$$\begin{aligned} L - H_D(X) &= \sum_i p_i \log_D (p_i D^{l_i}) &(3.14)\\ &= (\ln D)^{-1} \sum_i p_i \ln (p_i D^{l_i}) &(3.15)\\ &\geq (\ln D)^{-1} \sum_i p_i \left(1 - \frac{1}{p_i D^{l_i}}\right) &(3.16)\end{aligned}$$

$$= (\ln D)^{-1}\left[\sum_i p_i - \sum_i D^{-l_i}\right] \quad (3.17)$$

$$\geq (\ln D)^{-1}(1-1) \quad (3.18)$$

$$= 0, \quad (3.19)$$

where we have invoked the fundamental inequality in (3.16) and the Kraft inequality in (3.18). This proves (3.11). In order for this lower bound to be tight, both (3.16) and (3.18) have to be tight simultaneously. Now (3.16) is tight if and only if $p_i D^{l_i} = 1$, or $l_i = -\log_D p_i$ for all i. If this holds, we have

$$\sum_i D^{-l_i} = \sum_i p_i = 1, \quad (3.20)$$

i.e., (3.18) is also tight. This completes the proof of the theorem. □

The entropy bound can be regarded as a generalization of Theorem 2.43, as is seen from the following corollary.

COROLLARY 3.6 $H(X) \leq \log|\mathcal{X}|$.

Proof Considering encoding each outcome of a random variable X by a distinct symbol in $\{1, 2, \cdots, |\mathcal{X}|\}$. This is obviously a $|\mathcal{X}|$-ary uniquely decodable code with expected length 1. Then by the entropy bound, we have

$$H_{|\mathcal{X}|}(X) \leq 1, \quad (3.21)$$

which becomes

$$H(X) \leq \log|\mathcal{X}| \quad (3.22)$$

when the base of the logarithm is not specified. □

Motivated by the entropy bound, we now introduce the *redundancy* of a uniquely decodable code.

DEFINITION 3.7 *The redundancy R of a D-ary uniquely decodable code is the difference between the expected length of the code and the entropy of the source.*

We see from the entropy bound that the redundancy of a uniquely decodable code is always nonnegative.

3.2 PREFIX CODES

3.2.1 DEFINITION AND EXISTENCE

DEFINITION 3.8 *A code is called a prefix-free code if no codeword is a prefix of any other codeword. For brevity, a prefix-free code will be referred to as a prefix code.*

EXAMPLE 3.9 *The code C in Example 3.3 is not a prefix code because the codeword 0 is a prefix of the codeword 01, and the codeword 1 is a prefix of the codeword 10. It can easily be checked that the following code C' is a prefix code.*

x	$C'(x)$
A	0
B	10
C	110
D	1111

A D-ary tree is a graphical representation of a collection of finite sequences of D-ary symbols. In a D-ary tree, each node has at most D children. If a node has at least one child, it is called an *internal node*, otherwise it is called a *leaf*. The children of an internal node are labeled by the D symbols in the code alphabet.

A D-ary prefix code can be represented by a D-ary tree with the leaves of the tree being the codewords. Such a tree is called the *code tree* for the prefix code. Figure 3.1 shows the code tree for the prefix code C' in Example 3.9.

As we have mentioned in Section 3.1, once a sequence of codewords are concatenated, the boundaries of the codewords are no longer explicit. Prefix codes have the desirable property that the end of a codeword can be recognized instantaneously so that it is not necessary to make reference to the future codewords during the decoding process. For example, for the source sequence $BCDAC\cdots$, the code C' in Example 3.9 produces the code sequence $1011011110110\cdots$. Based on this binary sequence, the decoder can reconstruct the source sequence as follows. The first bit 1 cannot form the first codeword because 1 is not a valid codeword. The first two bits 10 must form the first codeword because it is a valid codeword and it is not the prefix of any other codeword. The same procedure is repeated to locate the end of the next

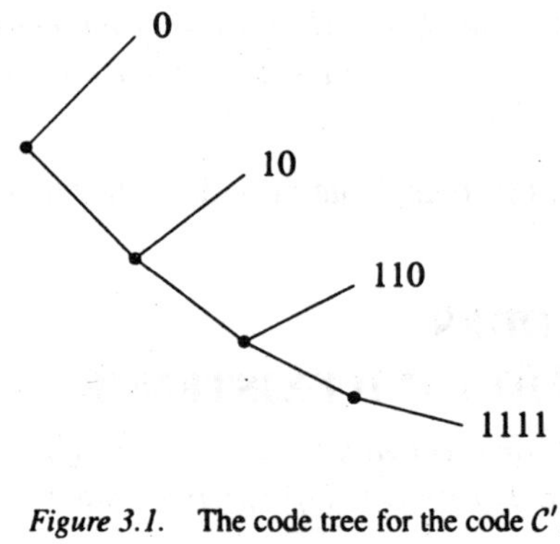

Figure 3.1. The code tree for the code C'.

codeword, and the code sequence is parsed as $10, 110, 1111, 0, 110, \cdots$. Then the source sequence $BCDAC\cdots$ can be reconstructed correctly.

Since a prefix code can always be decoded correctly, it is a uniquely decodable code. Therefore, by Theorem 3.4, the codeword lengths of a prefix code also satisfies the Kraft inequality. In the next theorem, we show that the Kraft inequality fully characterizes the existence of a prefix code.

THEOREM 3.10 *There exists a D-ary prefix code with codeword lengths l_1, $l_2, \cdots, l_m$ if and only if the Kraft inequality*

$$\sum_{k=1}^{m} D^{-l_k} \leq 1 \tag{3.23}$$

is satisfied.

Proof We only need to prove the existence of a D-ary prefix code with codeword lengths $l_1, l_2, \cdots, l_m$ if these lengths satisfy the Kraft inequality. Without loss of generality, assume that $l_1 \leq l_2 \leq \cdots \leq l_m$.

Consider all the D-ary sequences of lengths less than or equal to l_m and regard them as the nodes of the full D-ary tree of depth l_m. We will refer to a sequence of length l as a node of *order* l. Our strategy is to choose nodes as codewords in nondecreasing order of the codeword lengths. Specifically, we choose a node of order l_1 as the first codeword, then a node of order l_2 as the second codeword, so on and so forth, such that each newly chosen codeword is not prefixed by any of the previously chosen codewords. If we can successfully choose all the m codewords, then the resultant set of codewords forms a prefix code with the desired set of lengths.

There are $D^{l_1} > 1$ (since $l_1 \geq 1$) nodes of order l_1 which can be chosen as the first codeword. Thus choosing the first codeword is always possible. Assume that the first i codewords have been chosen successfully, where $1 \leq i \leq m-1$, and we want to choose a node of order l_{i+1} as the $(i+1)$st codeword such that it is not prefixed by any of the previously chosen codewords. In other words, the $(i+1)$st node to be chosen cannot be a descendant of any of the previously chosen codewords. Observe that for $1 \leq j \leq i$, the codeword with length l_j has $D^{l_{i+1}-l_j}$ descendents of order l_{i+1}. Since all the previously chosen codewords are not prefeces of each other, their descendents of order l_{i+1} do not overlap. Therefore, upon noting that the total number of nodes of order l_{i+1} is $D^{l_{i+1}}$, the number of nodes which can be chosen as the $(i+1)$st codeword is

$$D^{l_{i+1}} - D^{l_{i+1}-l_1} - \cdots - D^{l_{i+1}-l_i}. \tag{3.24}$$

If $l_1, l_2, \cdots, l_m$ satisfy the Kraft inequality, we have

$$D^{-l_1} + \cdots + D^{-l_i} + D^{-l_{i+1}} \leq 1. \tag{3.25}$$

Multiplying by $D^{l_{i+1}}$ and rearranging the terms, we have

$$D^{l_{i+1}} - D^{l_{i+1}-l_1} - \cdots - D^{l_{i+1}-l_i} \geq 1. \tag{3.26}$$

The left hand side is the number of nodes which can be chosen as the $(i+1)$st codeword as given in (3.24). Therefore, it is possible to choose the $(i+1)$st codeword. Thus we have shown the existence of a prefix code with codeword lengths $l_1, l_2, \cdots, l_m$, completing the proof. □

A probability distribution $\{p_i\}$ such that for all i, $p_i = D^{-t_i}$, where t_i is a positive integer, is called a *D-adic* distribution. When $D = 2$, $\{p_i\}$ is called a *dyadic* distribution. From Theorem 3.5 and the above theorem, we can obtain the following result as a corollary.

COROLLARY 3.11 *There exists a D-ary prefix code which achieves the entropy bound for a distribution* $\{p_i\}$ *if and only if* $\{p_i\}$ *is D-adic.*

Proof Consider a D-ary prefix code which achieves the entropy bound for a distribution $\{p_i\}$. Let l_i be the length of the codeword assigned to the probability p_i. By Theorem 3.5, for all i, $l_i = -\log_D p_i$, or $p_i = D^{-l_i}$. Thus $\{p_i\}$ is D-adic.

Conversely, suppose $\{p_i\}$ is D-adic, and let $p_i = D^{-t_i}$ for all i. Let $l_i = t_i$ for all i. Then by the Kraft inequality, there exists a prefix code with codeword lengths $\{l_i\}$, because

$$\sum_i D^{-l_i} = \sum_i D^{-t_i} = \sum_i p_i = 1. \tag{3.27}$$

Assigning the codeword with length l_i to the probability p_i for all i, we see from Theorem 3.5 that this code achieves the entropy bound. □

3.2.2 HUFFMAN CODES

As we have mentioned, the efficiency of a uniquely decodable code is measured by its expected length. Thus for a given source X, we are naturally interested in prefix codes which have the minimum expected length. Such codes, called optimal codes, can be constructed by the *Huffman procedure*, and these codes are referred to as *Huffman codes*. In general, there exists more than one optimal code for a source, and some optimal codes cannot be constructed by the Huffman procedure.

For simplicity, we first discuss binary Huffman codes. A binary prefix code for a source X with distribution $\{p_i\}$ is represented by a binary code tree, with each leaf in the code tree corresponding to a codeword. The Huffman procedure is to form a code tree such that the expected length is minimum. The procedure is described by a very simple rule:

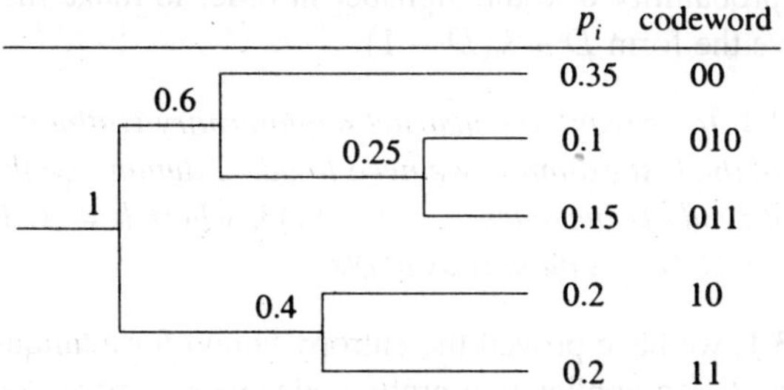

Figure 3.2. The Huffman procedure.

Keep merging the two smallest probability masses until one probability mass (i.e., 1) is left.

The merging of two probability masses corresponds to the formation of an internal node of the code tree. We now illustrate the Huffman procedure by the following example.

EXAMPLE 3.12 *Let X be the source with $\mathcal{X} = \{A, B, C, D, E\}$, and the probabilities are 0.35, 0.1, 0.15, 0.2, 0.2, respectively. The Huffman procedure is shown in Figure 3.2. In the first step, we merge probability masses 0.1 and 0.15 into a probability mass 0.25. In the second step, we merge probability masses 0.2 and 0.2 into a probability mass 0.4. In the third step, we merge probability masses 0.35 and 0.25 into a probability mass 0.6. Finally, we merge probability masses 0.6 and 0.4 into a probability mass 1. A code tree is then formed. Upon assigning 0 and 1 (in any convenient way) to each pair of branches at an internal node, we obtain the codeword assigned to each source symbol.*

In the Huffman procedure, sometimes there is more than one choice of merging the two smallest probability masses. We can take any one of these choices without affecting the optimality of the code eventually obtained.

For an alphabet of size m, it takes $m - 1$ steps to complete the Huffman procedure for constructing a binary code, because we merge two probability masses in each step. In the resulting code tree, there are m leaves and $m - 1$ internal nodes.

In the Huffman procedure for constructing a D-ary code, the smallest D probability masses are merged in each step. If the resulting code tree is formed in $k + 1$ steps, where $k \geq 0$, then there will be $k + 1$ internal nodes and $D + k(D - 1)$ leaves, where each leaf corresponds to a source symbol in the alphabet. If the alphabet size m has the form $D + k(D - 1)$, then we can apply the Huffman procedure directly. Otherwise, we need to add a few dummy symbols with probability 0 to the alphabet in order to make the total number of symbols have the form $D + k(D - 1)$.

EXAMPLE 3.13 *If we want to construct a quaternary Huffman code ($D = 4$) for the source in the last example, we need to add 2 dummy symbols so that the total number of symbols becomes $7 = 4 + (1)3$, where $k = 1$. In general, we need to add at most $D - 2$ dummy symbols.*

In Section 3.1, we have proved the entropy bound for a uniquely decodable code. This bound also applies to a prefix code since a prefix code is uniquely decodable. In particular, it applies to a Huffman code, which is a prefix code by construction. Thus the expected length of a Huffman code is at least the entropy of the source. In Example 3.12, the entropy $H(X)$ is 2.202 bits, while the expected length of the Huffman code is

$$0.35(2) + 0.1(3) + 0.15(3) + 0.2(2) + 0.2(2) = 2.25. \tag{3.28}$$

We now turn to proving the optimality of a Huffman code. For simplicity, we will only prove the optimality of a binary Huffman code. Extension of the proof to the general case is straightforward.

Without loss of generality, assume that

$$p_1 \geq p_2 \geq \cdots \geq p_m. \tag{3.29}$$

Denote the codeword assigned to p_i by c_i, and denote its length by l_i. To prove that a Huffman code is actually optimal, we make the following observations.

LEMMA 3.14 *In an optimal code, shorter codewords are assigned to larger probabilities.*

Proof Consider $1 \leq i < j \leq m$ such that $p_i > p_j$. Assume that in a code, the codewords c_i and c_j are such that $l_i > l_j$, i.e., a shorter codeword is assigned to a smaller probability. Then by exchanging c_i and c_j, the expected length of the code is changed by

$$(p_i l_j + p_j l_i) - (p_i l_i + p_j l_j) = (p_i - p_j)(l_j - l_i) < 0 \tag{3.30}$$

since $p_i > p_j$ and $l_i > l_j$. In other words, the code can be improved and therefore is not optimal. The lemma is proved. □

LEMMA 3.15 *There exists an optimal code in which the codewords assigned to the two smallest probabilities are siblings, i.e., the two codewords have the same length and they differ only in the last symbol.*

Proof The reader is encouraged to trace the steps in this proof by drawing a code tree. Consider any optimal code. From the last lemma, the codeword c_m

assigned to p_m has the longest length. Then the sibling of c_m cannot be the prefix of another codeword.

We claim that the sibling of c_m must be a codeword. To see this, assume that it is not a codeword (and it is not the prefix of another codeword). Then we can replace c_m by its parent to improve the code because the length of the codeword assigned to p_m is reduced by 1, while all the other codewords remain unchanged. This is a contradiction to the assumption that the code is optimal. Therefore, the sibling of c_m must be a codeword.

If the sibling of c_m is assigned to p_{m-1}, then the code already has the desired property, i.e., the codewords assigned to the two smallest probabilities are siblings. If not, assume that the sibling of c_m is assigned to p_i, where $i < m - 1$. Since $p_i \geq p_{m-1}$, $l_{m-1} \geq l_i = l_m$. On the other hand, by Lemma 3.14, l_{m-1} is always less than or equal to l_m, which implies that $l_{m-1} = l_m = l_i$. Then we can exchange the codewords for p_i and p_{m-1} without changing the expected length of the code (i.e., the code remains optimal) to obtain the desired code. The lemma is proved. □

Suppose c_i and c_j are siblings in a code tree. Then $l_i = l_j$. If we replace c_i and c_j by a common codeword at their parent, call it c_{ij}, then we obtain a reduced code tree, and the probability of c_{ij} is $p_i + p_j$. Accordingly, the probability set becomes a reduced probability set with p_i and p_j replaced by a probability $p_i + p_j$. Let L and L' be the expected lengths of the original code and the reduced code, respectively. Then

$$L - L' = (p_i l_i + p_j l_j) - (p_i + p_j)(l_i - 1) \quad (3.31)$$

$$= (p_i l_i + p_j l_i) - (p_i + p_j)(l_i - 1) \quad (3.32)$$

$$= p_i + p_j, \quad (3.33)$$

which implies

$$L = L' + (p_i + p_j). \quad (3.34)$$

This relation says that the difference between the expected length of the original code and the expected length of the reduced code depends only on the values of the two probabilities merged but not on the structure of the reduced code tree.

THEOREM 3.16 *The Huffman procedure produces an optimal prefix code.*

Proof Consider an optimal code in which c_m and c_{m-1} are siblings. Such an optimal code exists by Lemma 3.15. Let $\{p'_i\}$ be the reduced probability set obtained from $\{p_i\}$ by merging p_m and p_{m-1}. From (3.34), we see that L' is the expected length of an optimal code for $\{p'_i\}$ if and only if L is the expected length of an optimal code for $\{p_i\}$. Therefore, if we can find an optimal code for $\{p'_i\}$, we can use it to construct an optimal code for $\{p_i\}$. Note that by merging p_m and p_{m-1}, the size of the problem, namely the total number of probability masses, is reduced by one. To find an optimal code for $\{p'_i\}$, we again merge the two smallest probability in $\{p'_i\}$. This is repeated until the size of the problem is eventually reduced to 2, which we know that an optimal code has two codewords of length 1. In the last step of the Huffman procedure, two probability masses are merged, which corresponds to the formation of a code with two codewords of length 1. Thus the Huffman procedure indeed produces an optimal code. □

We have seen that the expected length of a Huffman code is lower bounded by the entropy of the source. On the other hand, it would be desirable to obtain an upper bound in terms of the entropy of the source. This is given in the next theorem.

THEOREM 3.17 *The expected length of a Huffman code, denoted by L_{Huff}, satisfies*

$$L_{\text{Huff}} < H_D(X) + 1. \tag{3.35}$$

This bound is the tightest among all the upper bounds on L_{Huff} which depend only on the source entropy.

Proof We will construct a prefix code with expected length less than $H(X)+1$. Then, because a Huffman code is an optimal prefix code, its expected length L_{Huff} is upper bounded by $H(X)+1$.

Consider constructing a prefix code with codeword lengths $\{l_i\}$, where

$$l_i = \lceil -\log_D p_i \rceil. \tag{3.36}$$

Then

$$-\log_D p_i \le l_i < -\log_D p_i + 1, \tag{3.37}$$

or

$$p_i \ge D^{-l_i} > D^{-1} p_i. \tag{3.38}$$

Thus

$$\sum_i D^{-l_i} \le \sum_i p_i = 1, \tag{3.39}$$

i.e., $\{l_i\}$ satisfies the Kraft inequality, which implies that it is possible to construct a prefix code with codeword lengths $\{l_i\}$.

It remains to show that L, the expected length of this code, is less than $H(X)+1$. Toward this end, consider

$$L = \sum_i p_i l_i \tag{3.40}$$

$$< \sum_i p_i(-\log_D p_i + 1) \tag{3.41}$$

$$= -\sum_i p_i \log_D p_i + \sum_i p_i \tag{3.42}$$

$$= H(X) + 1, \tag{3.43}$$

where (3.41) follows from the upper bound in (3.37). Thus we conclude that

$$L_{\text{Huff}} \le L < H(X) + 1. \tag{3.44}$$

To see that this upper bound is the tightest possible, we have to show that there exists a sequence of distributions P_k such that L_{Huff} approaches $H(X)+1$ as $k \to \infty$. This can be done by considering the sequence of D-ary distributions

$$P_k = \left\{1 - \frac{D-1}{k}, \frac{1}{k}, \cdots, \frac{1}{k}\right\}, \tag{3.45}$$

where $k \ge D$. The Huffman code for each P_k consists of D codewords of length 1. Thus L_{Huff} is equal to 1 for all k. As $k \to \infty$, $H(X) \to 0$, and hence L_{Huff} approaches $H(X) + 1$. The theorem is proved. □

The code constructed in the above proof is known as the *Shannon code*. The idea is that in order for the code to be near-optimal, we should choose l_i close to $-\log p_i$ for all i. When $\{p_i\}$ is D-adic, l_i can be chosen to be exactly $-\log p_i$ because the latter are integers. In this case, the entropy bound is tight.

From the entropy bound and the above theorem, we have

$$H(X) \le L_{\text{Huff}} < H(X) + 1. \tag{3.46}$$

Now suppose we use a Huffman code to encode $X_1, X_2, \cdots, X_n$ which are n i.i.d. copies of X. Let us denote the length of this Huffman code by L^n_{Huff}. Then (3.46) becomes

$$nH(X) \le L^n_{\text{Huff}} < nH(X) + 1. \tag{3.47}$$

Dividing by n, we obtain

$$H(X) \le \frac{1}{n} L^n_{\text{Huff}} < H(X) + \frac{1}{n}. \tag{3.48}$$

As $n \to \infty$, the upper bound approaches the lower bound. Therefore, $n^{-1}L^n_{\text{Huff}}$, the coding rate of the code, namely the average number of code symbols needed to encode a source symbol, approaches $H(X)$ as $n \to \infty$. But of course, as n becomes large, constructing a Huffman code becomes very complicated. Nevertheless, this result indicates that entropy is a fundamental measure of information.

3.3 REDUNDANCY OF PREFIX CODES

The entropy bound for a uniquely decodable code has been proved in Section 3.1. In this section, we present an alternative proof specifically for prefix codes which offers much insight into the redundancy of such codes.

Let X be a source random variable with probability distribution

$$\{p_1, p_2, \cdots, p_m\}, \tag{3.49}$$

where $m \geq 2$. A D-ary prefix code for X can be represented by a D-ary code tree with m leaves, where each leaf corresponds to a codeword. We denote the leaf corresponding to p_i by c_i and the order of c_i by l_i, and assume that the alphabet is

$$\{0, 1, \cdots, D-1\}. \tag{3.50}$$

Let $\mathcal{I}$ be the index set of all the internal nodes (including the root) in the code tree.

Instead of matching codewords by brute force, we can use the code tree of a prefix code for more efficient decoding. To decode a codeword, we trace the path specified by the codeword from the root of the code tree until it terminates at the leaf corresponding to that codeword. Let q_k be the probability of reaching an internal node $k \in \mathcal{I}$ during the decoding process. The probability q_k is called the *reaching probability* of internal node k. Evidently, q_k is equal to the sum of the probabilities of all the leaves descending from node k.

Let $\tilde{p}_{k,j}$ be the probability that the jth branch of node k is taken during the decoding process. The probabilities $\tilde{p}_{k,j}, 0 \leq j \leq D-1$, are called the *branching probabilities* of node k, and

$$q_k = \sum_j \tilde{p}_{k,j}. \tag{3.51}$$

Once node k is reached, the *conditional branching distribution* is

$$\left\{ \frac{\tilde{p}_{k,0}}{q_k}, \frac{\tilde{p}_{k,1}}{q_k}, \cdots, \frac{\tilde{p}_{k,D-1}}{q_k} \right\}. \tag{3.52}$$

Then define the *conditional entropy* of node k by

$$h_k = H_D\left(\left\{ \frac{\tilde{p}_{k,0}}{q_k}, \frac{\tilde{p}_{k,1}}{q_k}, \cdots, \frac{\tilde{p}_{k,D-1}}{q_k} \right\} \right), \tag{3.53}$$

where with a slight abuse of notation, we have used $H_D(\cdot)$ to denote the entropy in the base D of the conditional branching distribution in the parenthesis. By Theorem 2.43, $h_k \leq 1$. The following lemma relates the entropy of X with the structure of the code tree.

LEMMA 3.18 $H_D(X) = \sum_{k \in \mathcal{I}} q_k h_k$.

Proof We prove the lemma by induction on the number of internal nodes of the code tree. If there is only one internal node, it must be the root of the tree. Then the lemma is trivially true upon observing that the reaching probability of the root is equal to 1.

Assume the lemma is true for all code trees with n internal nodes. Now consider a code tree with $n+1$ internal nodes. Let k be an internal node such that k is the parent of a leaf c with maximum order. Each sibling of c may or may not be a leaf. If it is not a leaf, then it cannot be the ascendent of another leaf because we assume that c is a leaf with maximum order. Now consider revealing the outcome of X in two steps. In the first step, if the outcome of X is not a leaf descending from node k, we identify the outcome exactly, otherwise we identify the outcome to be a child of node k. We call this random variable V. If we do not identify the outcome exactly in the first step, which happens with probability q_k, we further identify in the second step which of the children (child) of node k the outcome is (there is only one child of node k which can be the outcome if all the siblings of c are not leaves). We call this random variable W. If the second step is not necessary, we assume that W takes a constant value with probability 1. Then $X = (V, W)$.

The outcome of V can be represented by a code tree with n internal nodes which is obtained by pruning the original code tree at node k. Then by the induction hypothesis,

$$H(V) = \sum_{k' \in \mathcal{I} \backslash \{k\}} q_{k'} h_{k'}. \tag{3.54}$$

By the chain rule for entropy, we have

$$\begin{aligned} H(X) &= H(V) + H(W|V) && (3.55) \\ &= \sum_{k' \in \mathcal{I} \backslash \{k\}} q_{k'} h_{k'} + (1 - q_k) \cdot 0 + q_k h_k && (3.56) \\ &= \sum_{k' \in \mathcal{I}} q_{k'} h_{k'}. && (3.57) \end{aligned}$$

The lemma is proved. □

The next lemma expresses the expected length L of a prefix code in terms of the reaching probabilities of the internal nodes of the code tree.

LEMMA 3.19 $L = \sum_{k \in \mathcal{I}} q_k$.

Proof Define

$$a_{ki} = \begin{cases} 1 & \text{if leaf } c_i \text{ is a descendent of internal node k} \\ 0 & \text{otherwise.} \end{cases} \tag{3.58}$$

Then

$$l_i = \sum_{k \in \mathcal{I}} a_{ki}, \tag{3.59}$$

because there are exactly l_i internal nodes of which c_i is a descendent if the order of c_i is l_i. On the other hand,

$$q_k = \sum_i a_{ki} p_i. \tag{3.60}$$

Then

$$\begin{aligned} L &= \sum_i p_i l_i && (3.61) \\ &= \sum_i p_i \sum_{k \in \mathcal{I}} a_{ki} && (3.62) \\ &= \sum_{k \in \mathcal{I}} \sum_i p_i a_{ki} && (3.63) \\ &= \sum_{k \in \mathcal{I}} q_k, && (3.64) \end{aligned}$$

proving the lemma. □

Define the *local redundancy* of an internal node k by

$$r_k = q_k(1 - h_k). \tag{3.65}$$

This quantity is local to node k in the sense that it depends only on the branching probabilities of node k, and it vanishes if and only if $\tilde{p}_{k,j} = q_k/D$ for all j, i.e., if and only if the node is *balanced.* Note that $r_k \geq 0$ because $h_k \leq 1$.

The next theorem says that the redundancy R of a prefix code is equal to the sum of the local redundancies of all the internal nodes of the code tree.

THEOREM 3.20 (LOCAL REDUNDANCY THEOREM) *Let L be the expected length of a D-ary prefix code for a source random variable X, and R be the redundancy of the code. Then*

$$R = \sum_{k \in \mathcal{I}} r_k. \tag{3.66}$$

Proof By Lemmas 3.18 and 3.19, we have

$$\begin{aligned} R &= L - H_D(X) && (3.67) \\ &= \sum_{k \in \mathcal{I}} q_k - \sum_k q_k h_k && (3.68) \end{aligned}$$

$$= \sum_{k \in \mathcal{I}} q_k(1 - h_k) \tag{3.69}$$

$$= \sum_{k \in \mathcal{I}} r_k. \tag{3.70}$$

The theorem is proved. □

We now present an slightly different version of the entropy bound.

COROLLARY 3.21 (ENTROPY BOUND) *Let R be the redundancy of a prefix code. Then $R \geq 0$ with equality if and only if all the internal nodes in the code tree are balanced.*

Proof Since $r_k \geq 0$ for all k, it is evident from the local redundancy theorem that $R \geq 0$. Moreover $R = 0$ if and only if $r_k = 0$ for all k, which means that all the internal nodes in the code tree are balanced. □

Remark Before the entropy bound was stated in Theorem 3.5, we gave the intuitive explanation that the entropy bound results from the fact that a D-ary symbol can carry at most one D-it of information. Therefore, when the entropy bound is tight, each code symbol has to carry exactly one D-it of information. Now consider revealing a random codeword one symbol after another. The above corollary states that in order for the entropy bound to be tight, all the internal nodes in the code tree must be balanced. That is, as long as the codeword is not completed, the next code symbol to be revealed always carries one D-it of information because it distributes uniformly on the alphabet. This is consistent with the intuitive explanation we gave for the entropy bound.

EXAMPLE 3.22 *The local redundancy theorem allows us to lower bound the redundancy of a prefix code based on partial knowledge on the structure of the code tree. More specifically,*

$$R \geq \sum_{k \in \mathcal{I}'} r_k \tag{3.71}$$

for any subset $\mathcal{I}'$ of $\mathcal{I}$.

Let p_{m-1}, p_m be the two smallest probabilities in the source distribution. In constructing a binary Huffman code, p_{m-1} and p_m are merged. Then the redundancy of a Huffman code is lower bounded by

$$(p_{m-1} + p_m) \left[1 - H_2\left(\left\{\frac{p_{m-1}}{p_{m-1} + p_m}, \frac{p_m}{p_{m-1} + p_m}\right\}\right)\right], \tag{3.72}$$

the local redundancy of the parent of the two leaves corresponding to p_{m-1} and p_m. See Yeung [229] for progressive lower and upper bounds on the redundancy of a Huffman code.

PROBLEMS

1. Construct a binary Huffman code for the distribution $\{0.25, 0.05, 0.1, 0.13, 0.2, 0.12, 0.08, 0.07\}$.

2. Construct a ternary Huffman code for the source distribution in Problem 1.

3. Show that a Huffman code is an optimal uniquely decodable code for a given source distribution.

4. Construct an optimal binary prefix code for the source distribution in Problem 1 such that all the codewords have even lengths.

5. Prove directly that the codeword lengths of a prefix code satisfy the Kraft inequality without using Theorem 3.4.

6. Prove that if $p_1 > 0.4$, then the shortest codeword of a binary Huffman code has length equal to 1. Then prove that the redundancy of such a Huffman code is lower bounded by $1 - h_b(p_1)$. (Johnsen [110].)

7. *Suffix codes* A code is a suffix code if no codeword is a suffix of any other codeword. Show that a suffix code is uniquely decodable.

8. *Fix-free codes* A code is a fix-free code if it is both a prefix code and a suffix code. Let $l_1, l_2, \cdots, l_m$ be m positive integers. Prove that if

$$\sum_{k=1}^{m} 2^{-l_k} \leq \frac{1}{2},$$

then there exists a binary fix-free code with codeword lengths $l_1, l_2, \cdots, l_m$. (Ahlswede *et al.* [5].)

9. *Random coding for prefix codes* Construct a binary prefix code with codeword lengths $l_1 \leq l_2 \leq \cdots \leq l_m$ as follows. For each $1 \leq k \leq m$, the codeword with length l_k is chosen independently from the set of all 2^{l_k} possible binary strings with length l_k according the uniform distribution. Let $P_m(good)$ be the probability that the code so constructed is a prefix code.

   a) Prove that $P_2(good) = (1 - 2^{-l_1})^+$, where

$$(x)^+ = \begin{cases} x & \text{if } x \geq 0 \\ 0 & \text{if } x < 0. \end{cases}$$

   b) Prove by induction on m that

$$P_m(good) = \prod_{k=1}^{m} \left(1 - \sum_{j=1}^{k-1} s^{-l_j}\right)^+.$$

c) Observe that there exists a prefix code with codeword lengths $l_1, l_2, \cdots, l_m$ if and only if $P_m(good) > 0$. Show that $P_m(good) > 0$ is equivalent to the Kraft inequality.

By using this random coding method, one can derive the Kraft inequality without knowing the inequality ahead of time. (Ye and Yeung [225].)

10. Let X be a source random variable. Suppose a certain probability mass p_k in the distribution of X is given. Let

$$l_j = \begin{cases} \lceil -\log p_j \rceil & \text{if } j = k \\ \lceil -\log(p_j + x_j) \rceil & \text{if } j \neq k, \end{cases}$$

where

$$x_j = p_j \left(\frac{p_k - 2^{-\lceil -\log p_k \rceil}}{1 - p_k} \right)$$

for all $j \neq k$.

a) Show that $1 \leq l_j \leq \lceil -\log p_j \rceil$ for all j.

b) Show that $\{l_j\}$ satisfies the Kraft inequality.

c) Obtain an upper bound on L_{Huff} in terms of $H(X)$ and p_k which is tighter than $H(X) + 1$. This shows that when partial knowledge about the source distribution in addition to the source entropy is available, tighter upper bounds on L_{Huff} can be obtained.

(Ye and Yeung [226].)

HISTORICAL NOTES

The foundation for the material in this chapter can be found in Shannon's original paper [184]. The Kraft inequality for uniquely decodable codes was first proved by McMillan [151]. The proof given here is due to Karush [113]. The Huffman coding procedure was devised and proved to be optimal by Huffman [103]. The same procedure was devised independently by Zimmerman [240]. Linder *et al.* [134] have proved the existence of an optimal prefix code for an infinite source alphabet which can be constructed from Huffman codes for truncations of the source distribution. The local redundancy theorem is due to Yeung [229]. A comprehensive survey of code trees for lossless data compression can be found in Abrahams [1].

Chapter 4

WEAK TYPICALITY

In the last chapter, we have discussed the significance of entropy in the context of zero-error data compression. In this chapter and the next, we explore entropy in terms of the asymptotic behavior of i.i.d. sequences. Specifically, two versions of the *asymptotic equipartition property* (AEP), namely the weak AEP and the strong AEP, are discussed. The role of these AEP's in information theory is analogous to the role of the weak law of large numbers in probability theory. In this chapter, the weak AEP and its relation with the source coding theorem are discussed. All the logarithms are in the base 2 unless otherwise specified.

4.1 THE WEAK AEP

We consider an information source $\{X_k, k \geq 1\}$ where X_k are i.i.d. with distribution $p(x)$. We use X to denote the generic random variable and $H(X)$ to denote the common entropy for all X_k, where $H(X) < \infty$. Let $\mathbf{X} = (X_1, X_2, \cdots, X_n)$. Since X_k are i.i.d.,

$$p(\mathbf{X}) = p(X_1)p(X_2)\cdots p(X_n). \tag{4.1}$$

Note that $p(\mathbf{X})$ is a random variable because it is a function of the random variables $X_1, X_2, \cdots, X_n$. We now prove an asymptotic property of $p(\mathbf{X})$ called the *weak asymptotic equipartition property* (weak AEP).

THEOREM 4.1 (WEAK AEP I)

$$-\frac{1}{n}\log p(\mathbf{X}) \to H(X) \tag{4.2}$$

in probability as $n \to \infty$, i.e., for any $\epsilon > 0$, for n sufficiently large,

$$\Pr\left\{\left|-\frac{1}{n}\log p(\mathbf{X}) - H(X)\right| \leq \epsilon\right\} > 1 - \epsilon. \tag{4.3}$$

Proof Since $X_1, X_2, \cdots, X_n$ are i.i.d., by (4.1),

$$-\frac{1}{n}\log p(\mathbf{X}) = -\frac{1}{n}\sum_{k=1}^{n}\log p(X_k). \tag{4.4}$$

The random variables $\log p(X_k)$ are also i.i.d. Then by the weak law of large numbers, the right hand side of (4.4) tends to

$$-E\log p(X) = H(X), \tag{4.5}$$

in probability, proving the theorem. □

The weak AEP is nothing more than a straightforward application of the weak law of large numbers. However, as we will see shortly, this property has significant implications.

DEFINITION 4.2 *The weakly typical set $W^n_{[X]\epsilon}$ with respect to $p(x)$ is the set of sequences $\mathbf{x} = (x_1, x_2, \cdots, x_n) \in \mathcal{X}^n$ such that*

$$\left|-\frac{1}{n}\log p(\mathbf{x}) - H(X)\right| \le \epsilon, \tag{4.6}$$

or equivalently,

$$H(X) - \epsilon \le -\frac{1}{n}\log p(\mathbf{x}) \le H(X) + \epsilon, \tag{4.7}$$

where ϵ is an arbitrarily small positive real number. The sequences in $W^n_{[X]\epsilon}$ are called weakly ϵ-typical sequences.

The quantity

$$-\frac{1}{n}\log p(\mathbf{x}) = -\frac{1}{n}\sum_{k=1}^{n}\log p(x_k) \tag{4.8}$$

is called the *empirical entropy* of the sequence $\mathbf{x}$. The empirical entropy of a weakly typical sequence is close to the true entropy $H(X)$. The important properties of the set $W^n_{[X]\epsilon}$ are summarized in the next theorem which we will see is equivalent to the weak AEP.

THEOREM 4.3 (WEAK AEP II) *The following hold for any $\epsilon > 0$:*

1) If $\mathbf{x} \in W^n_{[X]\epsilon}$, then

$$2^{-n(H(X)+\epsilon)} \le p(\mathbf{x}) \le 2^{-n(H(X)-\epsilon)}. \tag{4.9}$$

2) For n sufficiently large,

$$\Pr\{\mathbf{X} \in W^n_{[X]\epsilon}\} > 1 - \epsilon. \tag{4.10}$$

3) For n sufficiently large,

$$(1-\epsilon)2^{n(H(X)-\epsilon)} \leq |W^n_{[X]\epsilon}| \leq 2^{n(H(X)+\epsilon)}. \tag{4.11}$$

Proof Property 1 follows immediately from the definition of $W^n_{[X]\epsilon}$ in (4.7). Property 2 is equivalent to Theorem 4.1. To prove Property 3, we use the lower bound in (4.9) and consider

$$|W^n_{[X]\epsilon}|2^{-n(H(X)+\epsilon)} \leq \Pr\{W^n_{[X]\epsilon}\} \leq 1, \tag{4.12}$$

which implies

$$|W^n_{[X]\epsilon}| \leq 2^{n(H(X)+\epsilon)}. \tag{4.13}$$

Note that this upper bound holds for any $n \geq 1$. On the other hand, using the upper bound in (4.9) and Theorem 4.1, for n sufficiently large, we have

$$1 - \epsilon \leq \Pr\{W^n_{[X]\epsilon}\} \leq |W^n_{[X]\epsilon}|2^{-n(H(X)-\epsilon)}. \tag{4.14}$$

Then

$$|W^n_{[X]\epsilon}| \geq (1-\epsilon)2^{n(H(X)-\epsilon)}. \tag{4.15}$$

Combining (4.13) and (4.15) gives Property 3. The theorem is proved. □

Remark Theorem 4.3 is a consequence of Theorem 4.1. However, Property 2 in Theorem 4.3 is equivalent to Theorem 4.1. Therefore, Theorem 4.1 and Theorem 4.3 are equivalent, and they will both be referred to as the weak AEP.

The weak AEP has the following interpretation. Suppose $\mathbf{X} = (X_1, X_2, \cdots, X_n)$ is drawn i.i.d. according to $p(x)$, where n is large. After the sequence is drawn, we ask what the probability of occurrence of the sequence is. The weak AEP says that the probability of occurrence of the sequence drawn is close to $2^{-nH(X)}$ with very high probability. Such a sequence is called a weakly typical sequence. Moreover, the total number of weakly typical sequences is approximately equal to $2^{nH(X)}$. The weak AEP, however, does not say that most of the sequences in $\mathcal{X}^n$ are weakly typical. In fact, the number of weakly typical sequences is in general insignificant compared with the total number of sequences, because

$$\frac{|W^n_{[X]\delta}|}{|\mathcal{X}|^n} \approx \frac{2^{nH(X)}}{2^{n\log|\mathcal{X}|}} = 2^{-n(\log|\mathcal{X}|-H(X))} \to 0 \tag{4.16}$$

as $n \to \infty$ as long as $H(X)$ is strictly less than $\log|\mathcal{X}|$. The idea is that, although the size of the weakly typical set may be insignificant compared with the size of the set of all sequences, the former has almost all the probability.

When n is large, one can almost think of the sequence $\mathbf{X}$ as being obtained by choosing a sequence from the weakly typical set according to the uniform distribution. Very often, we concentrate on the properties of typical sequences because any property which is proved to be true for typical sequences will then be true with high probability. This in turn determines the average behavior of a large sample.

Remark The most likely sequence is in general not weakly typical although the probability of the weakly typical set is close to 1 when n is large. For example, for X_k i.i.d. with $p(0) = 0.1$ and $p(1) = 0.9$, $(1, 1, \cdots, 1)$ is the most likely sequence, but it is not weakly typical because its empirical entropy is not close to the true entropy. The idea is that as $n \to \infty$, the probability of every sequence, including that of the most likely sequence, tends to 0. Therefore, it is not necessary for a weakly typical set to include the most likely sequence in order to possess a probability close to 1.

4.2 THE SOURCE CODING THEOREM

To encode a random sequence $\mathbf{X} = (X_1, X_2, \cdots, X_n)$ drawn i.i.d. according to $p(x)$ by a *block code*, we construct a one-to-one mapping from a subset $\mathcal{A}$ of $\mathcal{X}^n$ to an index set

$$\mathcal{I} = \{1, 2, \cdots, M\}, \tag{4.17}$$

where $|\mathcal{A}| = M \leq |\mathcal{X}|^n$. We do not have to assume that $|\mathcal{X}|$ is finite. The indices in $\mathcal{I}$ are called *codewords*, and the integer n is called the *block length* of the code. If a sequence $\mathbf{x} \in \mathcal{A}$ occurs, the encoder outputs the corresponding codeword which is specified by approximately $\log M$ bits. If a sequence $\mathbf{x} \notin \mathcal{A}$ occurs, the encoder outputs the constant codeword 1. In either case, the codeword output by the encoder is decoded to the sequence in $\mathcal{A}$ corresponding to that codeword by the decoder. If a sequence $\mathbf{x} \in \mathcal{A}$ occurs, then $\mathbf{x}$ is decoded correctly by the decoder. If a sequence $\mathbf{x} \notin \mathcal{A}$ occurs, then $\mathbf{x}$ is not decoded correctly by the decoder. For such a code, its performance is measured by the coding rate defined as $n^{-1} \log M$ (in bits per source symbol), and the probability of error is given by

$$P_e = \Pr\{\mathbf{X} \notin \mathcal{A}\}. \tag{4.18}$$

If the code is not allowed to make any error, i.e., $P_e = 0$, it is clear that M must be taken to be $|\mathcal{X}|^n$, or $\mathcal{A} = \mathcal{X}^n$. In that case, the coding rate is equal to $\log|\mathcal{X}|$. However, if we allow P_e to be any small quantity, Shannon [184] showed that there exists a block code whose coding rate is arbitrarily close to $H(X)$ when n is sufficiently large. This is the direct part of Shannon's *source coding theorem*, and in this sense the source sequence $\mathbf{X}$ is said to be reconstructed *almost perfectly*.

We now prove the direct part of the source coding theorem by constructing a desired code. First, we fix $\epsilon > 0$ and take

$$\mathcal{A} = W^n_{[X]\epsilon} \tag{4.19}$$

and

$$M = |\mathcal{A}|. \tag{4.20}$$

For sufficiently large n, by the weak AEP,

$$(1-\epsilon)2^{n(H(X)-\epsilon)} \leq M = |\mathcal{A}| = |W^n_{[X]\epsilon}| \leq 2^{n(H(X)+\epsilon)}. \tag{4.21}$$

Therefore, the coding rate $n^{-1}\log M$ satisfies

$$\frac{1}{n}\log(1-\epsilon) + H(X) - \epsilon \leq \frac{1}{n}\log M \leq H(X) + \epsilon. \tag{4.22}$$

Also by the weak AEP,

$$P_e = \Pr\{\mathbf{X} \notin \mathcal{A}\} = \Pr\{\mathbf{X} \notin W^n_{[X]\epsilon}\} < \epsilon. \tag{4.23}$$

Letting $\epsilon \to 0$, the coding rate tends to $H(X)$, while P_e tends to 0. This proves the direct part of the source coding theorem.

The converse part of the source coding theorem says that if we use a block code with block length n and coding rate less than $H(X) - \zeta$, where $\zeta > 0$ does not change with n, then $P_e \to 1$ as $n \to \infty$. To prove this, consider any code with block length n and coding rate less than $H(X) - \zeta$, so that M, the total number of codewords, is at most $2^{n(H(X)-\zeta)}$. We can use some of these codewords for the typical sequences $\mathbf{x} \in W^n_{[X]\epsilon}$, and some for the non-typical sequences $\mathbf{x} \notin W^n_{[X]\epsilon}$. The total probability of the typical sequences covered by the code, by the weak AEP, is upper bounded by

$$2^{n(H(X)-\zeta)}2^{-n(H(X)-\epsilon)} = 2^{-n(\zeta-\epsilon)}. \tag{4.24}$$

Therefore, the total probability covered by the code is upper bounded by

$$2^{-n(\zeta-\epsilon)} + \Pr\{\mathbf{X} \notin W^n_{[X]\epsilon}\} < 2^{-n(\zeta-\epsilon)} + \epsilon \tag{4.25}$$

for n sufficiently large, again by the weak AEP. This probability is equal to $1 - P_e$ because P_e is the probability that the source sequence $\mathbf{X}$ is not covered by the code. Thus

$$1 - P_e < 2^{-n(\zeta-\epsilon)} + \epsilon, \tag{4.26}$$

or

$$P_e > 1 - (2^{-n(\zeta-\epsilon)} + \epsilon). \tag{4.27}$$

This inequality holds when n is sufficiently large for any $\epsilon > 0$, in particular for $\epsilon < \zeta$. Then for any $\epsilon < \zeta$, $P_e > 1 - 2\epsilon$ when n is sufficiently large. Hence, $P_e \to 1$ as $n \to \infty$ and then $\epsilon \to 0$. This proves the converse part of the source coding theorem.

4.3 EFFICIENT SOURCE CODING

THEOREM 4.4 *Let $\mathbf{Y} = (Y_1, Y_2, \cdots, Y_m)$ be a random binary sequence of length m. Then $H(\mathbf{Y}) \leq m$ with equality if and only if Y_i are drawn i.i.d. according to the uniform distribution on $\{0, 1\}$.*

Proof By the independence bound for entropy,

$$H(\mathbf{Y}) \leq \sum_{i=1}^{m} H(Y_i) \tag{4.28}$$

with equality if and only if Y_i are mutually independent. By Theorem 2.43,

$$H(Y_i) \leq \log 2 = 1 \tag{4.29}$$

with equality if and only if Y_i distributes uniformly on $\{0, 1\}$. Combining (4.28) and (4.29), we have

$$H(\mathbf{Y}) \leq \sum_{i=1}^{m} H(Y_i) \leq m, \tag{4.30}$$

where this upper bound is tight if and only if Y_i are mutually independent and each of them distributes uniformly on $\{0, 1\}$. The theorem is proved. □

Let $\mathbf{Y} = (Y_1, Y_2, \cdots, Y_n)$ be a sequence of length n such that Y_i are drawn i.i.d. according to the uniform distribution on $\{0, 1\}$, and let Y denote the generic random variable. Then $H(Y) = 1$. According to the source coding theorem, for almost perfect reconstruction of $\mathbf{Y}$, the coding rate of the source code must be at least 1. It turns out that in this case it is possible to use a source code with coding rate exactly equal to 1 while the source sequence $\mathbf{Y}$ can be reconstructed with zero error. This can be done by simply encoding all the 2^n possible binary sequences of length n, i.e., by taking $M = 2^n$. Then the coding rate is given by

$$n^{-1} \log M = n^{-1} \log 2^n = 1. \tag{4.31}$$

Since each symbol in $\mathbf{Y}$ is a bit and the rate of the best possible code describing $\mathbf{Y}$ is 1 bit per symbol, $Y_1, Y_2, \cdots, Y_n$ are called *raw bits*, with the connotation that they are incompressible.

It turns out that the whole idea of efficient source coding by a block code is to describe the information source by a binary sequence consisting of "almost raw" bits. Consider a sequence of block codes which encode $\mathbf{X} = (X_1, X_2, \cdots, X_n)$ into $\mathbf{Y} = (Y_1, Y_2, \cdots, Y_m)$, where X_k are i.i.d. with generic random variable X, $\mathbf{Y}$ is a binary sequence with length

$$m \approx nH(X), \tag{4.32}$$

and $n \to \infty$. For simplicity, we assume that the common alphabet $\mathcal{X}$ is finite. Let $\hat{\mathbf{X}} \in \mathcal{X}^n$ be the reconstruction of $\mathbf{X}$ by the decoder and P_e be the probability of error, i.e.,

$$P_e = \Pr\{\mathbf{X} \neq \hat{\mathbf{X}}\}. \tag{4.33}$$

Further assume $P_e \to 0$ as $n \to \infty$. We will show that $\mathbf{Y}$ consists of almost raw bits.

By Fano's inequality,

$$H(\mathbf{X}|\hat{\mathbf{X}}) \leq 1 + P_e \log |\mathcal{X}|^n = 1 + nP_e \log |\mathcal{X}|. \tag{4.34}$$

Since $\hat{\mathbf{X}}$ is a function of $\mathbf{Y}$,

$$H(\mathbf{Y}) = H(\mathbf{Y}, \hat{\mathbf{X}}) \geq H(\hat{\mathbf{X}}). \tag{4.35}$$

It follows that

$$\begin{aligned} H(\mathbf{Y}) &\geq H(\hat{\mathbf{X}}) && (4.36)\\ &\geq I(\mathbf{X}; \hat{\mathbf{X}}) && (4.37)\\ &= H(\mathbf{X}) - H(\mathbf{X}|\hat{\mathbf{X}}) && (4.38)\\ &\geq nH(X) - (1 + nP_e \log |\mathcal{X}|) && (4.39)\\ &= n(H(X) - P_e \log |\mathcal{X}|) - 1. && (4.40) \end{aligned}$$

On the other hand, by Theorem 4.4,

$$H(\mathbf{Y}) \leq m. \tag{4.41}$$

Combining (4.40) and (4.41), we have

$$n(H(X) - P_e \log |\mathcal{X}|) - 1 \leq H(\mathbf{Y}) \leq m. \tag{4.42}$$

Since $P_e \to 0$ as $n \to \infty$, the above lower bound on $H(\mathbf{Y})$ is approximately equal to

$$nH(X) \approx m \tag{4.43}$$

when n is large (cf. (4.32)). Therefore,

$$H(\mathbf{Y}) \approx m. \tag{4.44}$$

In light of Theorem 4.4, $\mathbf{Y}$ almost attains the maximum possible entropy. In this sense, we say that $\mathbf{Y}$ consists of almost raw bits.

4.4 THE SHANNON-MCMILLAN-BREIMAN THEOREM

For an i.i.d. information source $\{X_k\}$ with generic random variable X and generic distribution $p(x)$, the weak AEP states that

$$-\frac{1}{n}\log p(\mathbf{X}) \to H(X) \tag{4.45}$$

in probability as $n \to \infty$, where $\mathbf{X} = (X_1, X_2, \cdots, X_n)$. Here $H(X)$ is the entropy of the generic random variables X as well as the entropy rate of the source $\{X_k\}$.

In Section 2.9, we showed that the entropy rate H of a source $\{X_k\}$ exists if the source is stationary. The *Shannon-McMillan-Breiman theorem* states that if $\{X_k\}$ is also *ergodic*, then

$$\Pr\left\{-\lim_{n\to\infty}\frac{1}{n}\log\Pr\{\mathbf{X}\} = H\right\} = 1. \tag{4.46}$$

This means that if $\{X_k\}$ is stationary and ergodic, then $-\frac{1}{n}\log\Pr\{\mathbf{X}\}$ not only almost always converges, but it also almost always converges to H. For this reason, the Shannon-McMillan-Breiman theorem is also referred to as the weak AEP for ergodic stationary sources.

The formal definition of an ergodic source and the statement of the Shannon-McMillan-Breiman theorem require the use of measure theory which is beyond the scope of this book. We point out that the event in (4.46) involves an infinite collection of random variables which cannot be described by a joint distribution except in very special cases. Without measure theory, the probability of this event in general cannot be properly defined. However, this does not prevent us from developing some appreciation of the Shannon-McMillan-Breiman theorem.

Let $\mathcal{X}$ be the common alphabet for a stationary source $\{X_k\}$. Roughly speaking, a stationary source $\{X_k\}$ is ergodic if the time average exhibited by a single realization of the source is equal to the ensemble average with probability 1. More specifically, for any $k_1, k_2, \cdots, k_m$,

$$\Pr\left\{\lim_{n\to\infty}\frac{1}{n}\sum_{l=0}^{n-1} f(X_{k_1+l}, X_{k_2+l}, \cdots, X_{k_m+l}) = Ef(X_{k_1}, X_{k_2}, \cdots, X_{k_m})\right\} = 1, \tag{4.47}$$

where f is a function defined on $\mathcal{X}^m$ which satisfies suitable conditions. For the special case that $\{X_k\}$ satisfies

$$\Pr\left\{\lim_{n\to\infty}\frac{1}{n}\sum_{l=1}^{n} X_l = EX_k\right\} = 1, \tag{4.48}$$

we say that $\{X_k\}$ is *mean ergodic*.

EXAMPLE 4.5 *The i.i.d. source $\{X_k\}$ is mean ergodic under suitable conditions because the strong law of the large numbers states that (4.48) is satisfied.*

EXAMPLE 4.6 *Consider the source $\{X_k\}$ defined as follows. Let Z be a binary random variable uniformly distributed on $\{0, 1\}$. For all k, let $X_k = Z$. Then*

$$\Pr\left\{\lim_{n\to\infty} \frac{1}{n}\sum_{l=1}^{n} X_l = 0\right\} = \frac{1}{2} \tag{4.49}$$

and

$$\Pr\left\{\lim_{n\to\infty} \frac{1}{n}\sum_{l=1}^{n} X_l = 1\right\} = \frac{1}{2}. \tag{4.50}$$

Since $EX_k = \frac{1}{2}$,

$$\Pr\left\{\lim_{n\to\infty} \frac{1}{n}\sum_{l=1}^{n} X_l = EX_k\right\} = 0. \tag{4.51}$$

Therefore, $\{X_k\}$ is not mean ergodic and hence not ergodic.

If an information source $\{X_k\}$ is stationary and ergodic, by the Shannon-McMillan-Breiman theorem,

$$-\frac{1}{n}\log \Pr\{\mathbf{X}\} \approx H \tag{4.52}$$

when n is large. That is, with probability close to 1, the probability of the sequence $\mathbf{X}$ which occurs is approximately equal to 2^{-nH}. Then by means of arguments similar to the proof of Theorem 4.3, we see that there exist approximately 2^{nH} sequences in $\mathcal{X}^n$ whose probabilities are approximately equal to 2^{-nH}, and the total probability of these sequences is almost 1. Therefore, by encoding these sequences with approximately nH bits, the source sequence $\mathbf{X}$ can be recovered with an arbitrarily small probability of error when the block length n is sufficiently large. This is a generalization of the direct part of the source coding theorem which gives a physical meaning to the entropy rate of an ergodic stationary sources. We remark that if a source is stationary but not ergodic, although the entropy rate always exists, it may not carry any physical meaning.

As an example, by regarding printed English as a stationary ergodic process, Shannon [186] estimated by a guessing game that its entropy rate is about 1.3 bits per letter. Cover and King [50] described a gambling estimate of the entropy rate of printed English which gives 1.34 bits per letter. These results show that it is not possible to describe printed English accurately by using less than about 1.3 bits per letter.

PROBLEMS

1. Show that for any $\epsilon > 0$, $W^n_{[X]\epsilon}$ is nonempty for sufficiently large n.

2. *The source coding theorem with a general block code* In proving the converse of the source coding theorem, we assume that each codeword in $\mathcal{I}$ corresponds to a unique sequence in $\mathcal{X}^n$. More generally, a block code with block length n is defined by an encoding function $f : \mathcal{X}^n \to \mathcal{I}$ and a decoding function $g : \mathcal{I} \to \mathcal{X}^n$. Prove that $P_e \to 1$ as $n \to \infty$ even if we are allowed to use a general block code.

3. Following Problem 2, we further assume that we can use a block code with probabilistic encoding and decoding. For such a code, encoding is defined by a transition matrix F from $\mathcal{X}^n$ to $\mathcal{I}$ and decoding is defined by a transition matrix G from $\mathcal{I}$ to $\mathcal{X}^n$. Prove that $P_e \to 1$ as $n \to \infty$ even if we are allowed to use such a code.

4. In the discussion in Section 4.3, we made the assumption that the common alphabet $\mathcal{X}$ is finite. Can you draw the same conclusion when $\mathcal{X}$ is infinite but $H(X)$ is finite? Hint: use Problem 2.

5. Let p and q be two probability distributions on the same alphabet $\mathcal{X}$ such that $H(p) \neq H(q)$. Show that there exists an $\epsilon > 0$ such that
$$p^n\left(\left\{x^n \in \mathcal{X}^n : \left|-\frac{1}{n}\log p^n(x^n) - H(q)\right| < \epsilon\right\}\right) \to 0$$
as $n \to \infty$. Give an example that $p \neq q$ but the above convergence does not hold.

6. Let p and q be two probability distributions on the same alphabet $\mathcal{X}$ with the same support.

   a) Prove that for any $\delta > 0$,
   $$p^n\left(\left\{x^n \in \mathcal{X}^n : \left|-\frac{1}{n}\log q^n(x^n) - (H(p) + D(p\|q))\right| < \delta\right\}\right) \to 1$$
   as $n \to \infty$.

   b) Prove that for any $\delta > 0$,
   $$\left|\left\{x^n \in \mathcal{X}^n : \left|-\frac{1}{n}\log q^n(x^n) - (H(p) + D(p\|q))\right| < \delta\right\}\right| \le 2^{n(H(p)+D(p\|q)+\delta)}.$$

7. *Universal source coding* Let $\mathcal{F} = \{\{X_k^{(s)}, k \geq 1\} : s \in \mathcal{S}\}$ be a family of i.i.d. information sources indexed by a finite set $\mathcal{S}$ with a common alphabet $\mathcal{X}$. Define
$$\bar{H} = \max_{s\in\mathcal{S}} H(X^{(s)})$$
where $X^{(s)}$ is the generic random variable for $\{X_k^{(s)}, k \geq 1\}$, and
$$A_\epsilon^n(\mathcal{S}) = \bigcup_{s\in\mathcal{S}} W^n_{[X^{(s)}]\epsilon},$$

where $\epsilon > 0$.

a) Prove that for all $s \in \mathcal{S}$,

$$\Pr\{\mathbf{X}^{(s)} \in A_\epsilon^n(\mathcal{S})\} \to 1$$

as $n \to \infty$, where $\mathbf{X}^{(s)} = (X_1^{(s)}, X_2^{(s)}, \cdots, X_n^{(s)})$.

b) Prove that for any $\epsilon' > \epsilon$,

$$|A_\epsilon^n(\mathcal{S})| \leq 2^{n(\bar{H}+\epsilon')}$$

for sufficiently large n.

c) Suppose we know that an information source is in the family $\mathcal{F}$ but we do not know which one it is. Devise a compression scheme for the information source such that it is asymptotically optimal for every possible source in $\mathcal{F}$.

8. Let $\{X_k, k \geq 1\}$ be an i.i.d. information source with generic random variable X and alphabet $\mathcal{X}$. Assume

$$\sum_x p(x)[\log p(x)]^2 < \infty$$

and define

$$Z_n = -\frac{\log p(\mathbf{X})}{\sqrt{n}} - \sqrt{n}H(X)$$

for $n = 1, 2, \cdots$. Prove that $Z_n \to Z$ in distribution, where Z is a Gaussian random variable with mean 0 and variance $\sum_x p(x)[\log p(x)]^2 - H(X)^2$.

HISTORICAL NOTES

The weak asymptotic equipartition property (AEP), which is instrumental in proving the source coding theorem, was first proved by Shannon in his original paper [184]. In this paper, he also stated that this property can be extended to an ergodic stationary source. Subsequently, McMillan [150] and Breiman [36] proved this property for an ergodic stationary source with a finite alphabet. Chung [46] extended the theme to a countable alphabet.

Chapter 5

STRONG TYPICALITY

Weak typicality requires that the empirical entropy of a sequence is close to the true entropy. In this chapter, we introduce a stronger notion of typicality which requires that the relative frequency of each possible outcome is close to the corresponding probability. As we will see later, strong typicality is more powerful and flexible than weak typicality as a tool for theorem proving for memoryless problems. However, strong typicality can be used only for random variables with finite alphabets. Throughout this chapter, typicality refers to strong typicality and all the logarithms are in the base 2 unless otherwise specified.

5.1 STRONG AEP

We consider an information source $\{X_k, k \geq 1\}$ where X_k are i.i.d. with distribution $p(x)$. We use X to denote the generic random variable and $H(X)$ to denote the common entropy for all X_k, where $H(X) < \infty$. Let $\mathbf{X} = (X_1, X_2, \cdots, X_n)$.

DEFINITION 5.1 *The strongly typical set $T^n_{[X]\delta}$ with respect to $p(x)$ is the set of sequences $\mathbf{x} = (x_1, x_2, \cdots, x_n) \in \mathcal{X}^n$ such that $N(x;\mathbf{x}) = 0$ for $x \notin S_X$, and*

$$\sum_x \left| \frac{1}{n} N(x;\mathbf{x}) - p(x) \right| \leq \delta, \tag{5.1}$$

where $N(x;\mathbf{x})$ is the number of occurrences of x in the sequence $\mathbf{x}$, and δ is an arbitrarily small positive real number. The sequences in $T^n_{[X]\delta}$ are called strongly δ-typical sequences.

Throughout this chapter, we adopt the convention that all the summations, products, unions, etc, are taken over the corresponding supports unless otherwise specified. The strongly typical set $T^n_{[X]\delta}$ shares similar properties with its weakly typical counterpart, which is summarized as the *strong asymptotic equipartition property* (strong AEP) below. The interpretation of the strong AEP is similar to that of the weak AEP.

THEOREM 5.2 (STRONG AEP) *In the following, η is a small positive quantity such that $\eta \to 0$ as $\delta \to 0$.*

1) If $\mathbf{x} \in T^n_{[X]\delta}$, then

$$2^{-n(H(X)+\eta)} \le p(\mathbf{x}) \le 2^{-n(H(X)-\eta)}. \tag{5.2}$$

2) For n sufficiently large,

$$\Pr\{\mathbf{X} \in T^n_{[X]\delta}\} > 1 - \delta. \tag{5.3}$$

3) For n sufficiently large,

$$(1-\delta)2^{n(H(X)-\eta)} \le |T^n_{[X]\delta}| \le 2^{n(H(X)+\eta)}. \tag{5.4}$$

Proof To prove Property 1, for $\mathbf{x} \in T^n_{[X]\delta}$, we write

$$p(\mathbf{x}) = \prod_x p(x)^{N(x;\mathbf{x})}. \tag{5.5}$$

Then

$$\begin{aligned}
&\log p(\mathbf{x}) \\
&= \sum_x N(x;\mathbf{x}) \log p(x) && (5.6)\\
&= \sum_x (N(x;\mathbf{x}) - np(x) + np(x)) \log p(x) && (5.7)\\
&= n\sum_x p(x)\log p(x) - n\sum_x \left(\frac{1}{n}N(x;\mathbf{x}) - p(x)\right)(-\log p(x)) && (5.8)\\
&= -n\left[H(X) + \sum_x \left(\frac{1}{n}N(x;\mathbf{x}) - p(x)\right)(-\log p(x))\right]. && (5.9)
\end{aligned}$$

Since $\mathbf{x} \in T^n_{[X]\delta}$,

$$\sum_x \left|\frac{1}{n}N(x;\mathbf{x}) - p(x)\right| \le \delta, \tag{5.10}$$

which implies

$$
\begin{aligned}
&\left|\sum_x \left(\frac{1}{n}N(x;\mathbf{x}) - p(x)\right)(-\log p(x))\right| \\
&\quad\le \sum_x \left|\frac{1}{n}N(x;\mathbf{x}) - p(x)\right|(-\log p(x)) && (5.11)\\
&\quad\le -\log\left(\min_x p(x)\right)\sum_x \left|\frac{1}{n}N(x;\mathbf{x}) - p(x)\right| && (5.12)\\
&\quad\le -\delta \log\left(\min_x p(x)\right) && (5.13)\\
&\quad= \eta, && (5.14)
\end{aligned}
$$

where

$$\eta = -\delta \log\left(\min_x p(x)\right) > 0. \tag{5.15}$$

Therefore,

$$-\eta \le \sum_x \left(\frac{1}{n}N(x;\mathbf{x}) - p(x)\right)(-\log p(x)) \le \eta. \tag{5.16}$$

It then follows from (5.9) that

$$-n(H(X)+\eta) \le \log p(\mathbf{x}) \le -n(H(X)-\eta), \tag{5.17}$$

or

$$2^{-n(H(X)+\eta)} \le p(\mathbf{x}) \le 2^{-n(H(X)-\eta)}, \tag{5.18}$$

where $\eta \to 0$ as $\delta \to 0$, proving Property 1.

To prove Property 2, we write

$$N(x;\mathbf{X}) = \sum_{k=1}^{n} B_k(x), \tag{5.19}$$

where

$$B_k(x) = \begin{cases} 1 & \text{if } X_k = x \\ 0 & \text{if } X_k \ne x. \end{cases} \tag{5.20}$$

Then $B_k(x), k = 1, 2, \cdots, n$ are i.i.d. random variables with

$$\Pr\{B_k(x) = 1\} = p(x) \tag{5.21}$$

and

$$\Pr\{B_k(x) = 0\} = 1 - p(x). \tag{5.22}$$

Note that

$$EB_k(x) = (1 - p(x)) \cdot 0 + p(x) \cdot 1 = p(x). \tag{5.23}$$

By the weak law of large numbers, for any $\delta > 0$ and for any $x \in \mathcal{X}$,

$$\Pr\left\{\left|\frac{1}{n}\sum_{k=1}^{n} B_k(x) - p(x)\right| > \frac{\delta}{|\mathcal{X}|}\right\} < \frac{\delta}{|\mathcal{X}|} \tag{5.24}$$

for n sufficiently large. Then

$$\begin{aligned}
&\Pr\left\{\left|\frac{1}{n}N(x;\mathbf{X}) - p(x)\right| > \frac{\delta}{|\mathcal{X}|} \text{ for some } x\right\} \\
&= \Pr\left\{\left|\frac{1}{n}\sum_{k=1}^{n} B_k(x) - p(x)\right| > \frac{\delta}{|\mathcal{X}|} \text{ for some } x\right\} &(5.25)\\
&= \Pr\left\{\bigcup_x \left\{\left|\frac{1}{n}\sum_{k=1}^{n} B_k(x) - p(x)\right| > \frac{\delta}{|\mathcal{X}|}\right\}\right\} &(5.26)\\
&\le \sum_x \Pr\left\{\left|\frac{1}{n}\sum_{k=1}^{n} B_k(x) - p(x)\right| > \frac{\delta}{|\mathcal{X}|}\right\} &(5.27)\\
&< \sum_x \frac{\delta}{|\mathcal{X}|} &(5.28)\\
&= \delta, &(5.29)
\end{aligned}$$

where we have used the union bound[1] to obtain (5.27). Since

$$\sum_x \left|\frac{1}{n}N(x;\mathbf{x}) - p(x)\right| > \delta \tag{5.30}$$

implies

$$\left|\frac{1}{n}N(x;\mathbf{x}) - p(x)\right| > \frac{\delta}{|\mathcal{X}|} \quad \text{for some } x \in \mathcal{X}, \tag{5.31}$$

we have

$$\begin{aligned}
&\Pr\left\{\mathbf{X} \in T^n_{[X]\delta}\right\} \\
&= \Pr\left\{\sum_x \left|\frac{1}{n}N(x;\mathbf{X}) - p(x)\right| \le \delta\right\} &(5.32)\\
&= 1 - \Pr\left\{\sum_x \left|\frac{1}{n}N(x;\mathbf{X}) - p(x)\right| > \delta\right\} &(5.33)\\
&\ge 1 - \Pr\left\{\left|\frac{1}{n}N(x;\mathbf{X}) - p(x)\right| > \frac{\delta}{|\mathcal{X}|} \text{ for some } x \in \mathcal{X}\right\} &(5.34)\\
&> 1 - \delta, &(5.35)
\end{aligned}$$

[1] The union bound refers to $\Pr\{A \cup B\} \le \Pr\{A\} + \Pr\{B\}$.

proving Property 2.

Finally, Property 3 follows from Property 1 and Property 2 in exactly the same way as in Theorem 4.3, so the proof is omitted. □

Remark Analogous to weak typicality, we note that the upper bound on $|T^n_{[X]\delta}|$ in Property 3 holds for all $n \geq 1$, and for any $\delta > 0$, there exists at least one strongly typical sequence when n is sufficiently large. See Problem 1 in Chapter 4.

In the rest of the section, we prove an enhancement of Property 2 of the strong AEP which gives an exponential bound on the probability of obtaining a non-typical vector[2]. This result, however, will not be used until Chapter 15.

THEOREM 5.3 *For sufficiently large n, there exists $\varphi(\delta) > 0$ such that*

$$\Pr\{\mathbf{X} \notin T^n_{[X]\delta}\} < 2^{-n\varphi(\delta)}. \tag{5.36}$$

The proof of this theorem is based on the Chernoff bound [45] which we prove in the next lemma.

LEMMA 5.4 (CHERNOFF BOUND) *Let Y be a real random variable and s be any nonnegative real number. Then for any real number a,*

$$\log \Pr\{Y \geq a\} \leq -sa + \log E\left[2^{sY}\right] \tag{5.37}$$

and

$$\log \Pr\{Y \leq a\} \leq sa + \log E\left[2^{-sY}\right]. \tag{5.38}$$

Proof Let

$$u(y) = \begin{cases} 1 & \text{if } y \geq 0 \\ 0 & \text{if } y < 0. \end{cases} \tag{5.39}$$

Then for any $s \geq 0$,

$$u(y-a) \leq 2^{s(y-a)}. \tag{5.40}$$

This is illustrated in Fig. 5.1. Then

$$E[u(Y-a)] \leq E\left[2^{s(Y-a)}\right] = 2^{-sa} E\left[2^{sY}\right]. \tag{5.41}$$

---

[2] This result is due to Ning Cai and Raymond W. Yeung. An alternative proof based on Pinsker's inequality (Theorem 2.33) and the method of types has been given by Prakash Narayan (private communication).

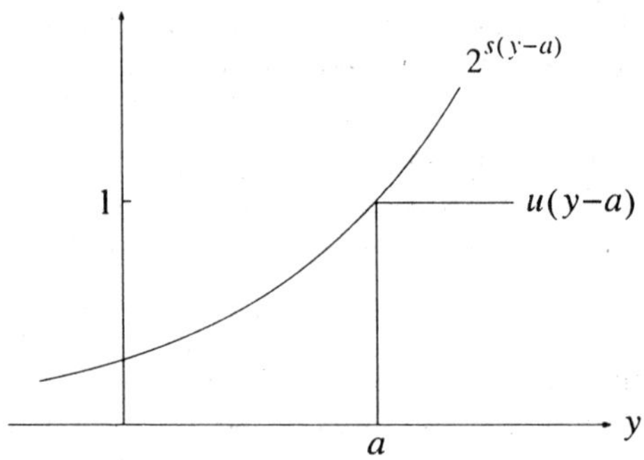

Figure 5.1. An illustration of $u(y-a) \le 2^{s(y-a)}$.

Since

$$E[u(Y-a)] = \Pr\{Y \ge a\} \cdot 1 + \Pr\{Y < a\} \cdot 0 = \Pr\{Y \ge a\}, \tag{5.42}$$

we see that

$$\Pr\{Y \ge a\} \le 2^{-sa} E\left[2^{sY}\right] = 2^{-sa + \log E\left[2^{sY}\right]}. \tag{5.43}$$

Then (5.37) is obtained by taking logarithm in the base 2. Upon replacing Y by $-Y$ and a by $-a$ in (5.37), (5.38) is obtained. The lemma is proved. □

Proof of Theorem 5.3 We will follow the notation in the proof of Theorem 5.2. Consider $x \in \mathcal{X}$ such that $p(x) > 0$. Applying (5.37), we have

$$\begin{aligned}
&\log \Pr\left\{\sum_{k=1}^{n} B_k(x) \ge n\left(p(x)+\delta\right)\right\} \\
&\le -sn\left(p(x)+\delta\right) + \log E\left[2^{s\sum_{k=1}^{n} B_k(x)}\right] && (5.44)\\
&\overset{a)}{=} -sn\left(p(x)+\delta\right) + \log\left(\prod_{k=1}^{n} E\left[2^{sB_k(x)}\right]\right) && (5.45)\\
&\overset{b)}{=} -sn\left(p(x)+\delta\right) + n\log(1-p(x)+p(x)2^s) && (5.46)\\
&\overset{c)}{\le} -sn\left(p(x)+\delta\right) + n(\ln 2)^{-1}(-p(x)+p(x)2^s) && (5.47)\\
&= -n\left[s\left(p(x)+\delta\right) + (\ln 2)^{-1}p(x)(1-2^s)\right], && (5.48)
\end{aligned}$$

where

a) follows because $B_k(x)$ are mutually independent;

b) is a direct evaluation of the expectation from the definition of $B_k(x)$ in (5.20);

c) follows from the fundamental inequality $\ln a \leq a - 1$.

In (5.48), upon defining

$$\beta_x(s, \delta) = s\,(p(x) + \delta) + (\ln 2)^{-1} p(x)(1 - 2^s), \tag{5.49}$$

we have

$$\log \Pr \left\{ \sum_{k=1}^{n} B_k(x) \geq n\,(p(x) + \delta) \right\} \leq -n\beta_x(s, \delta), \tag{5.50}$$

or

$$\Pr \left\{ \sum_{k=1}^{n} B_k(x) \geq n\,(p(x) + \delta) \right\} \leq 2^{-n\beta_x(s,\delta)}. \tag{5.51}$$

It is readily seen that

$$\beta_x(0, \delta) = 0. \tag{5.52}$$

Regarding δ as fixed and differentiate with respect to s, we have

$$\beta_x'(s, \delta) = p(x)(1 - 2^s) + \delta. \tag{5.53}$$

Then

$$\beta_x'(0, \delta) = \delta > 0 \tag{5.54}$$

and it is readily verified that

$$\beta_x'(s, \delta) \geq 0 \tag{5.55}$$

for

$$0 \leq s \leq \log \left(1 + \frac{\delta}{p(x)} \right). \tag{5.56}$$

Therefore, we conclude that $\beta_x(s, \delta)$ is strictly positive for

$$0 < s \leq \log \left(1 + \frac{\delta}{p(x)} \right). \tag{5.57}$$

On the other hand, by applying (5.38), we can obtain in the same fashion the bound

$$\log \Pr \left\{ \sum_{k=1}^{n} B_k(x) \leq n\,(p(x) - \delta) \right\} \leq -n\sigma_x(s, \delta), \tag{5.58}$$

or

$$\Pr \left\{ \sum_{k=1}^{n} B_k(x) \leq n\,(p(x) - \delta) \right\} \leq 2^{-n\sigma_x(s,\delta)}, \tag{5.59}$$

where

$$\sigma_x(s,\delta) = -s\,(p(x)-\delta) + (\ln 2)^{-1}p(x)(1-2^{-s}). \tag{5.60}$$

Then

$$\sigma_x(0,\delta) = 0, \tag{5.61}$$

and

$$\sigma_x'(s,\delta) = p(x)(2^{-s}-1)+\delta, \tag{5.62}$$

which is nonnegative for

$$0 \le s \le -\log\left(1-\frac{\delta}{p(x)}\right). \tag{5.63}$$

In particular,

$$\sigma_x'(0,\delta) = \delta > 0. \tag{5.64}$$

Therefore, we conclude that $\sigma_x(s,\delta)$ is strictly positive for

$$0 < s \le -\log\left(1-\frac{\delta}{p(x)}\right). \tag{5.65}$$

By choosing s satisfying

$$0 < s \le \min\left[\log\left(1+\frac{\delta}{p(x)}\right), -\log\left(1-\frac{\delta}{p(x)}\right)\right], \tag{5.66}$$

both $\beta_x(s,\delta)$ and $\sigma_x(s,\delta)$ are strictly positive. From (5.51) and (5.59), we have

$$\begin{aligned}
&\Pr\left\{\left|\frac{1}{n}\sum_{k=1}^{n} B_k(x) - p(x)\right| \ge \delta\right\} \\
&= \Pr\left\{\left|\sum_{k=1}^{n} B_k(x) - np(x)\right| \ge n\delta\right\} &(5.67)\\
&\le \Pr\left\{\sum_{k=1}^{n} B_k(x) \ge n\,(p(x)+\delta)\right\} \\
&\quad +\Pr\left\{\sum_{k=1}^{n} B_k(x) \le n\,(p(x)-\delta)\right\} &(5.68)\\
&\le 2^{-n\beta_x(s,\delta)} + 2^{-n\sigma_x(s,\delta)} &(5.69)\\
&\le 2\cdot 2^{-n\min(\beta_x(s,\delta),\sigma_x(s,\delta))} &(5.70)\\
&= 2^{-n\left[\min(\beta_x(s,\delta),\sigma_x(s,\delta))-\frac{1}{n}\right]} &(5.71)\\
&= 2^{-n\varphi_x(\delta)}, &(5.72)
\end{aligned}$$

where

$$\varphi_x(\delta) = \min(\beta_x(s,\delta), \sigma_x(s,\delta)) - \frac{1}{n}. \tag{5.73}$$

Then $\varphi_x(\delta)$ is strictly positive for sufficiently large n because both $\beta_x(s,\delta)$ and $\sigma_x(s,\delta)$ are strictly positive.

Finally, consider

$$\begin{aligned}
&\Pr\{\mathbf{X} \in T^n_{[X]\delta}\} \\
&= \Pr\left\{\sum_x \left|\frac{1}{n}N(x;\mathbf{X}) - p(x)\right| \le \delta\right\} && (5.74)\\
&\ge \Pr\left\{\left|\frac{1}{n}N(x;\mathbf{X}) - p(x)\right| \le \frac{\delta}{|\mathcal{X}|} \text{ for all } x \in \mathcal{X}\right\} && (5.75)\\
&= 1 - \Pr\left\{\left|\frac{1}{n}N(x;\mathbf{X}) - p(x)\right| > \frac{\delta}{|\mathcal{X}|} \text{ for some } x \in \mathcal{X}\right\} && (5.76)\\
&\ge 1 - \sum_x \Pr\left\{\left|\frac{1}{n}N(x;\mathbf{X}) - p(x)\right| > \frac{\delta}{|\mathcal{X}|}\right\} && (5.77)\\
&= 1 - \sum_x \Pr\left\{\left|\frac{1}{n}\sum_{k=1}^n B_k(x) - p(x)\right| > \frac{\delta}{|\mathcal{X}|}\right\} && (5.78)\\
&= 1 - \sum_{x:p(x)>0} \Pr\left\{\left|\frac{1}{n}\sum_{k=1}^n B_k(x) - p(x)\right| > \frac{\delta}{|\mathcal{X}|}\right\} && (5.79)\\
&\ge 1 - \sum_{x:p(x)>0} 2^{-n\varphi_x\left(\frac{\delta}{|\mathcal{X}|}\right)}, && (5.80)
\end{aligned}$$

where the last step follows from (5.72). Define

$$\varphi(\delta) = \frac{1}{2}\left[\min_{x:p(x)>0} \varphi_x\left(\frac{\delta}{|\mathcal{X}|}\right)\right]. \tag{5.81}$$

Then for sufficiently large n,

$$\Pr\{\mathbf{X} \in T^n_{[X]\delta}\} > 1 - 2^{-n\varphi(\delta)}, \tag{5.82}$$

or

$$\Pr\{\mathbf{X} \notin T^n_{[X]\delta}\} < 2^{-n\varphi(\delta)}, \tag{5.83}$$

where $\varphi(\delta)$ is strictly positive. The theorem is proved. □

5.2 STRONG TYPICALITY VERSUS WEAK TYPICALITY

As we have mentioned at the beginning of the chapter, strong typicality is more powerful and flexible than weak typicality as a tool for theorem proving for memoryless problems, but it can be used only for random variables with finite alphabets. We will prove in the next proposition that strong typicality is stronger than weak typicality in the sense that the former implies the latter.

PROPOSITION 5.5 *For any* $\mathbf{x} \in \mathcal{X}^n$, *if* $\mathbf{x} \in T^n_{[X]\delta}$, *then* $\mathbf{x} \in W^n_{[X]\eta}$, *where* $\eta \to 0$ *as* $\delta \to 0$.

Proof By Property 1 of strong AEP (Theorem 5.2), if $\mathbf{x} \in T^n_{[X]\delta}$, then

$$2^{-n(H(X)+\eta)} \leq p(\mathbf{x}) \leq 2^{-n(H(X)-\eta)}, \tag{5.84}$$

or

$$H(X) - \eta \leq -\frac{1}{n} \log p(\mathbf{x}) \leq H(X) + \eta, \tag{5.85}$$

where $\eta \to 0$ as $\delta \to 0$. Then $\mathbf{x} \in W^n_{[X]\eta}$ by Definition 4.2. The proposition is proved. □

We have proved in this proposition that strong typicality implies weak typicality, but the converse is not true. This idea can be explained without any detailed analysis. Let X be distributed with p such that $p(0) = 0.5$, $p(1) = 0.25$, and $p(2) = 0.25$. Consider a sequence $\mathbf{x}$ of length n and let $q(i)$ be the relative frequency of occurrence of symbol i in $\mathbf{x}$, i.e., $\frac{1}{n}N(i;\mathbf{x})$, where $i = 0, 1, 2$. In order for the sequence $\mathbf{x}$ to be weakly typical, we need

$$\begin{aligned} & -\frac{1}{n} \log p(\mathbf{x}) \\ = \;& -q(0) \log 0.5 - q(1) \log 0.25 - q(2) \log 0.25 && (5.86) \\ \approx \;& H(X) && (5.87) \\ = \;& -(0.5) \log 0.5 - (0.25) \log 0.25 - (0.25) \log 0.25. && (5.88) \end{aligned}$$

Obviously, this can be satisfied by choosing $q(i) = p(i)$ for all i. But alternatively, we can choose $q(0) = 0.5$, $q(1) = 0.5$, and $q(2) = 0$. With such a choice of $\{q(i)\}$, the sequence $\mathbf{x}$ is weakly typical with respect to p but obviously not strongly typical with respect to p, because the relative frequency of occurrence of each symbol i is $q(i)$, which is not close to $p(i)$ for $i = 1, 2$.

Therefore, we conclude that strong typicality is indeed stronger than weak typicality. However, as we have pointed out at the beginning of the chapter, strong typicality can only be used for random variables with finite alphabets.

5.3 JOINT TYPICALITY

In this section, we discuss strong joint typicality with respect to a bivariate distribution. Generalization to a multivariate distribution is straightforward.

Consider a bivariate information source $\{(X_k, Y_k), k \geq 1\}$ where (X_k, Y_k) are i.i.d. with distribution $p(x, y)$. We use (X, Y) to denote the pair of generic random variables.

DEFINITION 5.6 *The strongly jointly typical set $T^n_{[XY]\delta}$ with respect to $p(x, y)$ is the set of $(\mathbf{x}, \mathbf{y}) \in \mathcal{X}^n \times \mathcal{Y}^n$ such that $N(x, y; \mathbf{x}, \mathbf{y}) = 0$ for $(x, y) \notin S_{XY}$, and*

$$\sum_x \sum_y \left| \frac{1}{n} N(x, y; \mathbf{x}, \mathbf{y}) - p(x, y) \right| \leq \delta, \tag{5.89}$$

where $N(x, y; \mathbf{x}, \mathbf{y})$ is the number of occurrences of (x, y) in the pair of sequences $(\mathbf{x}, \mathbf{y})$, and δ is an arbitrarily small positive real number. A pair of sequences $(\mathbf{x}, \mathbf{y})$ is called strongly jointly δ-typical if it is in $T^n_{[XY]\delta}$.

Strong typicality satisfies the following *consistency* property.

THEOREM 5.7 (CONSISTENCY) *If $(\mathbf{x}, \mathbf{y}) \in T^n_{[XY]\delta}$, then $\mathbf{x} \in T^n_{[X]\delta}$ and $\mathbf{y} \in T^n_{[Y]\delta}$.*

Proof If $(\mathbf{x}, \mathbf{y}) \in T^n_{[XY]\delta}$, then

$$\sum_x \sum_y \left| \frac{1}{n} N(x, y; \mathbf{x}, \mathbf{y}) - p(x, y) \right| \leq \delta. \tag{5.90}$$

Upon observing that

$$N(x; \mathbf{x}) = \sum_y N(x, y; \mathbf{x}, \mathbf{y}), \tag{5.91}$$

we have

$$\begin{aligned}
&\sum_x \left| \frac{1}{n} N(x; \mathbf{x}) - p(x) \right| \\
&= \sum_x \left| \frac{1}{n} \sum_y N(x, y; \mathbf{x}, \mathbf{y}) - \sum_y p(x, y) \right| && (5.92)\\
&= \sum_x \left| \sum_y \left(\frac{1}{n} N(x, y; \mathbf{x}, \mathbf{y}) - p(x, y) \right) \right| && (5.93)\\
&\leq \sum_x \sum_y \left| \frac{1}{n} N(x, y; \mathbf{x}, \mathbf{y}) - p(x, y) \right| && (5.94)\\
&\leq \delta. && (5.95)
\end{aligned}$$

Therefore, $\mathbf{x} \in T^n_{[X]\delta}$. Similarly, $\mathbf{y} \in T^n_{[Y]\delta}$. The theorem is proved. □

For a bivariate i.i.d. source $\{(X_k, Y_k)\}$, we have the *strong joint asymptotic equipartition property* (strong JAEP), which can readily be obtained by applying the strong AEP to the source $\{(X_k, Y_k)\}$.

THEOREM 5.8 (STRONG JAEP) *Let*

$$(\mathbf{X}, \mathbf{Y}) = ((X_1, Y_1), (X_2, Y_2), \cdots, (X_n, Y_n)), \tag{5.96}$$

where (X_i, Y_i) are i.i.d. with generic pair of random variables (X, Y). In the following, λ is a small positive quantity such that $\lambda \to 0$ as $\delta \to 0$.

1) If $(\mathbf{x}, \mathbf{y}) \in T^n_{[XY]\delta}$, then

$$2^{-n(H(X,Y)+\lambda)} \leq p(\mathbf{x}, \mathbf{y}) \leq 2^{-n(H(X,Y)-\lambda)}. \tag{5.97}$$

2) For n sufficiently large,

$$\Pr\{(\mathbf{X}, \mathbf{Y}) \in T^n_{[XY]\delta}\} > 1 - \delta. \tag{5.98}$$

3) For n sufficiently large,

$$(1 - \delta)2^{n(H(X,Y)-\lambda)} \leq |T^n_{[XY]\delta}| \leq 2^{n(H(X,Y)+\lambda)}. \tag{5.99}$$

From the strong JAEP, we can see the following. Since there are approximately $2^{nH(X,Y)}$ typical $(\mathbf{x}, \mathbf{y})$ pairs and approximately $2^{nH(X)}$ typical $\mathbf{x}$, for a typical $\mathbf{x}$, the number of $\mathbf{y}$ such that $(\mathbf{x}, \mathbf{y})$ is jointly typical is approximately

$$\frac{2^{nH(X,Y)}}{2^{nH(X)}} = 2^{nH(Y|X)} \tag{5.100}$$

on the average. The next theorem reveals that this is not only true on the average, but it is in fact true for every typical $\mathbf{x}$ as long as there exists at least one $\mathbf{y}$ such that $(\mathbf{x}, \mathbf{y})$ is jointly typical.

THEOREM 5.9 *For any $\mathbf{x} \in T^n_{[X]\delta}$, define*

$$T^n_{[Y|X]\delta}(\mathbf{x}) = \{\mathbf{y} \in T^n_{[Y]\delta} : (\mathbf{x}, \mathbf{y}) \in T^n_{[XY]\delta}\}. \tag{5.101}$$

If $|T^n_{[Y|X]\delta}(\mathbf{x})| \geq 1$, then

$$2^{n(H(Y|X)-\nu)} \leq |T^n_{[Y|X]\delta}(\mathbf{x})| \leq 2^{n(H(Y|X)+\nu)}, \tag{5.102}$$

where $\nu \to 0$ as $n \to \infty$ and $\delta \to 0$.

We first prove the following lemma which is along the line of Stirling's approximation [70].

LEMMA 5.10 *For any $n > 0$,*

$$n \ln n - n \leq \ln n! \leq (n+1)\ln(n+1) - n. \tag{5.103}$$

Proof First, we write

$$\ln n! = \ln 1 + \ln 2 + \cdots + \ln n. \tag{5.104}$$

Since $\ln x$ is a monotonically increasing function of x, we have

$$\int_{k-1}^{k} \ln x \, dx < \ln k < \int_{k}^{k+1} \ln x \, dx. \tag{5.105}$$

Summing over $1 \leq k \leq n$, we have

$$\int_{0}^{n} \ln x \, dx < \ln n! < \int_{1}^{n+1} \ln x \, dx, \tag{5.106}$$

or

$$n \ln n - n < \ln n! < (n+1)\ln(n+1) - n. \tag{5.107}$$

The lemma is proved. □

Proof of Theorem 5.9 Let δ be a small positive real number and n be a large positive integer to be specified later. Fix an $\mathbf{x} \in T^n_{[X]\delta}$, so that

$$\sum_x \left| \frac{1}{n} N(X;\mathbf{x}) - p(x) \right| \leq \delta. \tag{5.108}$$

This implies that for all $x \in \mathcal{X}$,

$$\left| \frac{1}{n} N(X;\mathbf{x}) - p(x) \right| \leq \delta, \tag{5.109}$$

or

$$p(x) - \delta \leq \frac{1}{n} N(x;\mathbf{x}) \leq p(x) + \delta. \tag{5.110}$$

Assume that $|T^n_{[Y|X]\delta}(\mathbf{x})| \geq 1$, and let

$$\{K(x,y), (x,y) \in \mathcal{X} \times \mathcal{Y}\} \tag{5.111}$$

be any set of nonnegative integers such that

1.
$$\sum_y K(x,y) = N(x;\mathbf{x}) \tag{5.112}$$

for all $x \in \mathcal{X}$, and

2. for any $\mathbf{y} \in \mathcal{Y}^n$, if
$$N(x, y; \mathbf{x}, \mathbf{y}) = K(x, y) \tag{5.113}$$
for all $(x, y) \in \mathcal{X} \times \mathcal{Y}$, then $(\mathbf{x}, \mathbf{y}) \in T^n_{[XY]\delta}$.

Then from (5.89), $\{K(x, y)\}$ satisfy
$$\sum_x \sum_y \left| \frac{1}{n} K(x, y) - p(x, y) \right| \le \delta, \tag{5.114}$$
which implies that for all $(x, y) \in \mathcal{X} \times \mathcal{Y}$,
$$\left| \frac{1}{n} K(x, y) - p(x, y) \right| \le \delta, \tag{5.115}$$
or
$$p(x, y) - \delta \le \frac{1}{n} K(x, y) \le p(x, y) + \delta. \tag{5.116}$$
Such a set $\{K(x, y)\}$ exists because $T^n_{[Y|X]\delta}(\mathbf{x})$ is assumed to be nonempty. Straightforward combinatorics reveals that the number of $\mathbf{y}$ which satisfy the constraints in (5.113) is equal to
$$M(K) = \prod_x \frac{N(x; \mathbf{x})!}{\prod_y K(x, y)!}. \tag{5.117}$$
We will show that for large n,
$$n^{-1} \log M(K) \approx H(Y|X). \tag{5.118}$$
When this is done, the proof will almost be complete.

Using Lemma 5.10, we can lower bound $\ln M(K)$ as follows.
$$\begin{aligned} &\ln M(K) \\ &\ge \sum_x \Big\{ N(x; \mathbf{x}) \ln N(x; \mathbf{x}) - N(x; \mathbf{x}) \\ &\quad - \sum_y [(K(x, y) + 1) \ln(K(x, y) + 1) - K(x, y)] \Big\} \qquad (5.119) \\ &\stackrel{a)}{=} \sum_x \Big[N(x; \mathbf{x}) \ln N(x; \mathbf{x}) \\ &\quad - \sum_y (K(x, y) + 1) \ln(K(x, y) + 1) \Big] \qquad (5.120) \end{aligned}$$

$$\stackrel{b)}{\geq} \sum_x \{N(x;\mathbf{x}) \ln(n\,(p(x)-\delta))$$

$$-\sum_y (K(x,y)+1) \ln\left[n\left(p(x,y)+\delta+\frac{1}{n}\right)\right]\right\}. \quad (5.121)$$

In the above, a) follows from (5.112), and b) is obtained by applying the lower bound on $n^{-1}N(x;\mathbf{x})$ in (5.110) and the upper bound on $n^{-1}K(x,y)$ in (5.116). Also from (5.112), the coefficient of $\ln n$ in (5.121) is given by

$$\sum_x \left[N(x;\mathbf{x}) - \sum_y (K(x,y)+1)\right] = -|\mathcal{X}||\mathcal{Y}|. \quad (5.122)$$

Let δ be sufficiently small and n be sufficiently large so that

$$0 < p(x) - \delta < 1 \quad (5.123)$$

and

$$p(x,y) + \delta + \frac{1}{n} < 1 \quad (5.124)$$

for all x and y. Then in (5.121), both the logarithms

$$\ln(p(x) - \delta) \quad (5.125)$$

and

$$\ln\left(p(x,y) + \delta + \frac{1}{n}\right) \quad (5.126)$$

are negative. Note that the logarithm in (5.125) is well-defined by virtue of (5.123). Rearranging the terms in (5.121), applying the upper bound in (5.110) and the lower bound[3] in (5.116), and dividing by n, we have

$$\begin{aligned} & n^{-1}\ln M(K) \\ \geq\ & \sum_x (p(x)+\delta)\ln(p(x)-\delta) - \sum_x\sum_y \left(p(x,y) - \delta + \frac{1}{n}\right) \\ & \times \ln\left(p(x,y)+\delta+\frac{1}{n}\right) - \frac{|\mathcal{X}||\mathcal{Y}|\ln n}{n} && (5.127) \\ =\ & -H_e(X) + H_e(X,Y) + L_l(n,\delta) && (5.128) \\ =\ & H_e(Y|X) + L_l(n,\delta), && (5.129) \end{aligned}$$

[3]For the degenerate case when $p(x,y) = 1$ for some x and y, $p(x,y) + \delta + \frac{1}{n} > 1$, and the logarithm in (5.126) is in fact positive. Then the upper bound instead of the lower bound should be applied. The details are omitted.

where $L_l(n, \delta)$ denotes a function of n and δ which tends to 0 as $n \to \infty$ and $\delta \to 0$. Using a similar method, we can obtain a corresponding upper bound as follows.

$$\begin{aligned}
\ln M(K) & \\
\leq & \sum_x \Big\{ (N(x;\mathbf{x}) + 1)\ln(N(x;\mathbf{x}) + 1) - N(x;\mathbf{x}) \\
& - \sum_y [K(x,y)\ln K(x,y) - K(x,y)] \Big\} && (5.130) \\
= & \sum_x \Big[(N(x;\mathbf{x}) + 1)\ln(N(x;\mathbf{x}) + 1) \\
& - \sum_y K(x,y)\ln K(x,y) \Big] && (5.131) \\
\leq & \sum_x \Big\{ (N(x;\mathbf{x}) + 1)\ln\left[n\left(p(x) + \delta + \frac{1}{n}\right)\right] \\
& - \sum_y K(x,y)\ln\left[n\left(p(x,y) - \delta\right)\right] \Big\}. && (5.132)
\end{aligned}$$

In the above, the coefficient of $\ln n$ is given by

$$\sum_x \left[(N(x;\mathbf{x}) + 1) - \sum_y K(x,y) \right] = |\mathcal{X}|. \tag{5.133}$$

We let δ be sufficiently small and n be sufficiently large so that

$$p(x) + \delta + \frac{1}{n} < 1 \tag{5.134}$$

and

$$0 < p(x,y) - \delta < 1 \tag{5.135}$$

for all x and y. Then in (5.132), both the logarithms

$$\ln\left(p(x) + \delta + \frac{1}{n}\right) \tag{5.136}$$

and

$$\ln(p(x,y) - \delta) \tag{5.137}$$

are negative, and the logarithm in (5.137) is well-defined by virtue of (5.135). Rearranging the terms in (5.132), applying the lower bound[4] in (5.110) and the

---

[4]If $p(x) = 1$, $p(x) + \delta + \frac{1}{n} > 0$, and the logarithm in (5.136) is in fact positive. Then the upper bound instead of the lower bound should be applied. The details are omitted.

upper bound in (5.116), and dividing by n, we have

$$\begin{aligned} & n^{-1}\ln M(K) \\ \leq\ & \sum_x \left(p(x) - \delta + \frac{1}{n}\right) \ln \left(p(x) + \delta + \frac{1}{n}\right) \\ & - \sum_x \sum_y (p(x,y) + \delta) \ln (p(x,y) - \delta) + \frac{|\mathcal{X}| \ln n}{n} \qquad (5.138) \\ =\ & -H_e(X) + H_e(X,Y) + L_u(n,\delta) \qquad (5.139) \\ =\ & H_e(Y|X) + L_u(n,\delta), \qquad (5.140) \end{aligned}$$

where $L_u(n,\delta)$ denotes a function of n and δ which tends to 0 as $n \to \infty$ and $\delta \to 0$.

Now from (5.129) and (5.140) and changing the base of the logarithm to 2, we have

$$H(Y|X) + L_l(n,\delta) \leq n^{-1} \log M(K) \leq H(Y|X) + L_u(n,\delta). \qquad (5.141)$$

From the lower bound above, we immediately have

$$n^{-1} \log |T^n_{[Y|X]\delta}(\mathbf{x})| \geq H(Y|X) + L_l(n,\delta). \qquad (5.142)$$

To upper bound $|T^n_{[Y|X]\delta}(\mathbf{x})|$, we have to consider that there are in general more than one possible set of $\{K(x,y)\}$. Toward this end, by observing that $K(x,y)$ takes values between 0 and n, the total number of possible sets of $\{K(x,y)\}$ is at most

$$(n+1)^{|\mathcal{X}||\mathcal{Y}|}. \qquad (5.143)$$

This is a very crude bound but it is good enough for our purpose. Then from the upper bound in (5.141), we obtain

$$n^{-1} \log |T^n_{[Y|X]\delta}(\mathbf{x})| \leq n^{-1} \log(n+1)^{|\mathcal{X}||\mathcal{Y}|} + H(Y|X) + L_u(n,\delta). \quad (5.144)$$

Since the first term of the above upper bound tends to 0 as $n \to \infty$, combining (5.142) and (5.144), it is possible to choose a ν (depending on n and δ) which tends to 0 as $n \to \infty$ and $\delta \to 0$ such that

$$|n^{-1} \log |T_{[Y|X]\delta^n(\mathbf{x})}| - H(Y|X)| \leq \nu, \qquad (5.145)$$

or

$$2^{n(H(Y|X)-\nu)} \leq |T^n_{[Y|X]\delta}(\mathbf{x})| \leq 2^{n(H(Y|X)+\nu)}. \qquad (5.146)$$

The theorem is proved. □

The above theorem says that for any typical $\mathbf{x}$, as long as there is one typical $\mathbf{y}$ such that $(\mathbf{x},\mathbf{y})$ is jointly typical, there are approximately $2^{nH(Y|X)}$ $\mathbf{y}$ such

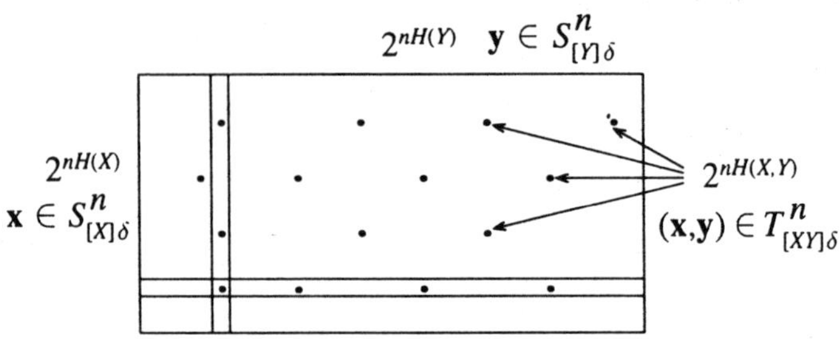

Figure 5.2. A two-dimensional strong joint typicality array.

that $(\mathbf{x}, \mathbf{y})$ is jointly typical. This theorem has the following corollary that the number of such typical $\mathbf{x}$ grows with n at almost the same rate as the total number of typical $\mathbf{x}$.

COROLLARY 5.11 *For a joint distribution $p(x, y)$ on $\mathcal{X} \times \mathcal{Y}$, let $S^n_{[X]\delta}$ be the set of all sequences $\mathbf{x} \in T^n_{[X]\delta}$ such that $T^n_{[Y|X]\delta}(\mathbf{x})$ is nonempty. Then*

$$|S^n_{[X]\delta}| \geq (1 - \delta)2^{n(H(X)-\psi)}, \tag{5.147}$$

where $\psi \to 0$ as $n \to \infty$ and $\delta \to 0$.

Proof By the consistency of strong typicality (Theorem 5.7), if $(\mathbf{x}, \mathbf{y}) \in T^n_{[XY]\delta}$, then $\mathbf{x} \in T^n_{[X]\delta}$. In particular, $\mathbf{x} \in S^n_{[X]\delta}$. Then

$$T^n_{[XY]\delta} = \bigcup_{\mathbf{x} \in S^n_{[X]\delta}} \{(\mathbf{x}, \mathbf{y}) : \mathbf{y} \in T^n_{[Y|X]\delta}(\mathbf{x})\}. \tag{5.148}$$

Using the lower bound on $|T^n_{[XY]\delta}|$ in Theorem 5.8 and the upper bound on $|T^n_{[Y|X]\delta}(\mathbf{x})|$ in the last theorem, we have

$$(1 - \delta)2^{n(H(X,Y)-\lambda)} \leq |T^n_{[XY]\delta}| \leq |S^n_{[X]\delta}|2^{n(H(Y|X)+\nu)} \tag{5.149}$$

which implies

$$|S^n_{[X]\delta}| \geq (1 - \delta)2^{n(H(X)-(\lambda+\nu))}. \tag{5.150}$$

The theorem is proved upon letting $\psi = \lambda + \nu$. □

We have established a rich set of structural properties for strong typicality with respect to a bivariate distribution $p(x, y)$, which is summarized in the two-dimensional *strong joint typicality array* in Figure 5.2. In this array, the rows and the columns are the typical sequences $\mathbf{x} \in S^n_{[X]\delta}$ and $\mathbf{y} \in S^n_{[Y]\delta}$,

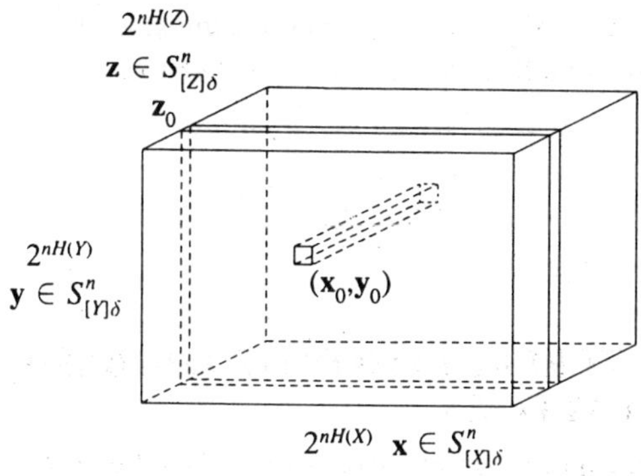

Figure 5.3. A three-dimensional strong joint typicality array.

respectively. The total number of rows and columns are approximately equal to $2^{nH(X)}$ and $2^{nH(Y)}$, respectively. An entry indexed by $(\mathbf{x}, \mathbf{y})$ receives a dot if $(\mathbf{x}, \mathbf{y})$ is strongly jointly typical. The total number of dots is approximately equal to $2^{nH(X,Y)}$. The number of dots in each row is approximately equal to $2^{nH(Y|X)}$, while the number of dots in each column is approximately equal to $2^{nH(X|Y)}$.

For reasons which will become clear in Chapter 16, the strong joint typicality array in Figure 5.2 is said to exhibit an *asymptotic quasi-uniform* structure. By a two-dimensional asymptotic quasi-uniform structure, we mean that in the array all the columns have approximately the same number of dots, and all the rows have approximately the same number of dots. The strong joint typicality array for a multivariate distribution continues to exhibit an asymptotic quasi-uniform structure. The three-dimensional strong joint typicality array with respect to a distribution $p(x, y, z)$ is illustrated in Figure 5.3. As before, an entry $(\mathbf{x}, \mathbf{y}, \mathbf{z})$ receives a dot if $(\mathbf{x}, \mathbf{y}, \mathbf{z})$ is strongly jointly typical. This is not shown in the figure otherwise it will be very confusing. The total number of dots in the whole array is approximately equal to $2^{nH(X,Y,Z)}$. These dots are distributed in the array such that all the planes parallel to each other have approximately the same number of dots, and all the cylinders parallel to each other have approximately the same number of dots. More specifically, the total number of dots on the plane for any fixed $\mathbf{z}_0 \in S^n_{[Z]\delta}$ (as shown) is approximately equal to $2^{nH(X,Y|Z)}$, and the total number of dots in the cylinder for any fixed $(\mathbf{x}_0, \mathbf{y}_0)$ pair in $S^n_{[XY]\delta}$ (as shown) is approximately equal to $2^{nH(Z|X,Y)}$, so on and so forth.

We see from the strong AEP and Corollary 5.11 that $S^n_{[X]\delta}$ and $T^n_{[X]\delta}$ grow with n at approximately the same rate. We end this section by stating in the next proposition that $S^n_{[X]\delta}$ indeed contains almost all the probability when n is large. The proof is left as an exercise (see Problem 5).

PROPOSITION 5.12 *With respect to a joint distribution $p(x,y)$ on $\mathcal{X} \times \mathcal{Y}$, for any $\delta > 0$,*

$$\Pr\{\mathbf{X} \in S^n_{[X]\delta}\} > 1 - \delta \tag{5.151}$$

for n sufficiently large.

5.4 AN INTERPRETATION OF THE BASIC INEQUALITIES

The asymptotic quasi-uniform structure exhibited in a strong joint typicality array discussed in the last section is extremely important in information theory. Later in the book, we will see how this structure is involved in proving results such as the channel coding theorem and the rate-distortion theorem. In this section, we show how the basic inequalities can be revealed by examining this structure. It has further been shown by Chan [41] that all unconstrained information inequalities can be obtained from this structure, thus giving a physical meaning to these inequalities.

Consider random variables X, Y, and Z and a fixed $\mathbf{z} \in S^n_{[Z]\delta}$, so that $T^n_{[XY|Z]\delta}(\mathbf{z})$ is nonempty. By the consistency of strong typicality, if $(\mathbf{x}, \mathbf{y}, \mathbf{z}) \in T^n_{[XYZ]\delta}$, then $(\mathbf{x}, \mathbf{z}) \in T^n_{[XZ]\delta}$ and $(\mathbf{y}, \mathbf{z}) \in T^n_{[YZ]\delta}$, or $\mathbf{x} \in T^n_{[X|Z]\delta}(\mathbf{z})$ and $\mathbf{y} \in T^n_{[Y|Z]\delta}(\mathbf{z})$, respectively. Thus

$$T^n_{[XY|Z]\delta}(\mathbf{z}) \subset T^n_{[X|Z]\delta}(\mathbf{z}) \times T^n_{[Y|Z]\delta}(\mathbf{z}), \tag{5.152}$$

which implies

$$|T^n_{[XY|Z]\delta}(\mathbf{z})| \le |T^n_{[X|Z]\delta}(\mathbf{z})||T^n_{[Y|Z]\delta}(\mathbf{z})|. \tag{5.153}$$

Applying the lower bound in Theorem 5.9 to $T^n_{[XY|Z]\delta}(\mathbf{z})$ and the upper bound to $T^n_{[X|Z]\delta}(\mathbf{z})$ and $T^n_{[Y|Z]\delta}(\mathbf{z})$, we have

$$2^{n(H(X,Y|Z)-\zeta)} \le 2^{n(H(X|Z)+\gamma)} 2^{n(H(Y|Z)+\phi)}, \tag{5.154}$$

where $\zeta, \gamma, \phi \to 0$ as $n \to \infty$ and $\delta \to 0$. Taking logarithm to the base 2 and dividing by n, we obtain

$$H(X,Y|Z) \le H(X|Z) + H(Y|Z) \tag{5.155}$$

upon letting $n \to \infty$ and $\delta \to 0$. This inequality is equivalent to

$$I(X;Y|Z) \ge 0. \tag{5.156}$$

Thus we have proved the nonnegativity of conditional mutual information. Since all Shannon's information measures are special cases of conditional mutual information, we have proved the nonnegativity of all Shannon's information measures, namely the basic inequalities.

PROBLEMS

1. Show that $(\mathbf{x}, \mathbf{y}) \in T^n_{[X,Y]\delta}$ and $(\mathbf{y}, \mathbf{z}) \in T^n_{[Y,Z]\delta}$ do not imply $(\mathbf{x}, \mathbf{z}) \in T^n_{[X,Z]\delta}$.

2. Let $\mathbf{X} = (X_1, X_2, \cdots, X_n)$, where X_k are i.i.d. with generic random variable X. Prove that

$$\Pr\{\mathbf{X} \in T^n_{[X]\delta}\} \geq 1 - \frac{|\mathcal{X}|^3}{n\delta^2}$$

for any n and $\delta > 0$. This shows that $\Pr\{\mathbf{X} \in T^n_{[X]\delta}\} \to 1$ as $\delta \to 0$ and $n \to \infty$ if $\sqrt{n}\delta \to \infty$.

3. Prove that for a discrete random variable X with an infinite alphabet, Property 2 of the strong AEP holds, while Properties 1 and 3 do not hold.

4. Consider a sequence $\mathbf{x} = (x_1, x_2, \cdots, x_n)$ where x_i is in some finite alphabet $\mathcal{X}$ for all i. The type $P_{\mathbf{x}}$ of the sequence $\mathbf{x}$ is the empirical probability distribution over $\mathcal{X}$ in $\mathbf{x}$, i.e., $P_{\mathbf{x}}(x) = N(x; \mathbf{x})/n$ for all $x \in \mathcal{X}$. Prove that there are a total of $\binom{n+|\mathcal{X}|-1}{n}$ distinct types $P_{\mathbf{x}}$. Hint: There are $\binom{a+b-1}{a}$ ways to distribute a identical balls in b boxes.

5. a) Show that for a joint distribution $p(x, y)$, for any $\delta > 0$, there exists $0 < \delta' < \delta$ such that $T^n_{[X]\delta'} \subset S^n_{[X]\delta}$ for sufficiently large n. Hint: Take δ' to be as small as necessary.

   b) Prove Proposition 5.12.

6. Let $\mathcal{P}(\mathcal{X})$ be the set of all probability distributions over a finite alphabet $\mathcal{X}$. Find a polynomial $Q(n)$ such that for any integer n, there exists a subset $\mathcal{P}_n(\mathcal{X})$ of $\mathcal{P}(\mathcal{X})$ such that

   a) $|\mathcal{P}_n(\mathcal{X})| \leq Q(n)$;

   b) for all $P \in \mathcal{P}(\mathcal{X})$, there exists $P_n \in \mathcal{P}_n(\mathcal{X})$ such that

$$|P_n(x) - P(x)| < \frac{1}{n}$$

   for all $x \in \mathcal{X}$.

   Hint: Let $\mathcal{P}_n(\mathcal{X})$ be the set of all probability distributions over $\mathcal{X}$ such that all the probability masses can be expressed as fractions with denominator n.

7. Let p be any probability distribution over a finite set $\mathcal{X}$ and η be a real number in $(0, 1)$. Prove that for any subset A of $\mathcal{X}^n$ with $p^n(A) \geq \eta$,

$$|A \cap T^n_{[X]\delta}| \geq 2^{n(H(p)-\delta')},$$

where $\delta' \to 0$ as $\delta \to 0$ and $n \to \infty$.

HISTORICAL NOTES

Berger [21] introduced strong typicality which was further developed into the method of types in the book by Csiszár and Körner [55]. The treatment of the subject here has a level between the two, while avoiding the heavy notation in [55]. The interpretation of the basic inequalities in Section 5.4 is a preamble to the relation between entropy and groups to be discussed in Chapter 16.

Chapter 6

THE I-MEASURE

In Chapter 2, we have shown the relationship between Shannon's information measures for two random variables by the diagram in Figure 2.2. For convenience, Figure 2.2 is reproduced in Figure 6.1 with the random variables X and Y replaced by X_1 and X_2, respectively. This diagram suggests that Shannon's information measures for any $n \geq 2$ random variables may have a set-theoretic structure.

In this chapter, we develop a theory which establishes a one-to-one correspondence between Shannon's information measures and set theory in full generality. With this correspondence, manipulations of Shannon's information measures can be viewed as set operations, thus allowing the rich suite of tools in set theory to be used in information theory. Moreover, the structure of Shannon's information measures can easily be visualized by means of an

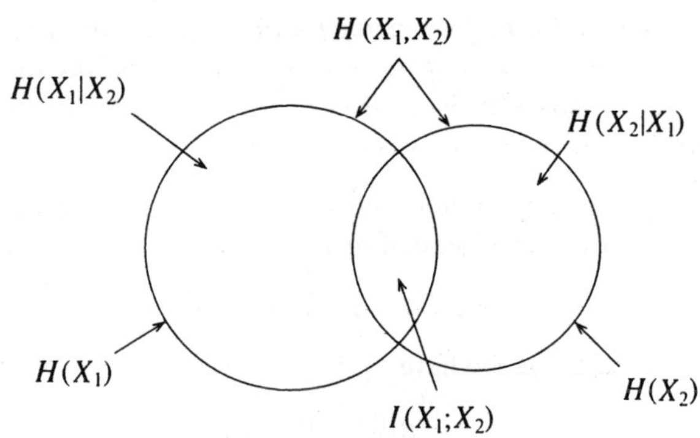

Figure 6.1. Relationship between entropies and mutual information for two random variables.

information diagram if four or fewer random variables are involved. The use of information diagrams simplifies many difficult proofs in information theory problems. More importantly, these results, which may be difficult to discover in the first place, can easily be obtained by inspection of an information diagram.

The main concepts to be used in this chapter are from measure theory. However, it is not necessary for the reader to know measure theory to read this chapter.

6.1 PRELIMINARIES

In this section, we introduce a few basic concepts in measure theory which will be used subsequently. These concepts will be illustrated by simple examples.

DEFINITION 6.1 *The field $\mathcal{F}_n$ generated by sets $\tilde{X}_1, \tilde{X}_2, \cdots, \tilde{X}_n$ is the collection of sets which can be obtained by any sequence of usual set operations (union, intersection, complement, and difference) on $\tilde{X}_1, \tilde{X}_2, \cdots, \tilde{X}_n$.*

DEFINITION 6.2 *The atoms of $\mathcal{F}_n$ are sets of the form $\cap_{i=1}^n Y_i$, where Y_i is either $\tilde{X}_i$ or $\tilde{X}_i^c$, the complement of $\tilde{X}_i$.*

There are 2^n atoms and 2^{2^n} sets in $\mathcal{F}_n$. Evidently, all the atoms in $\mathcal{F}_n$ are disjoint, and each set in $\mathcal{F}_n$ can be expressed uniquely as the union of a subset of the atoms of $\mathcal{F}_n$[1]. We assume that the sets $\tilde{X}_1, \tilde{X}_2, \cdots, \tilde{X}_n$ intersect with each other generically, i.e., all the atoms of $\mathcal{F}_n$ are nonempty unless otherwise specified.

EXAMPLE 6.3 *The sets $\tilde{X}_1$ and $\tilde{X}_2$ generate the field $\mathcal{F}_2$. The atoms of $\mathcal{F}_2$ are*

$$\tilde{X}_1 \cap \tilde{X}_2, \tilde{X}_1^c \cap \tilde{X}_2, \tilde{X}_1 \cap \tilde{X}_2^c, \tilde{X}_1^c \cap \tilde{X}_2^c, \tag{6.1}$$

which are represented by the four distinct regions in the Venn diagram in Figure 6.2. The field $\mathcal{F}_2$ consists of the unions of subsets of the atoms in (6.1). There are a total of 16 sets in $\mathcal{F}_2$, which are precisely all the sets which can be obtained from $\tilde{X}_1$ and $\tilde{X}_2$ by the usual set operations.

DEFINITION 6.4 *A real function μ defined on $\mathcal{F}_n$ is called a signed measure if it is set-additive, i.e., for disjoint A and B in $\mathcal{F}_n$,*

$$\mu(A \cup B) = \mu(A) + \mu(B). \tag{6.2}$$

For a signed measure μ, we have

$$\mu(\emptyset) = 0, \tag{6.3}$$

[1] We adopt the convention that the union of the empty subset of the atoms of $\mathcal{F}_n$ is the empty set.

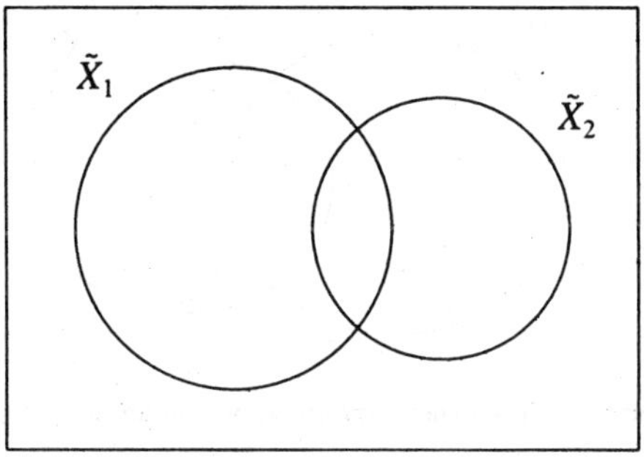

Figure 6.2. The Venn diagram for $\tilde{X}_1$ and $\tilde{X}_2$.

which can be seen as follows. For any A in $\mathcal{F}_n$,

$$\mu(A) = \mu(A \cup \emptyset) = \mu(A) + \mu(\emptyset) \tag{6.4}$$

by set-additivity because A and $\emptyset$ are disjoint, which implies (6.3).

A signed measure μ on $\mathcal{F}_n$ is completely specified by its values on the atoms of $\mathcal{F}_n$. The values of μ on the other sets in $\mathcal{F}_n$ can be obtained via set-additivity.

EXAMPLE 6.5 *A signed measure μ on $\mathcal{F}_2$ is completely specified by the values*

$$\mu(\tilde{X}_1 \cap \tilde{X}_2), \mu(\tilde{X}_1^c \cap \tilde{X}_2), \mu(\tilde{X}_1 \cap \tilde{X}_2^c), \mu(\tilde{X}_1^c \cap \tilde{X}_2^c). \tag{6.5}$$

The value of μ on $\tilde{X}_1$, for example, can be obtained as

$$\begin{aligned} \mu(\tilde{X}_1) &= \mu((\tilde{X}_1 \cap \tilde{X}_2) \cup (\tilde{X}_1 \cap \tilde{X}_2^c)) & (6.6)\\ &= \mu(\tilde{X}_1 \cap \tilde{X}_2) + \mu(\tilde{X}_1 \cap \tilde{X}_2^c). & (6.7) \end{aligned}$$

6.2 THE *I*-MEASURE FOR TWO RANDOM VARIABLES

To fix ideas, we first formulate in this section the one-to-one correspondence between Shannon's information measures and set theory for two random variables. For random variables X_1 and X_2, let $\tilde{X}_1$ and $\tilde{X}_2$ be sets corresponding to X_1 and X_2, respectively. The sets $\tilde{X}_1$ and $\tilde{X}_2$ generates the field $\mathcal{F}_2$ whose atoms are listed in (6.1). In our formulation, we set the universal set Ω to $\tilde{X}_1 \cup \tilde{X}_2$ for reasons which will become clear later. With this choice of Ω, the Venn diagram for $\tilde{X}_1$ and $\tilde{X}_2$ is represented by the diagram in Figure 6.3. For simplicity, the sets $\tilde{X}_1$ and $\tilde{X}_2$ are respectively labeled by X_1 and X_2 in the diagram. We call this the *information diagram* for the random variables X_1 and X_2. In this diagram, the universal set, which is the union of $\tilde{X}_1$ and $\tilde{X}_2$, is not shown explicitly just as in a usual Venn diagram. Note that with our choice

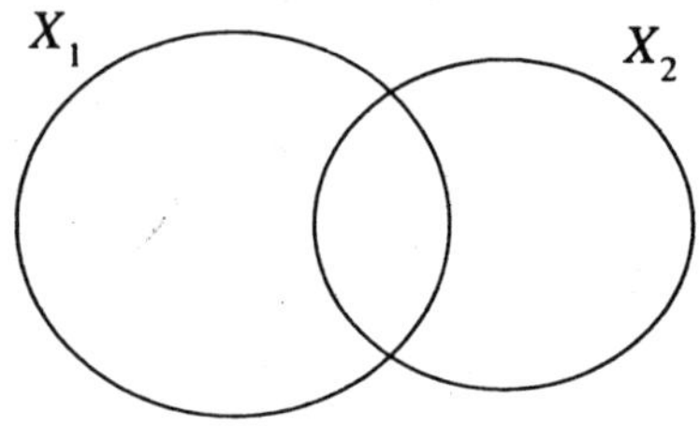

Figure 6.3. The generic information diagram for X_1 and X_2.

of the universal set, the atom $\tilde{X}_1^c \cap \tilde{X}_2^c$ degenerates to the empty set, because

$$\tilde{X}_1^c \cap \tilde{X}_2^c = (\tilde{X}_1 \cup \tilde{X}_2)^c = \Omega^c = \emptyset. \tag{6.8}$$

Thus this atom is not shown in the information diagram in Figure 6.3.

For random variables X_1 and X_2, the Shannon's information measures are

$$H(X_1), H(X_2), H(X_1|X_2), H(X_2|X_1), H(X_1,X_2), I(X_1;X_2). \tag{6.9}$$

Writing $A \cap B^c$ as $A - B$, we now define a signed measure[2] μ^* by

$$\mu^*(\tilde{X}_1 - \tilde{X}_2) = H(X_1|X_2) \tag{6.10}$$
$$\mu^*(\tilde{X}_2 - \tilde{X}_1) = H(X_2|X_1), \tag{6.11}$$

and

$$\mu^*(\tilde{X}_1 \cap \tilde{X}_2) = I(X_1;X_2). \tag{6.12}$$

These are the values of μ^* on the nonempty atoms of $\mathcal{F}_2$ (i.e., atoms of $\mathcal{F}_2$ other than $\tilde{X}_1^c \cap \tilde{X}_2^c$). The values of μ^* on the other sets in $\mathcal{F}_2$ can be obtained via set-additivity. In particular, the relations

$$\mu^*(\tilde{X}_1 \cup \tilde{X}_2) = H(X_1,X_2) \tag{6.13}$$
$$\mu^*(\tilde{X}_1) = H(X_1), \tag{6.14}$$

and

$$\mu^*(\tilde{X}_2) = H(X_2) \tag{6.15}$$

can readily be verified. For example, (6.13) is seen to be true by considering

$$\mu^*(\tilde{X}_1 \cup \tilde{X}_2)$$
$$= \mu^*(\tilde{X}_1 - \tilde{X}_2) + \mu^*(\tilde{X}_2 - \tilde{X}_1) + \mu^*(\tilde{X}_1 \cap \tilde{X}_2) \tag{6.16}$$
$$= H(X_1|X_2) + H(X_2|X_1) + I(X_1;X_2) \tag{6.17}$$
$$= H(X_1,X_2). \tag{6.18}$$

[2] It happens that μ^* defined here for $n = 2$ assumes only nonnegative values, but we will see in Section 6.4 that μ^* can assume negative values for $n \geq 3$.

The right hand sides of (6.10) to (6.15) are the six Shannon's information measures for X_1 and X_2 in (6.9). Now observe that (6.10) to (6.15) are consistent with how the Shannon's information measures on the right hand side are identified in Figure 6.1, with the left circle and the right circle representing the sets $\tilde{X}_1$ and $\tilde{X}_2$, respectively. Specifically, in each of these equations, the left hand side and the right hand side correspond to each other via the following substitution of symbols:

$$\begin{array}{rcl} H/I & \leftrightarrow & \mu^* \\ , & \leftrightarrow & \cup \\ ; & \leftrightarrow & \cap \\ | & \leftrightarrow & -. \end{array} \tag{6.19}$$

Note that we make no distinction between the symbols H and I in this substitution. Thus, for two random variables X_1 and X_2, Shannon's information measures can be regarded formally as a signed measure on $\mathcal{F}_2$. We will refer to μ^* as the *I-Measure* for the random variables X_1 and X_2[3].

Upon realizing that Shannon's information measures can be viewed as a signed measure, we can apply the rich family of operations in set theory to information theory. This explains why Figure 6.1 or Figure 6.3 represents the relationships among all Shannon's information measures for two random variables correctly. As an example, consider the following set identity which is readily identified in Figure 6.3:

$$\mu^*(\tilde{X}_1 \cup \tilde{X}_2) = \mu^*(\tilde{X}_1) + \mu^*(\tilde{X}_2) - \mu^*(\tilde{X}_1 \cap \tilde{X}_2) \tag{6.20}$$

This identity is a special case of the inclusion-exclusion formula in set theory. By means of the substitution of symbols in (6.19), we immediately obtain the information identity

$$H(X_1, X_2) = H(X_1) + H(X_2) - I(X_1; X_2). \tag{6.21}$$

We end this section with a remark. The value of μ^* on the atom $\tilde{X}_1^c \cap \tilde{X}_2^c$ has no apparent information-theoretic meaning. In our formulation, we set the universal set Ω to $\tilde{X}_1 \cup \tilde{X}_2$ so that the atom $\tilde{X}_1^c \cap \tilde{X}_2^c$ degenerates to the empty set. Then $\mu^*(\tilde{X}_1^c \cap \tilde{X}_2^c)$ naturally vanishes because μ^* is a measure, so that μ^* is completely specified by all Shannon's information measures involving the random variables X_1 and X_2.

[3]The reader should not confuse μ^* with the probability measure defining the random variables X_1 and X_2. The former, however, is determined by the latter.

6.3 CONSTRUCTION OF THE I-MEASURE μ^*

We have constructed the I-Measure for two random variables in the last section. We now construct the I-Measure for any $n \geq 2$ random variables.

Consider n random variables $X_1, X_2, \cdots, X_n$. For any random variable X, let $\tilde{X}$ be a set corresponding to X. Let

$$\mathcal{N}_n = \{1, 2, \cdots, n\}. \tag{6.22}$$

Define the universal set Ω to be the union of the sets $\tilde{X}_1, \tilde{X}_2, \cdots, \tilde{X}_n$, i.e.,

$$\Omega = \bigcup_{i \in \mathcal{N}_n} \tilde{X}_i. \tag{6.23}$$

We use $\mathcal{F}_n$ to denote the field generated by $\tilde{X}_1, \tilde{X}_2, \cdots, \tilde{X}_n$. The set

$$A_0 = \bigcap_{i \in \mathcal{N}_n} \tilde{X}_i^c \tag{6.24}$$

is called the *empty atom* of $\mathcal{F}_n$ because

$$\bigcap_{i \in \mathcal{N}_n} \tilde{X}_i^c = \left(\bigcup_{i \in \mathcal{N}_n} \tilde{X}_i \right)^c = \Omega^c = \emptyset. \tag{6.25}$$

All the atoms of $\mathcal{F}_n$ other than A_0 are called *nonempty atoms*.

Let $\mathcal{A}$ be the set of all nonempty atoms of $\mathcal{F}_n$. Then $|\mathcal{A}|$, the cardinality of $\mathcal{A}$, is equal to $2^n - 1$. A signed measure μ on $\mathcal{F}_n$ is completely specified by the values of μ on the nonempty atoms of $\mathcal{F}_n$.

To simplify notation, we will use X_G to denote $(X_i, i \in G)$ and $\tilde{X}_G$ to denote $\cup_{i \in G} \tilde{X}_i$ for any nonempty subset G of $\mathcal{N}_n$.

THEOREM 6.6 *Let*

$$\mathcal{B} = \left\{ \tilde{X}_G : G \text{ is a nonempty subset of } \mathcal{N}_n \right\}. \tag{6.26}$$

Then a signed measure μ on $\mathcal{F}_n$ is completely specified by $\{\mu(B), B \in \mathcal{B}\}$, which can be any set of real numbers.

Proof The number of elements in $\mathcal{B}$ is equal to the number of nonempty subsets of $\mathcal{N}_n$, which is $2^n - 1$. Thus $|\mathcal{A}| = |\mathcal{B}| = 2^n - 1$. Let $k = 2^n - 1$. Let $\mathbf{u}$ be a column k-vector of $\mu(A), A \in \mathcal{A}$, and $\mathbf{h}$ be a column k-vector of $\mu(B), B \in \mathcal{B}$. Since all the sets in $\mathcal{B}$ can expressed uniquely as the union of some nonempty atoms in $\mathcal{A}$, by the set-additivity of μ, for each $B \in \mathcal{B}$, $\mu(B)$ can be expressed uniquely as the sum of some components of $\mathbf{u}$. Thus

$$\mathbf{h} = \mathbf{C}_n \mathbf{u}, \tag{6.27}$$

where $\mathbf{C}_n$ is a *unique* $k \times k$ matrix. On the other hand, it can be shown (see Appendix 6.A) that for each $A \in \mathcal{A}$, $\mu(A)$ can be expressed as a linear combination of $\mu(B), B \in \mathcal{B}$ by applications, if necessary, of the following two identities:

$$\mu(A \cap B - C) = \mu(A - C) + \mu(B - C) - \mu(A \cup B - C) \tag{6.28}$$

$$\mu(A - B) = \mu(A \cup B) - \mu(B). \tag{6.29}$$

However, the existence of the said expression does not imply its uniqueness. Nevertheless, we can write

$$\mathbf{u} = \mathbf{D}_n \mathbf{h} \tag{6.30}$$

for some $k \times k$ matrix $\mathbf{D}_n$. Upon substituting (6.27) into (6.30), we obtain

$$\mathbf{u} = (\mathbf{D}_n \mathbf{C}_n)\mathbf{u}, \tag{6.31}$$

which implies that $\mathbf{D}_n$ is the inverse of $\mathbf{C}_n$ as (6.31) holds regardless of the choice of μ. Since $\mathbf{C}_n$ is unique, so is $\mathbf{D}_n$. Therefore, $\mu(A), A \in \mathcal{A}$ are uniquely determined once $\mu(B), B \in \mathcal{B}$ are specified. Hence, a signed measure μ on $\mathcal{F}_n$ is completely specified by $\{\mu(B), B \in \mathcal{B}\}$, which can be any set of real numbers. The theorem is proved. □

We now prove the following two lemmas which are related by the substitution of symbols in (6.19).

LEMMA 6.7

$$\mu(A \cap B - C) = \mu(A \cup C) + \mu(B \cup C) - \mu(A \cup B \cup C) - \mu(C). \tag{6.32}$$

Proof From (6.28) and (6.29), we have

$$\begin{aligned} &\mu(A \cap B - C) \\ &= \mu(A - C) + \mu(B - C) - \mu(A \cup B - C) && (6.33) \\ &= (\mu(A \cup C) - \mu(C)) + (\mu(B \cup C) - \mu(C)) \\ &\quad -(\mu(A \cup B \cup C) - \mu(C)) && (6.34) \\ &= \mu(A \cup C) + \mu(B \cup C) - \mu(A \cup B \cup C) - \mu(C). && (6.35) \end{aligned}$$

The lemma is proved. □

LEMMA 6.8

$$I(X;Y|Z) = H(X,Z) + H(Y,Z) - H(X,Y,Z) - H(Z). \tag{6.36}$$

Proof Consider

$$\begin{aligned} & I(X;Y|Z) \\ &= H(X|Z) - H(X|Y,Z) && (6.37)\\ &= H(X,Z) - H(Z) - (H(X,Y,Z) - H(Y,Z)) && (6.38)\\ &= H(X,Z) + H(Y,Z) - H(X,Y,Z) - H(Z). && (6.39)\end{aligned}$$

The lemma is proved. □

We now construct the I-Measure μ^* on $\mathcal{F}_n$ using Theorem 6.6 by defining

$$\mu^*(\tilde{X}_G) = H(X_G) \tag{6.40}$$

for all nonempty subsets G of $\mathcal{N}_n$. In order for μ^* to be meaningful, it has to be consistent with all Shannon's information measures (via the substitution of symbols in (6.19)). In that case, the following must hold for all (not necessarily disjoint) subsets G, G', G'' of $\mathcal{N}_n$ where G and G' are nonempty:

$$\mu^*(\tilde{X}_G \cap \tilde{X}_{G'} - \tilde{X}_{G''}) = I(X_G; X_{G'}|X_{G''}). \tag{6.41}$$

When $G'' = \emptyset$, (6.41) becomes

$$\mu^*(\tilde{X}_G \cap \tilde{X}_{G'}) = I(X_G; X_{G'}). \tag{6.42}$$

When $G = G'$, (6.41) becomes

$$\mu^*(\tilde{X}_G - \tilde{X}_{G''}) = H(X_G|X_{G''}). \tag{6.43}$$

When $G = G'$ and $G'' = \emptyset$, (6.41) becomes

$$\mu^*(\tilde{X}_G) = H(X_G). \tag{6.44}$$

Thus (6.41) covers all the four cases of Shannon's information measures, and it is the necessary and sufficient condition for μ^* to be consistent with all Shannon's information measures.

THEOREM 6.9 *μ^* is the unique signed measure on $\mathcal{F}_n$ which is consistent with all Shannon's information measures.*

Proof Consider

$$\begin{aligned} & \mu^*(\tilde{X}_G \cap \tilde{X}_{G'} - \tilde{X}_{G''}) \\ &= \mu^*(\tilde{X}_{G\cup G''}) + \mu^*(\tilde{X}_{G'\cup G''}) - \mu^*(\tilde{X}_{G\cup G'\cup G''}) - \mu^*(\tilde{X}_{G''}) && (6.45)\\ &= H(X_{G\cup G''}) + H(X_{G'\cup G''}) - H(X_{G\cup G'\cup G''}) - H(X_{G''}) && (6.46)\\ &= I(X_G; X_{G'}|X_{G''}), && (6.47)\end{aligned}$$

where (6.45) and (6.47) follow from Lemmas 6.7 and 6.8, respectively, and (6.46) follows from (6.40), the definition of μ^*. Thus we have proved (6.41), i.e., μ^* is consistent with all Shannon's information measures.

In order that μ^* is consistent with all Shannon's information measures, for all nonempty subsets G of $\mathcal{N}_n$, μ^* has to satisfy (6.44), which in fact is the definition of μ^* in (6.40). Therefore, μ^* is the unique signed measure on $\mathcal{F}_n$ which is consistent with all Shannon's information measures. □

6.4 μ^* CAN BE NEGATIVE

In the previous sections, we have been cautious in referring to the I-Measure μ^* as a signed measure instead of a measure[4]. In this section, we show that μ^* in fact can take negative values for $n \geq 3$.

For $n = 2$, the three nonempty atoms of $\mathcal{F}_2$ are

$$\tilde{X}_1 \cap \tilde{X}_2, \tilde{X}_1 - \tilde{X}_2, \tilde{X}_2 - \tilde{X}_1. \tag{6.48}$$

The values of μ^* on these atoms are respectively

$$I(X_1; X_2), H(X_1|X_2), H(X_2|X_1). \tag{6.49}$$

These quantities are Shannon's information measures and hence nonnegative by the basic inequalities. Therefore, μ^* is always nonnegative for $n = 2$.

For $n = 3$, the seven nonempty atoms of $\mathcal{F}_3$ are

$$\tilde{X}_i - \tilde{X}_{\{j,k\}}, \tilde{X}_i \cap \tilde{X}_j - \tilde{X}_k, \tilde{X}_1 \cap \tilde{X}_2 \cap \tilde{X}_3, \tag{6.50}$$

where $1 \leq i < j < k \leq 3$. The values of μ^* on the first two types of atoms are

$$\mu^*(\tilde{X}_i - \tilde{X}_{\{j,k\}}) = H(X_i|X_j, X_k) \tag{6.51}$$

and

$$\mu^*(\tilde{X}_i \cap \tilde{X}_j - \tilde{X}_k) = I(X_i; X_j|X_k), \tag{6.52}$$

respectively, which are Shannon's information measures and therefore nonnegative. However, $\mu^*(\tilde{X}_1 \cap \tilde{X}_2 \cap \tilde{X}_3)$ does not correspond to a Shannon's information measure. In the next example, we show that $\mu^*(\tilde{X}_1 \cap \tilde{X}_2 \cap \tilde{X}_3)$ can actually be negative.

EXAMPLE 6.10 *In this example, all entropies are in the base 2. Let X_1 and X_2 be independent binary random variables with*

$$\Pr\{X_i = 0\} = \Pr\{X_i = 1\} = 0.5, \tag{6.53}$$

[4] A measure can assume only nonnegative values.

$i = 1, 2$. Let

$$X_3 = (X_1 + X_2) \bmod 2. \tag{6.54}$$

It is easy to check that X_3 has the same marginal distribution as X_1 and X_2. Thus,

$$H(X_i) = 1 \tag{6.55}$$

for $i = 1, 2, 3$. Moreover, X_1, X_2, and X_3 are pairwise independent. Therefore,

$$H(X_i, X_j) = 2 \tag{6.56}$$

and

$$I(X_i; X_j) = 0 \tag{6.57}$$

for $1 \le i < j \le 3$. We further see from (6.54) that each random variable is a function of the other two random variables. Then by the chain rule for entropy, we have

$$\begin{aligned} H(X_1, X_2, X_3) &= H(X_1, X_2) + H(X_3|X_1, X_2) && (6.58) \\ &= 2 + 0 && (6.59) \\ &= 2. && (6.60) \end{aligned}$$

Now for $1 \le i < j < k \le 3$,

$$\begin{aligned} & I(X_i; X_j|X_k) \\ &= H(X_i, X_k) + H(X_j, X_k) - H(X_1, X_2, X_3) - H(X_k) && (6.61) \\ &= 2 + 2 - 2 - 1 && (6.62) \\ &= 1, && (6.63) \end{aligned}$$

where we have invoked Lemma 6.8. It then follows that

$$\begin{aligned} \mu^*(\tilde{X}_1 \cap \tilde{X}_2 \cap \tilde{X}_3) &= \mu^*(\tilde{X}_1 \cap \tilde{X}_2) - \mu^*(\tilde{X}_1 \cap \tilde{X}_2 - \tilde{X}_3) && (6.64) \\ &= I(X_1; X_2) - I(X_1; X_2|X_3) && (6.65) \\ &= 0 - 1 && (6.66) \\ &= -1. && (6.67) \end{aligned}$$

Thus μ^* takes a negative value on the atom $\tilde{X}_1 \cap X_2 \cap \tilde{X}_3$.

Motivated by the substitution of symbols in (6.19) for Shannon's information measures, we will write $\mu^*(\tilde{X}_1 \cap \tilde{X}_2 \cap \tilde{X}_3)$ as $I(X_1; X_2; X_3)$. In general, we will write

$$\mu^*(\tilde{X}_{G_1} \cap \tilde{X}_{G_2} \cap \cdots \cap \tilde{X}_{G_m} - \tilde{X}_F) \tag{6.68}$$

as

$$I(X_{G_1}; X_{G_2}; \cdots; X_{G_m}|X_F) \tag{6.69}$$

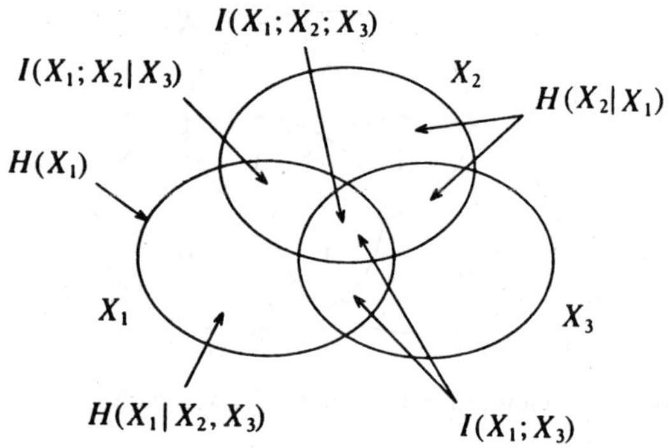

Figure 6.4. The generic information diagram for X_1, X_2, and X_3.

and refer to it as the ***mutual information*** between $X_{G_1}, X_{G_2}, \cdots, X_{G_m}$ conditioning on X_F. Then (6.64) in the above example can be written as

$$I(X_1;X_2;X_3) = I(X_1;X_2) - I(X_1;X_2|X_3). \tag{6.70}$$

For this example, $I(X_1;X_2;X_3) < 0$, which implies

$$I(X_1;X_2|X_3) > I(X_1;X_2). \tag{6.71}$$

Therefore, unlike entropy, the mutual information between two random variables can be increased by conditioning on a third random variable. Also, we note in (6.70) that although the expression on the right hand side is not symbolically symmetrical in X_1, X_2, and X_3, we see from the left hand side that it is in fact symmetrical in X_1, X_2, and X_3.

6.5 INFORMATION DIAGRAMS

We have established in Section 6.3 a one-to-one correspondence between Shannon's information measures and set theory. Therefore, it is valid to use an *information diagram*, which is a variation of a Venn diagram, to represent the relationship between Shannon's information measures.

For simplicity, a set $\tilde{X}_i$ will be labeled by X_i in an information diagram. We have seen the generic information diagram for $n = 2$ in Figure 6.3. A generic information diagram for $n = 3$ is shown in Figure 6.4. The information-theoretic labeling of the values of μ^* on some of the sets in $\mathcal{F}_3$ is shown in the diagram. As an example, the information diagram for the I-Measure for random variables X_1, X_2, and X_3 discussed in Example 6.10 is shown in Figure 6.5.

For $n \geq 4$, it is not possible to display an information diagram perfectly in two dimensions. In general, an information diagram for n random variables

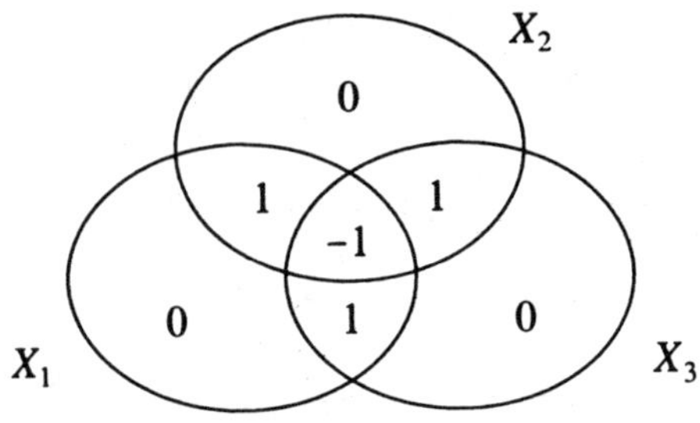

Figure 6.5. The information diagram for X_1, X_2, and X_3 in Example 6.10.

needs $n - 1$ dimensions to be displayed perfectly. Nevertheless, for $n = 4$, an information diagram can be displayed in two dimensions almost perfectly as shown in Figure 6.6. This information diagram is correct in that the region representing the set $\tilde{X}_4$ splits each atom in Figure 6.4 into two atoms. However, the adjacency of certain atoms are not displayed correctly. For example, the set $\tilde{X}_1 \cap \tilde{X}_2 \cap \tilde{X}_4^c$, which consists of the atoms $\tilde{X}_1 \cap \tilde{X}_2 \cap \tilde{X}_3 \cap \tilde{X}_4^c$ and $\tilde{X}_1 \cap \tilde{X}_2 \cap \tilde{X}_3^c \cap \tilde{X}_4^c$, is not represented by a connected region because the two atoms are not adjacent to each other.

When μ^* takes the value zero on an atom A of $\mathcal{F}_n$, we do not need to display the atom A in an information diagram because the atom A does not contribute to $\mu^*(B)$ for any set B containing the atom A. As we will see shortly, this can happen if certain Markov constraints are imposed on the random variables involved, and the information diagram can be simplified accordingly. In a generic information diagram (i.e., when there is no constraint on the random variables), however, all the atoms have to be displayed, as is implied by the next theorem.

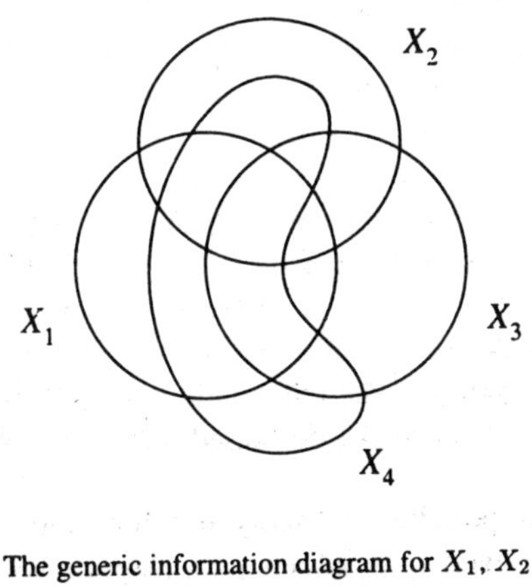

Figure 6.6. The generic information diagram for X_1, X_2, X_3, and X_4.

THEOREM 6.11 *If there is no constraint on* $X_1, X_2, \cdots, X_n$, *then* μ^* *can take any set of nonnegative values on the nonempty atoms of* $\mathcal{F}_n$.

Proof We will prove the theorem by constructing a μ^* which can take any set of nonnegative values on the nonempty atoms of $\mathcal{F}_n$. Recall that $\mathcal{A}$ is the set of all nonempty atoms of $\mathcal{F}_n$. Let $Y_A, A \in \mathcal{A}$ be mutually independent random variables. Now define the random variables $X_i, i = 1, 2, \cdots, n$ by

$$X_i = (Y_A : A \in \mathcal{A} \text{ and } A \subset \tilde{X}_i). \tag{6.72}$$

We determine the I-Measure μ^* for $X_1, X_2, \cdots, X_n$ so defined as follows. Since Y_A are mutually independent, for any nonempty subsets G of $\mathcal{N}_n$, we have

$$H(X_G) = \sum_{A \in \mathcal{A}: A \subset \tilde{X}_G} H(Y_A). \tag{6.73}$$

On the other hand,

$$H(X_G) = \mu^*(\tilde{X}_G) = \sum_{A \in \mathcal{A}: A \subset \tilde{X}_G} \mu^*(A). \tag{6.74}$$

Equating the right hand sides of (6.73) and (6.74), we have

$$\sum_{A \in \mathcal{A}: A \subset \tilde{X}_G} H(Y_A) = \sum_{A \in \mathcal{A}: A \subset \tilde{X}_G} \mu^*(A). \tag{6.75}$$

Evidently, we can make the above equality hold for all nonempty subsets G of $\mathcal{N}_n$ by taking

$$\mu^*(A) = H(Y_A) \tag{6.76}$$

for all $A \in \mathcal{A}$. By the uniqueness of μ^*, this is also the only possibility for μ^*. Since $H(Y_A)$ can take any nonnegative value by Corollary 2.44, μ^* can take any set of nonnegative values on the nonempty atoms of $\mathcal{F}_n$. The theorem is proved. □

In the rest of this section, we explore the structure of Shannon's information measures when $X_1 \to X_2 \to \cdots \to X_n$ forms a Markov chain. To start with, we consider $n = 3$, i.e., $X_1 \to X_2 \to X_3$ forms a Markov chain. Since

$$\mu^*(\tilde{X}_1 \cap \tilde{X}_2^c \cap \tilde{X}_3) = I(X_1; X_3 | X_2) = 0, \tag{6.77}$$

the atom $\tilde{X}_1 \cap \tilde{X}_2^c \cap \tilde{X}_3$ does not have to be displayed in an information diagram. As such, in constructing the information diagram, the regions representing the random variables X_1, X_2, and X_3 should overlap with each other such that the region corresponding to the atom $\tilde{X}_1 \cap \tilde{X}_2^c \cap \tilde{X}_3$ is empty, while the regions corresponding to all other nonempty atoms are nonempty. Figure 6.7 shows

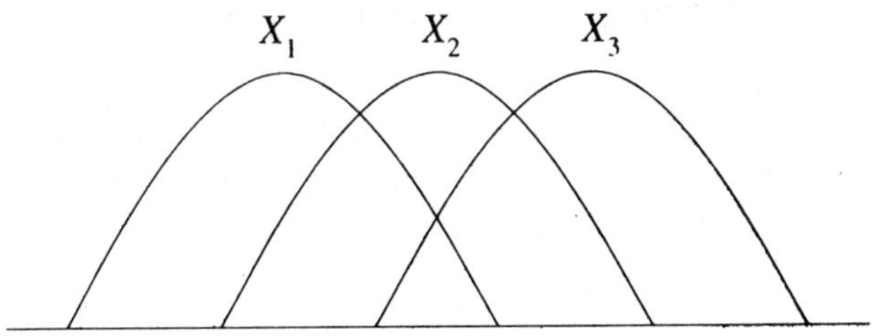

Figure 6.7. The information diagram for the Markov chain $X_1 \to X_2 \to X_3$.

such a construction, in which each random variable is represented by a *mountain*[5]. From Figure 6.7, we see that $\tilde{X}_1 \cap \tilde{X}_2 \cap \tilde{X}_3$, as the only atom on which μ^* may take a negative value, now becomes identical to the atom $\tilde{X}_1 \cap \tilde{X}_3$. Therefore, we have

$$
\begin{aligned}
I(X_1; X_2; X_3) &= \mu^*(\tilde{X}_1 \cap \tilde{X}_2 \cap \tilde{X}_3) && (6.78)\\
&= \mu^*(\tilde{X}_1 \cap \tilde{X}_3) && (6.79)\\
&= I(X_1; X_3) && (6.80)\\
&\geq 0. && (6.81)
\end{aligned}
$$

Hence, we conclude that when $X_1 \to X_2 \to X_3$ forms a Markov chain, μ^* is always nonnegative.

Next, we consider $n = 4$, i.e., $X_1 \to X_2 \to X_3 \to X_4$ forms a Markov chain. With reference to Figure 6.6, we first show that under this Markov constraint, μ^* always vanishes on certain nonempty atoms:

1. The Markov chain $X_1 \to X_2 \to X_3$ implies

$$I(X_1; X_3; X_4|X_2) + I(X_1; X_3|X_2, X_4) = I(X_1; X_3|X_2) = 0. \quad (6.82)$$

2. The Markov chain $X_1 \to X_2 \to X_4$ implies

$$I(X_1; X_3; X_4|X_2) + I(X_1; X_4|X_2, X_3) = I(X_1; X_4|X_2) = 0. \quad (6.83)$$

3. The Markov chain $X_1 \to X_3 \to X_4$ implies

$$I(X_1; X_2; X_4|X_3) + I(X_1; X_4|X_2, X_3) = I(X_1; X_4|X_3) = 0. \quad (6.84)$$

[5]This form of an information diagram for a Markov chain first appeared in Kawabata [114].

4. The Markov chain $X_2 \to X_3 \to X_4$ implies

$$I(X_1; X_2; X_4|X_3) + I(X_2; X_4|X_1, X_3) = I(X_2; X_4|X_3) = 0. \quad (6.85)$$

5. The Markov chain $(X_1, X_2) \to X_3 \to X_4$ implies

$$\begin{aligned} & I(X_1; X_2; X_4|X_3) + I(X_1; X_4|X_2, X_3) + I(X_2; X_4|X_1, X_3) \\ = \; & I(X_1, X_2; X_4|X_3) & (6.86) \\ = \; & 0. & (6.87) \end{aligned}$$

Now (6.82) and (6.83) imply

$$I(X_1; X_4|X_2, X_3) = I(X_1; X_3|X_2, X_4), \quad (6.88)$$

(6.84) and (6.88) imply

$$I(X_1; X_2; X_4|X_3) = -I(X_1; X_3|X_2, X_4), \quad (6.89)$$

and (6.85) and (6.89) imply

$$I(X_2; X_4|X_1, X_3) = I(X_1; X_3|X_2, X_4). \quad (6.90)$$

The terms on the left hand sides of (6.88), (6.89), and (6.90) are the three terms on the left hand side of (6.87). Then we substitute (6.88), (6.89), and (6.90) in (6.87) to obtain

$$\mu^*(\tilde{X}_1 \cap \tilde{X}_2^c \cap \tilde{X}_3 \cap \tilde{X}_4^c) = I(X_1; X_3|X_2, X_4) = 0. \quad (6.91)$$

From (6.82), (6.88), (6.89), and (6.90), (6.91) implies

$$\begin{aligned} \mu^*(\tilde{X}_1 \cap \tilde{X}_2^c \cap \tilde{X}_3 \cap \tilde{X}_4) &= I(X_1; X_3; X_4|X_2) = 0 & (6.92) \\ \mu^*(\tilde{X}_1 \cap \tilde{X}_2^c \cap \tilde{X}_3^c \cap \tilde{X}_4) &= I(X_1; X_4|X_2, X_3) = 0 & (6.93) \\ \mu^*(\tilde{X}_1 \cap \tilde{X}_2 \cap \tilde{X}_3^c \cap \tilde{X}_4) &= I(X_1; X_2; X_4|X_3) = 0 & (6.94) \\ \mu^*(\tilde{X}_1^c \cap \tilde{X}_2 \cap \tilde{X}_3^c \cap \tilde{X}_4) &= I(X_2; X_4|X_1, X_3) = 0. & (6.95) \end{aligned}$$

From (6.91) to (6.95), we see that μ^* always vanishes on the atoms

$$\begin{array}{c} \tilde{X}_1 \cap \tilde{X}_2^c \cap \tilde{X}_3 \cap \tilde{X}_4^c \\ \tilde{X}_1 \cap \tilde{X}_2^c \cap \tilde{X}_3 \cap \tilde{X}_4 \\ \tilde{X}_1 \cap \tilde{X}_2^c \cap \tilde{X}_3^c \cap \tilde{X}_4 \\ \tilde{X}_1 \cap \tilde{X}_2 \cap \tilde{X}_3^c \cap \tilde{X}_4 \\ \tilde{X}_1^c \cap \tilde{X}_2 \cap \tilde{X}_3^c \cap \tilde{X}_4 \end{array} \quad (6.96)$$

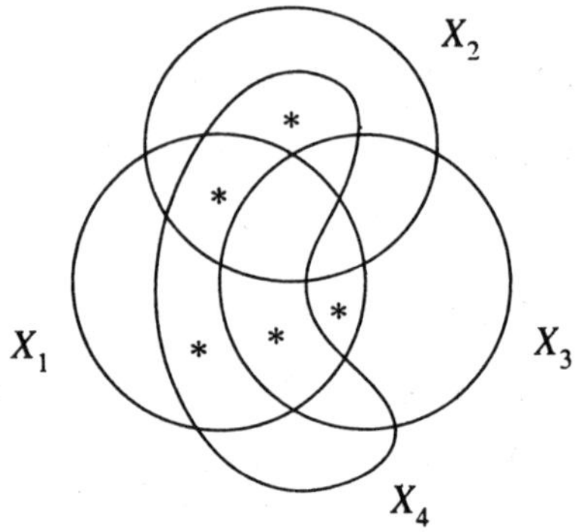

Figure 6.8. The atoms of $\mathcal{F}_4$ on which μ^* vanishes when $X_1 \to X_2 \to X_3 \to X_4$ forms a Markov chain.

of $\mathcal{F}_4$, which we mark by an asterisk in the information diagram in Figure 6.8. In fact, the reader can gain a lot of insight by letting $I(X_1; X_3|X_2, X_4) = a \geq 0$ in (6.82) and tracing the subsequent steps leading to the above conclusion in the information diagram in Figure 6.6.

It is not necessary to display the five atoms in (6.96) in an information diagram because μ^* always vanishes on these atoms. Therefore, in constructing the information diagram, the regions representing the random variables should overlap with each other such that the regions corresponding to these five nonempty atoms are empty, while the regions corresponding to the other ten nonempty atoms, namely

$$\begin{array}{c} \tilde{X}_1 \cap \tilde{X}_2^c \cap \tilde{X}_3^c \cap \tilde{X}_4^c \\ \tilde{X}_1 \cap \tilde{X}_2 \cap \tilde{X}_3^c \cap \tilde{X}_4^c \\ \tilde{X}_1 \cap \tilde{X}_2 \cap \tilde{X}_3 \cap \tilde{X}_4^c \\ \tilde{X}_1 \cap \tilde{X}_2 \cap \tilde{X}_3 \cap \tilde{X}_4 \\ \tilde{X}_1^c \cap \tilde{X}_2 \cap \tilde{X}_3^c \cap \tilde{X}_4^c \\ \tilde{X}_1^c \cap \tilde{X}_2 \cap \tilde{X}_3 \cap \tilde{X}_4^c \\ \tilde{X}_1^c \cap \tilde{X}_2 \cap \tilde{X}_3 \cap \tilde{X}_4 \\ \tilde{X}_1^c \cap \tilde{X}_2^c \cap \tilde{X}_3 \cap \tilde{X}_4^c \\ \tilde{X}_1^c \cap \tilde{X}_2^c \cap \tilde{X}_3 \cap \tilde{X}_4 \\ \tilde{X}_1^c \cap \tilde{X}_2^c \cap \tilde{X}_3^c \cap \tilde{X}_4, \end{array} \tag{6.97}$$

are nonempty. Figure 6.9 shows such a construction. The reader should compare the information diagrams in Figures 6.7 and 6.9 and observe that the latter is an extension of the former.

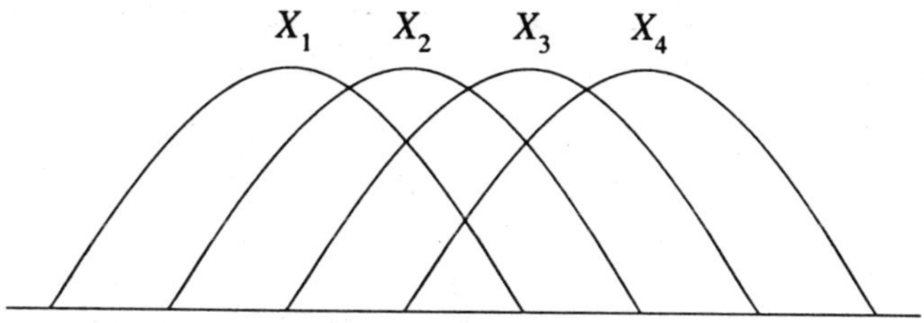

Figure 6.9. The information diagram for the Markov chain $X_1 \to X_2 \to X_3 \to X_4$.

From Figure 6.9, we see that the values of μ^* on the ten nonempty atoms in (6.97) are equivalent to

$$\begin{array}{l} H(X_1|X_2,X_3,X_4) \\ I(X_1;X_2|X_3,X_4) \\ I(X_1;X_3|X_4) \\ I(X_1;X_4) \\ H(X_2|X_1,X_3,X_4) \\ I(X_2;X_3|X_1;X_4) \\ I(X_2;X_4|X_1) \\ H(X_3|X_1,X_2,X_4) \\ I(X_3;X_4|X_1,X_2) \\ H(X_4|X_1,X_2,X_3), \end{array} \tag{6.98}$$

respectively[6]. Since these are all Shannon's information measures and thus nonnegative, we conclude that μ^* is always nonnegative.

When $X_1 \to X_2 \to \cdots \to X_n$ forms a Markov chain, for $n = 3$, there is only one nonempty atom, namely $\tilde{X}_1 \cap \tilde{X}_2^c \cap \tilde{X}_3$, on which μ^* always vanishes. This atom can be determined directly from the Markov constraint $I(X_1;X_3|X_2) = 0$. For $n = 4$, the five nonempty atoms on which μ^* always vanishes are listed in (6.96). The determination of these atoms, as we have seen, is not straightforward. We have also shown that for $n = 3$ and $n = 4$, μ^* is always nonnegative.

We will extend this theme in Chapter 7 to finite Markov random fields with Markov chains being a special case. For a Markov chain, the information diagram can always be displayed in two dimensions as in Figure 6.10, and μ^* is always nonnegative. These will be explained in Chapter 7.

[6] A formal proof will be given in Theorem 7.30.

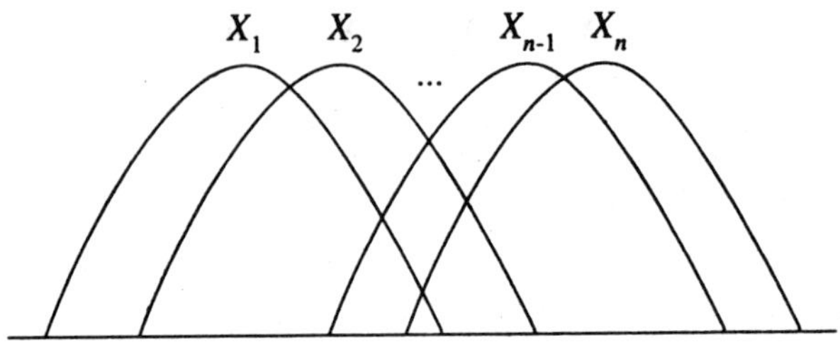

Figure 6.10. The information diagram for the Markov chain $X_1 \to X_2 \to \cdots \to X_n$.

6.6 EXAMPLES OF APPLICATIONS

In this section, we give a few examples of applications of information diagrams. These examples show how information diagrams can help solving information theory problems.

The use of an information diagram is highly intuitive. To obtain an information identity from an information diagram is WYSIWYG[7]. However, how to obtain an information inequality from an information diagram needs some explanation.

Very often, we use a Venn diagram to represent a measure μ which takes nonnegative values. If we see in the Venn diagram two sets A and B such that A is a subset of B, then we can immediately conclude that $\mu(A) \leq \mu(B)$ because

$$\mu(B) - \mu(A) = \mu(B - A) \geq 0. \tag{6.99}$$

However, an I-Measure μ^* can take negative values. Therefore, when we see in an information diagram that A is a subset of B, we cannot conclude from this fact alone that $\mu^*(A) \leq \mu^*(B)$ unless we know from the setup of the problem that μ^* is nonnegative. (For example, μ^* is nonnegative if the random variables involved form a Markov chain.) Instead, information inequalities can be obtained from an information diagram in conjunction with the basic inequalities. The following examples will illustrate how it works.

EXAMPLE 6.12 (CONCAVITY OF ENTROPY) *Let $X_1 \sim p_1(x)$ and $X_2 \sim p_2(x)$. Let*

$$X \sim p(x) = \lambda p_1(x) + \bar{\lambda} p_2(x), \tag{6.100}$$

where $0 \leq \lambda \leq 1$ and $\bar{\lambda} = 1 - \lambda$. We will show that

$$H(X) \geq \lambda H(X_1) + \bar{\lambda} H(X_2). \tag{6.101}$$

Consider the system in Figure 6.11 in which the position of the switch is determined by a random variable Z with

[7]What you see is what you get.

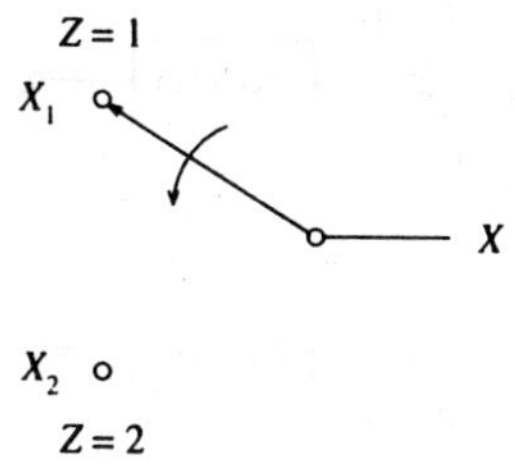

Figure 6.11. The schematic diagram for Example 6.12.

$$\Pr\{Z = 1\} = \lambda \quad and \quad \Pr\{Z = 2\} = \bar{\lambda}, \tag{6.102}$$

where Z is independent of X_1 and X_2. The switch takes position i if $Z = i$, $i = 1, 2$. Figure 6.12 shows the information diagram for X and Z. From the diagram, we see that $\tilde{X} - \tilde{Z}$ is a subset of $\tilde{X}$. Since μ^* is nonnegative for two random variables, we can conclude that

$$\mu^*(\tilde{X}) \geq \mu^*(\tilde{X} - \tilde{Z}), \tag{6.103}$$

which is equivalent to

$$H(X) \geq H(X|Z). \tag{6.104}$$

Then

$$\begin{aligned} H(X) &\geq H(X|Z) & (6.105)\\ &= \Pr\{Z = 1\}H(X|Z = 1) + \Pr\{Z = 2\}H(X|Z = 2) & (6.106)\\ &= \lambda H(X_1) + \bar{\lambda} H(X_2), & (6.107) \end{aligned}$$

proving (6.101). This shows that $H(X)$ is a concave functional of $p(x)$.

EXAMPLE 6.13 (CONVEXITY OF MUTUAL INFORMATION) *Let*

$$(X, Y) \sim p(x, y) = p(x)p(y|x). \tag{6.108}$$

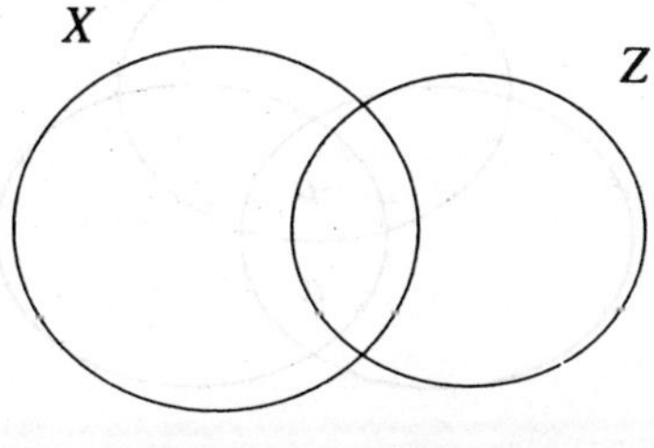

Figure 6.12. The information diagram for Example 6.12.

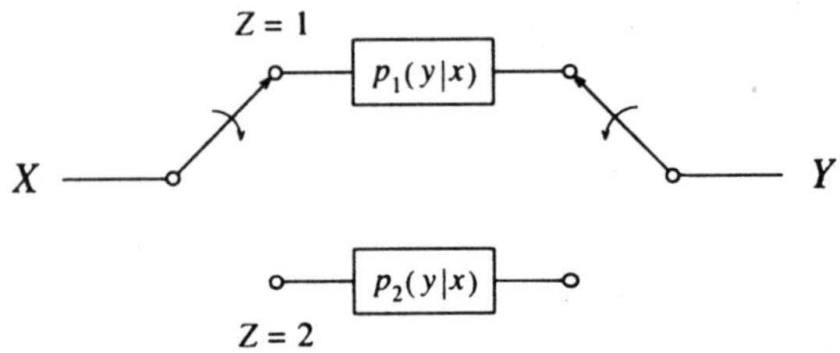

Figure 6.13. The schematic diagram for Example 6.14.

We will show that for fixed $p(x)$, $I(X;Y)$ is a convex functional of $p(y|x)$.

Let $p_1(y|x)$ and $p_2(y|x)$ be two transition matrices. Consider the system in Figure 6.13 in which the position of the switch is determined by a random variable Z as in the last example, where Z is independent of X, i.e.,

$$I(X;Z) = 0. \tag{6.109}$$

In the information diagram for X, Y, and Z in Figure 6.14, let

$$I(X;Z|Y) = a \geq 0. \tag{6.110}$$

Since $I(X;Z) = 0$, we see that

$$I(X;Y;Z) = -a, \tag{6.111}$$

because

$$I(X;Z) = I(X;Z|Y) + I(X;Y;Z). \tag{6.112}$$

Then

$$\begin{aligned} & I(X;Y) \\ = \; & I(X;Y|Z) + I(X;Y;Z) \end{aligned} \tag{6.113}$$

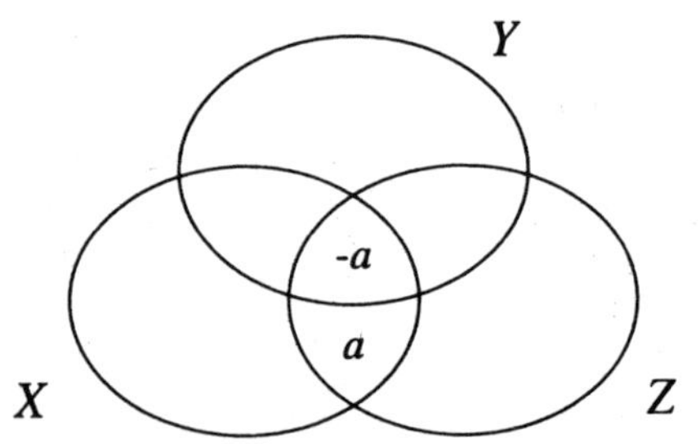

Figure 6.14. The information diagram for Example 6.14.

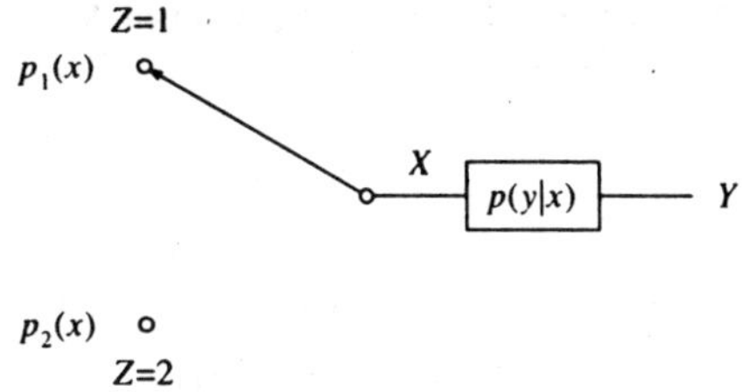

Figure 6.15. The schematic diagram for Example 6.14.

$$
\begin{aligned}
&= I(X;Y|Z) - a &(6.114)\\
&\le I(X;Y|Z) &(6.115)\\
&= \Pr\{Z=1\}I(X;Y|Z=1) + \Pr\{Z=2\}I(X;Y|Z=2) &(6.116)\\
&= \lambda I(p(x), p_1(y|x)) + \bar{\lambda} I(p(x), p_2(y|x)), &(6.117)
\end{aligned}
$$

where $I(p(x), p_i(y|x))$ denotes the mutual information between the input and output of a channel with input distribution $p(x)$ and transition matrix $p_i(y|x)$. This shows that for fixed $p(x)$, $I(X;Y)$ is a convex functional of $p(y|x)$.

EXAMPLE 6.14 (CONCAVITY OF MUTUAL INFORMATION) *Let*

$$(X,Y) \sim p(x,y) = p(x)p(y|x). \tag{6.118}$$

We will show that for fixed $p(y|x)$, $I(X;Y)$ is a concave functional of $p(x)$.

Consider the system in Figure 6.15, where the position of the switch is determined by a random variable Z as in the last example. In this system, when X is given, Y is independent of Z, or $Z \to X \to Y$ forms a Markov chain. Then μ^* is nonnegative, and the information diagram for X, Y, and Z is shown in Figure 6.16.

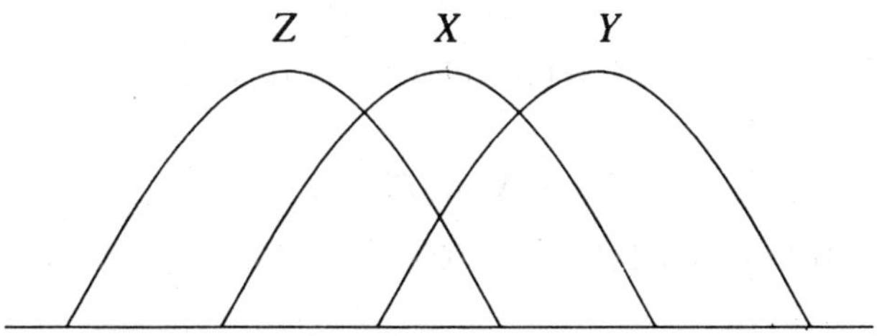

Figure 6.16. The information diagram for Example 6.14.

From Figure 6.16, since $\tilde{X} \cap \tilde{Y} - \tilde{Z}$ is a subset of $\tilde{X} \cap \tilde{Y}$ and μ^* is nonnegative, we immediately see that

$$\begin{aligned} I(X;Y) & \\ &\geq I(X;Y|Z) \qquad (6.119)\\ &= \Pr\{Z=1\}I(X;Y|Z=1) + \Pr\{Z=2\}I(X;Y|Z=2) \qquad (6.120)\\ &= \lambda I(p_1(x), p(y|x)) + \bar{\lambda} I(p_2(x), p(y|x)). \qquad (6.121)\end{aligned}$$

This shows that for fixed $p(y|x)$, $I(X;Y)$ is a concave functional of $p(x)$.

EXAMPLE 6.15 (IMPERFECT SECRECY THEOREM) *Let X be the plain text, Y be the cipher text, and Z be the key in a secret key cryptosystem. Since X can be recovered from Y and Z, we have*

$$H(X|Y,Z) = 0. \qquad (6.122)$$

We will show that this constraint implies

$$I(X;Y) \geq H(X) - H(Z). \qquad (6.123)$$

The quantity $I(X;Y)$ is a measure of the security level of the cryptosystem. In general, we want to make $I(X;Y)$ small so that the eavesdropper cannot obtain too much information about the plain text X by observing the cipher text Y. The inequality in (6.123) says that the system can attain a certain level of security only if $H(Z)$ (often called the key length) is sufficiently large. In particular, if perfect secrecy is required, i.e., $I(X;Y) = 0$, then $H(Z)$ must be at least equal to $H(X)$. This special case is known as Shannon's perfect secrecy theorem [185][8].

We now prove (6.123). Let

$$I(X;Y|Z) = a \geq 0 \qquad (6.124)$$

$$I(Y;Z|X) = b \geq 0 \qquad (6.125)$$

$$H(Z|X,Y) = c \geq 0, \qquad (6.126)$$

and

$$I(X;Y;Z) = d. \qquad (6.127)$$

(See Figure 6.17.) Since $I(Y;Z) \geq 0$,

---

[8] Shannon used a combinatorial argument to prove this theorem. An information-theoretic proof can be found in Massey [142].

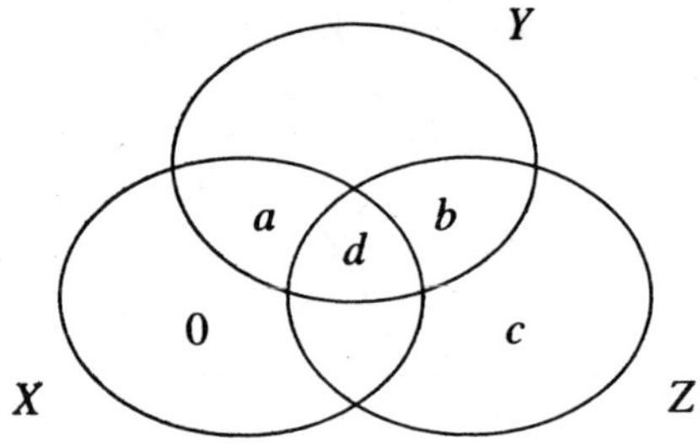

Figure 6.17. The information diagram for Example 6.15.

$$b + d \geq 0. \tag{6.128}$$

In comparing $H(X)$ with $H(Z)$, we do not have to consider $I(X;Z|Y)$ and $I(X;Y;Z)$ since they belong to both $H(X)$ and $H(Z)$. Then we see from Figure 6.17 that

$$H(X) - H(Z) = a - b - c. \tag{6.129}$$

Therefore,

$$\begin{aligned} I(X;Y) &= a + d && (6.130)\\ &\geq a - b && (6.131)\\ &\geq a - b - c && (6.132)\\ &= H(X) - H(Z), && (6.133) \end{aligned}$$

where (6.131) and (6.132) follow from (6.128) and (6.126), respectively, proving (6.123).

Note that in deriving our result, the assumptions that $H(Y|X,Z) = 0$, i.e., the cipher text is a function of the plain text and the key, and $I(X;Z) = 0$, i.e., the plain text and the key are independent, are not necessary.

EXAMPLE 6.16 *Figure 6.18 shows the information diagram for the Markov chain $X \to Y \to Z$. From this diagram, we can identify the following two information identities:*

$$\begin{aligned} I(X;Y) &= I(X;Y,Z) && (6.134)\\ H(X|Y) &= H(X|Y,Z). && (6.135) \end{aligned}$$

Since μ^* is nonnegative and $\tilde{X} \cap \tilde{Z}$ is a subset of $\tilde{X} \cap \tilde{Y}$, we have

$$I(X;Z) \leq I(X;Y), \tag{6.136}$$

which has already been obtained in Lemma 2.41. Similarly, we can also obtain

$$H(X|Y) \leq H(X|Z). \tag{6.137}$$

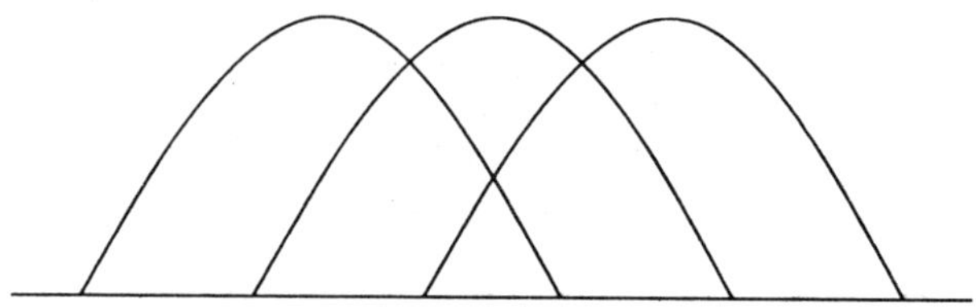

Figure 6.18. The information diagram for the Markov chain $X \to Y \to Z$.

EXAMPLE 6.17 (DATA PROCESSING THEOREM) *Figure 6.19 shows the information diagram for the Markov chain $X \to Y \to Z \to T$. Since μ^* is nonnegative and $\tilde{X} \cap \tilde{T}$ is a subset of $\tilde{Y} \cap \tilde{Z}$, we have*

$$I(X;T) \leq I(Y;Z), \tag{6.138}$$

which is the data processing theorem (Theorem 2.42).

We end this chapter by giving an application of the information diagram for a Markov chain with five random variables.

EXAMPLE 6.18 *In this example, we prove with the help of an information diagram that for five random variables X, Y, Z, T, and U such that $X \to Y \to Z \to T \to U$ forms a Markov chain,*

$$H(Y) + H(T) =$$

$$I(Z;X,Y,T,U) + I(X,Y;T,U) + H(Y|Z) + H(T|Z). \tag{6.139}$$

In the information diagram for X, Y, Z, T, and U in Figure 6.20, we first identify the atoms of $H(Y)$ and then the atoms of $H(T)$ by marking each of them by a dot. If an atom belongs to both $H(Y)$ and $H(T)$, it receives two dots. The

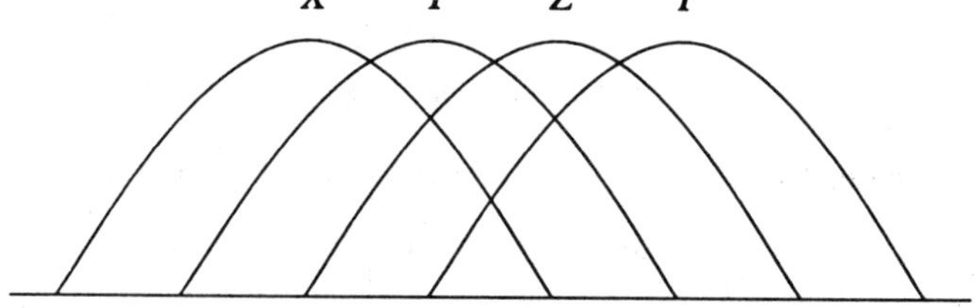

Figure 6.19. The information diagram for the Markov chain $X \to Y \to Z \to T$.

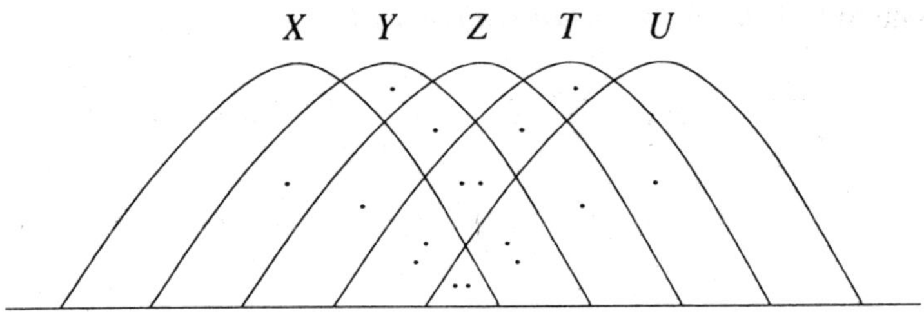

Figure 6.20. The atoms of $H(Y) + H(T)$.

resulting diagram represents

$$H(Y) + H(T). \tag{6.140}$$

By repeating the same procedure for

$$I(Z;X,Y,T,U) + I(X,Y;T,U) + H(Y|Z) + H(T|Z), \tag{6.141}$$

we obtain the information diagram in Figure 6.21. Comparing these two information diagrams, we find that they are identical. Hence, the information identity in (6.139) always holds conditioning on the Markov chain $X \to Y \to Z \to T \to U$. This identity is critical in proving an outer bound on the achievable coding rate region of the multiple descriptions problem in Fu et al. [76]. It is virtually impossible to discover this identity without the help of an information diagram!

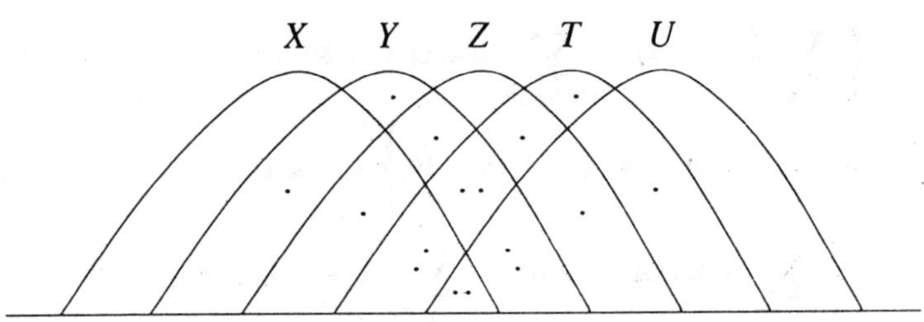

Figure 6.21. The atoms of $I(Z;X,Y,T,U) + I(X,Y;T,U) + H(Y|Z) + H(T|Z)$.

APPENDIX 6.A: A VARIATION OF THE INCLUSION-EXCLUSION FORMULA

In this appendix, we show that for each $A \in \mathcal{A}$, $\mu(A)$ can be expressed as a linear combination of $\mu(B), B \in \mathcal{B}$ via applications of (6.28) and (6.29). We first prove by using (6.28) the following variation of the inclusive-exclusive formula.

THEOREM 6-6.A.19 *For a set-additive function μ,*

$$\mu\left(\bigcap_{k=1}^{n} A_k - B\right) = \sum_{1\leq i\leq n} \mu(A_i - B) - \sum_{1\leq i<j\leq n} \mu(A_i \cup A_j - B)$$

$$+ \cdots + (-1)^{n+1}\mu(A_1 \cup A_2 \cup \cdots \cup A_n - B). \qquad (6.A.1)$$

Proof The theorem will be proved by induction on n. First, (6.A.1) is obviously true for $n = 1$. Assume (6.A.1) is true for some $n \geq 1$. Now consider

$$\mu\left(\bigcap_{k=1}^{n+1} A_k - B\right)$$

$$= \mu\left(\left(\bigcap_{k=1}^{n} A_k\right) \cap A_{n+1} - B\right) \qquad (6.A.2)$$

$$= \mu\left(\bigcap_{k=1}^{n} A_k - B\right) + \mu(A_{n+1} - B) - \mu\left(\left(\bigcap_{k=1}^{n} A_k\right) \cup A_{n+1} - B\right) \qquad (6.A.3)$$

$$= \left\{\sum_{1\leq i\leq n} \mu(A_i - B) - \sum_{1\leq i<j\leq n} \mu(A_i \cup A_j - B) + \cdots \right.$$

$$\left. + (-1)^{n+1}\mu(A_1 \cup A_2 \cup \cdots \cup A_n - B)\right\} + \mu(A_{n+1} - B)$$

$$-\mu\left(\bigcap_{k=1}^{n} (A_k \cup A_{n+1}) - B\right) \qquad (6.A.4)$$

$$= \left\{\sum_{1\leq i\leq n} \mu(A_i - B) - \sum_{1\leq i<j\leq n} \mu(A_i \cup A_j - B) + \cdots \right.$$

$$\left. + (-1)^{n+1}\mu(A_1 \cup A_2 \cup \cdots \cup A_n - B)\right\} + \mu(A_{n+1} - B)$$

$$-\left\{\sum_{1\leq i\leq n} \mu(A_i \cup A_{n+1} - B) - \sum_{1\leq i<j\leq n} \mu(A_i \cup A_j \cup A_{n+1} - B) \right.$$

$$\left. + \cdots + (-1)^{n+1}\mu(A_1 \cup A_2 \cup \cdots \cup A_n \cup A_{n+1} - B)\right\} \qquad (6.A.5)$$

$$= \sum_{1\leq i\leq n+1} \mu(A_i - B) - \sum_{1\leq i<j\leq n+1} \mu(A_i \cup A_j - B) + \cdots$$

$$+(-1)^{n+2}\mu(A_1 \cup A_2 \cup \cdots \cup A_{n+1} - B). \qquad (6.A.6)$$

In the above, (6.28) was used in obtaining (6.A.3), and the induction hypothesis was used in obtaining (6.A.4) and (6.A.5). The theorem is proved. □

Now a nonempty atom of $\mathcal{F}_n$ has the form

$$\bigcap_{i=1}^{n} Y_i, \tag{6.A.7}$$

where Y_i is either $\tilde{X}_i$ or $\tilde{X}_i^c$, and there exists at least one i such that $Y_i = \tilde{X}_i$. Then we can write the atom in (6.A.7) as

$$\bigcap_{i:Y_i=\tilde{X}_i} \tilde{X}_i - \left(\bigcup_{j:Y_j=\tilde{X}_j^c} \tilde{X}_j \right). \tag{6.A.8}$$

Note that the intersection above is always nonempty. Then using (6.A.1) and (6.29), we see that for each $A \in \mathcal{A}$, $\mu(A)$ can be expressed as a linear combination of $\mu(B)$, $B \in \mathcal{B}$.

PROBLEMS

1. Show that

$$I(X;Y;Z) = E \log \frac{p(X,Y)p(Y,Z)p(X,Z)}{p(X)p(Y)p(Z)p(X,Y,Z)}$$

and obtain a general formula for $I(X_1; X_2, ; \cdots ; X_n)$.

2. Show that $I(X;Y;Z)$ vanishes if at least one of the following conditions hold:

   a) X, Y, and Z are mutually independent;

   b) $X \to Y \to Z$ forms a Markov chain and X and Z are independent.

3. a) Verify that $I(X;Y;Z)$ vanishes for the distribution $p(x, y, z)$ given by

$$\begin{array}{lll} p(0,0,0) = 0.0625, & p(0,0,1) = 0.0772, & p(0,1,0) = 0.0625 \\ p(0,1,1) = 0.0625, & p(1,0,0) = 0.0625, & p(1,0,1) = 0.1103 \\ p(1,1,0) = 0.1875, & p(1,1,1) = 0.375. & \end{array}$$

   b) Verify that the distribution in part a) does not satisfy the conditions in Problem 2.

4. *Weak independence* X is weakly independent of Y if the rows of the transition matrix $[p(x|y)]$ are linearly dependent.

   a) Show that if X and Y are independent, then X is weakly independent of Y.

   b) Show that for random variables X and Y, there exists a random variable Z satisfying

      i) $X \to Y \to Z$

ii) X and Z are independent

iii) Y and Z are not independent

if and only if X is weakly independent of Y.

(Berger and Yeung [22].)

5. Prove that

   a) $I(X;Y;Z) \geq -\min\{I(X;Y|Z), I(Y;Z|X), I(X,Z|Y)\}$

   b) $I(X;Y;Z) \leq \min\{I(X;Y), I(Y;Z), I(X;Z)\}$.

6. a) Prove that if X and Y are independent, then $I(X,Y;Z) \geq I(X;Y|Z)$.

   b) Show that the inequality in part a) is not valid in general by giving a counterexample.

7. In Example 6.15, it was shown that $I(X;Y) \geq H(X) - H(Z)$, where X is the plain text, Y is the cipher text, and Z is the key in a secret key cryptosystem. Give an example of a secret key cryptosystem such that this inequality is tight.

8. *Secret sharing* For a given finite set $\mathcal{P}$ and a collection $\mathcal{A}$ of subsets of $\mathcal{P}$, a secret sharing scheme is a random variable S and a family of random variables $\{X_p : p \in \mathcal{P}\}$ such that for all $A \in \mathcal{A}$,

$$H(S|X_A) = 0,$$

and for all $B \notin \mathcal{A}$,

$$H(S|X_B) = H(S).$$

Here, S is the *secret* and $\mathcal{P}$ is the set of *participants* of the scheme. A participant p of the scheme possesses a *share* X_p of the secret. The set $\mathcal{A}$ specifies the *access structure* of the scheme: For a subset A of $\mathcal{P}$, by pooling their shares, if $A \in \mathcal{A}$, the participants in A can reconstruct S, otherwise they can know nothing about S.

   a) i) Prove that for $A, B \subset \mathcal{P}$, if $B \notin \mathcal{A}$ and $A \cup B \in \mathcal{A}$, then

$$H(X_A|X_B) = H(S) + H(X_A|X_B, S).$$

   ii) Prove that if $B \in \mathcal{A}$, then

$$H(X_A|X_B) = H(X_A|X_B, S).$$

   (Capocelli *et al.* [39].)

   b) Prove that for $A, B, C \subset \mathcal{P}$ such that $A \cup C \in \mathcal{A}$, $B \cup C \in \mathcal{A}$, and $C \notin \mathcal{A}$, then

$$I(X_A; X_B|X_C) \geq H(S).$$

(van Dijk [207].)

9. Consider four random variables X, Y, Z, and T which satisfy the following constraints: $H(T|X) = H(T)$, $H(T|X,Y) = 0$, $H(T|Y) = H(T)$, $H(Y|Z) = 0$, and $H(T|Z) = 0$. Prove that

   a) $H(T|X,Y,Z) = I(Z;T|X,Y) = 0$.
   b) $I(X;T|Y,Z) = I(X;Y;T|Z) = I(Y;T|X,Z) = 0$.
   c) $I(X;Z;T) = I(Y;Z;T) = 0$.
   d) $H(Y|X,Z,T) = I(X;Y|Z,T) = 0$.
   e) $I(X;Y;Z) \geq 0$.
   f) $I(X;Z) \geq H(T)$.

   The inequality in f) finds application in a secret sharing problem studied by Blundo *et al.* [33].

In the following, we use $X \perp Y|Z$ to denote that X and Y are independent given Z.

10. a) Prove that under the constraint that $X \to Y \to Z$ forms a Markov chain, $X \perp Y|Z$ and $X \perp Z$ imply $X \perp Y$.

    b) Prove that the implication in a) continues to be valid without the Markov chain constraint.

11. a) Show that $Y \perp Z|T$ does not imply $Y \perp Z|(X,T)$ by giving a counterexample.

    b) Prove that $Y \perp Z|T$ implies $Y \perp Z|(X,T)$ conditioning on $X \to Y \to Z \to T$.

12. Prove that for random variables X, Y, Z, and T,

$$\left.\begin{array}{l} X \perp Z|Y \\ (X,Y) \perp T|Z \\ Y \perp Z|T \\ Y \perp Z|X \\ X \perp T \end{array}\right\} \Rightarrow Y \perp Z.$$

    Hint: Observe that $X \perp Z|Y$ and $(X,Y) \perp T|Z$ are equivalent to $X \to Y \to Z \to T$ and use an information diagram.

13. Prove that

$$\left.\begin{array}{l} X \perp Y \\ X \perp Y|(Z,T) \\ Z \perp T|X \\ Z \perp T|Y \end{array}\right\} \Leftrightarrow \left\{\begin{array}{l} Z \perp T \\ Z \perp T|(X,Y) \\ X \perp Y|Z \\ X \perp Y|T. \end{array}\right.$$

    (Studený [202].)

HISTORICAL NOTES

The original work on the set-theoretic structure of Shannon's information measures is due to Hu [102]. It was established in this paper that every information identity implies a set identity via a substitution of symbols. This allows the tools for proving information identities to be used in proving set identities. Since the paper was published in Russian, it was largely unknown to the West until it was described in Csiszár and Körner [55]. Throughout the years, the use of Venn diagrams to represent the structure of Shannon's information measures for two or three random variables has been suggested by various authors, for example, Reza [170], Abramson [2], and Papoulis [159], but no formal justification was given until Yeung [228] introduced the I-Measure. Most of the examples in Section 6.6 were previously unpublished.

McGill [149] proposed a multiple mutual information for any number of random variables which is equivalent to the mutual information between two or more random variables discussed here. Properties of this quantity have been investigated by Kawabata [114] and Yeung [228].

Along a related direction, Han [92] viewed the linear combination of entropies as a vector space and developed a lattice-theoretic description of Shannon's information measures.

Chapter 7

MARKOV STRUCTURES

We have proved in Section 6.5 that if $X_1 \to X_2 \to X_3 \to X_4$ forms a Markov chain, the I-Measure μ^* always vanishes on the five atoms

$$
\begin{array}{l}
\tilde{X}_1 \cap \tilde{X}_2^c \cap \tilde{X}_3 \cap \tilde{X}_4^c \\
\tilde{X}_1 \cap \tilde{X}_2^c \cap \tilde{X}_3 \cap \tilde{X}_4 \\
\tilde{X}_1 \cap \tilde{X}_2^c \cap \tilde{X}_3^c \cap \tilde{X}_4 \\
\tilde{X}_1 \cap \tilde{X}_2 \cap \tilde{X}_3^c \cap \tilde{X}_4 \\
\tilde{X}_1^c \cap \tilde{X}_2 \cap \tilde{X}_3^c \cap \tilde{X}_4.
\end{array} \tag{7.1}
$$

Consequently, the I-Measure μ^* is completely specified by the values of μ^* on the other ten nonempty atoms of $\mathcal{F}_4$, and the information diagram for four random variables forming a Markov chain can be displayed in two dimensions as in Figure 6.11.

Figure 7.1 is a graph which represents the Markov chain $X_1 \to X_2 \to X_3 \to X_4$. The observant reader would notice that μ^* always vanishes on a nonempty atom A of $\mathcal{F}_4$ if and only if the graph in Figure 7.1 becomes disconnected upon removing all the vertices corresponding to the complemented set variables in A. For example, μ^* always vanishes on the atom $\tilde{X}_1 \cap \tilde{X}_2^c \cap \tilde{X}_3 \cap \tilde{X}_4^c$, and the graph in Figure 7.1 becomes disconnected upon removing vertices 2 and 4. On the other hand, μ^* does not necessarily vanish on the atom $\tilde{X}_1^c \cap \tilde{X}_2 \cap \tilde{X}_3 \cap \tilde{X}_4^c$, and the graph in Figure 7.1 remains connected

Figure 7.1. The graph representing the Markov chain $X_1 \to X_2 \to X_3 \to X_4$.

upon removing vertices 1 and 4. This observation will be explained in a more general setting in the subsequent sections.

The theory of I-Measure establishes a one-to-one correspondence between Shannon's information measures and set theory. Based on this theory, we develop in this chapter a set-theoretic characterization of a Markov structure called *full conditional mutual independence*. A Markov chain, and more generally a Markov random field, is a collection of full conditional mutual independencies. We will show that if a collection of random variables forms a Markov random field, then the structure of μ^* can be simplified. In particular, when the random variables form a Markov chain, μ^* exhibits a very simple structure so that the information diagram can be displayed in two dimensions regardless of the length of the Markov chain, and μ^* is always nonnegative. (See also Sections 6.5 and 6.6.)

The topics to be covered in this chapter are fundamental. Unfortunately, the proofs of the results are very heavy. The reader may skip this chapter without affecting further reading of the book.

7.1 CONDITIONAL MUTUAL INDEPENDENCE

In this section, we explore the effect of conditional mutual independence on the structure of the I-Measure μ^*. We begin with a simple example.

EXAMPLE 7.1 *Let X, Y, and Z be mutually independent random variables. Then*

$$I(X;Y) = I(X;Y;Z) + I(X;Y|Z) = 0. \tag{7.2}$$

Since $I(X;Y|Z) \geq 0$, we let

$$I(X;Y|Z) = a \geq 0, \tag{7.3}$$

so that

$$I(X;Y;Z) = -a. \tag{7.4}$$

Similarly,

$$I(Y;Z) = I(X;Y;Z) + I(Y;Z|X) = 0 \tag{7.5}$$

and

$$I(X;Z) = I(X;Y;Z) + I(X;Z|Y) = 0. \tag{7.6}$$

Then from (7.4), we obtain

$$I(Y;Z|X) = I(X;Z|Y) = a. \tag{7.7}$$

The relations (7.3), (7.4), and (7.7) are shown in the information diagram in Figure 7.2, which indicates that X, Y, and Z are pairwise independent.

We have proved in Theorem 2.39 that X, Y, and Z are mutually independent if and only if

$$H(X,Y,Z) = H(X) + H(Y) + H(Z). \tag{7.8}$$

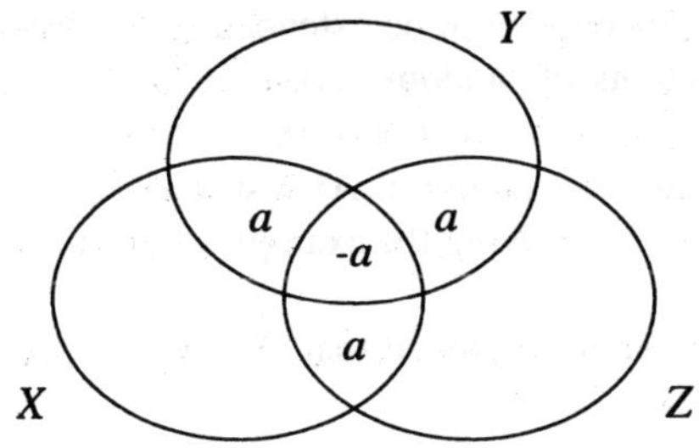

Figure 7.2. X, Y, and Z are pairwise independent.

By counting atoms in the information diagram, we see that

$$
\begin{aligned}
0 &= H(X) + H(Y) + H(Z) - H(X,Y,Z) && (7.9)\\
&= I(X;Y|Z) + I(Y;Z|X) + I(X;Z|Y) + 2I(X;Y;Z) && (7.10)\\
&= a. && (7.11)
\end{aligned}
$$

Thus $a = 0$, *which implies*

$$I(X;Y|Z), I(Y;Z|X), I(X;Z|Y), I(X;Y;Z) \tag{7.12}$$

are all equal to 0. Equivalently, μ^* *vanishes on*

$$\tilde{X} \cap \tilde{Y} - \tilde{Z}, \tilde{Y} \cap \tilde{Z} - \tilde{X}, \tilde{X} \cap \tilde{Z} - \tilde{Y}, \tilde{X} \cap \tilde{Y} \cap \tilde{Z}, \tag{7.13}$$

which are precisely the atoms in the intersection of any two of the set variables $\tilde{X}$, $\tilde{Y}$, *and* $\tilde{Z}$.

Conversely, if μ^* *vanishes on the sets in (7.13), then we see from (7.10) that (7.8) holds, i.e.,* X, Y, *and* Z *are mutually independent. Therefore,* X, Y, *and* Z *are mutually independent if and only if* μ^* *vanishes on the sets in (7.13). This is shown in the information diagram in Figure 7.3.*

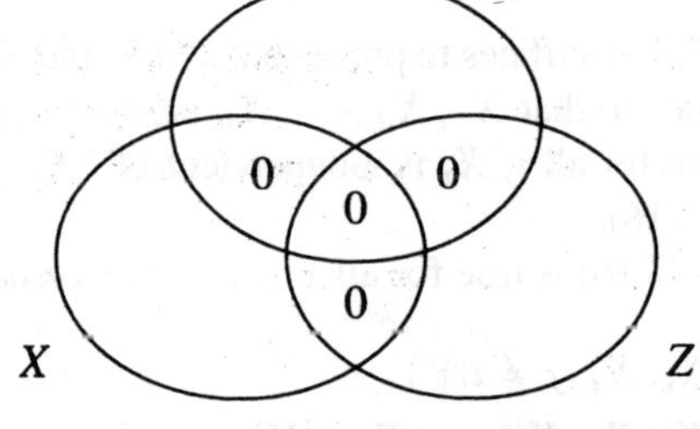

Figure 7.3. X, Y, and Z are mutually independent.

The theme of this example will be extended to conditional mutual independence among collections of random variables in Theorem 7.9, which is the main result in this section. In the rest of the section, we will develop the necessary tools for proving this theorem. At first reading, the reader should try to understand the results by studying the examples without getting into the details of the proofs.

In Theorem 2.39, we have proved that $X_1, X_2, \cdots, X_n$ are mutually independent if and only if

$$H(X_1, X_2, \cdots, X_n) = \sum_{i=1}^{n} H(X_i). \tag{7.14}$$

By conditioning on a random variable Y, one can readily prove the following.

THEOREM 7.2 *$X_1, X_2, \cdots, X_n$ are mutually independent conditioning on Y if and only if*

$$H(X_1, X_2, \cdots, X_n|Y) = \sum_{i=1}^{n} H(X_i|Y). \tag{7.15}$$

We now prove two alternative characterizations of conditional mutual independence.

THEOREM 7.3 *$X_1, X_2, \cdots, X_n$ are mutually independent conditioning on Y if and only if for all $1 \le i \le n$,*

$$I(X_i; X_j, j \neq i|Y) = 0, \tag{7.16}$$

i.e., X_i and $(X_j, j \neq i)$ are independent conditioning on Y.

Remark A conditional independency is a special case of a conditional mutual independency. However, this theorem says that a conditional mutual independency is equivalent to a set of conditional independencies.

Proof of Theorem 7.3 It suffices to prove that (7.15) and (7.16) are equivalent. Assume (7.15) is true, so that $X_1, X_2, \cdots, X_n$ are mutually independent conditioning on Y. Then for all i, X_i is independent of $(X_j, j \neq i)$ conditioning on Y. This proves (7.16).

Now assume that (7.16) is true for all $1 \le i \le n$. Consider

$$\begin{aligned} 0 &= I(X_i; X_j, j \neq i|Y) & (7.17)\\ &= I(X_i; X_1, X_2, \cdots, X_{i-1}|Y) \\ &\quad + I(X_i; X_{i+1}, \cdots, X_n|Y, X_1, X_2, \cdots, X_{i-1}). & (7.18) \end{aligned}$$

Since mutual information is always nonnegative, this implies

$$I(X_i; X_1, \cdots, X_{i-1}|Y) = 0, \tag{7.19}$$

or X_i and $(X_1, X_2, \cdots, X_{i-1})$ are independent conditioning on Y. Therefore, $X_1, X_2, \cdots, X_n$ are mutually independent conditioning on Y (see the proof of Theorem 2.39), proving (7.15). Hence, the theorem is proved. □

THEOREM 7.4 *$X_1, X_2, \cdots, X_n$ are mutually independent conditioning on Y if and only if*

$$H(X_1, X_2, \cdots, X_n|Y) = \sum_{i=1}^{n} H(X_i|Y, X_j, j \neq i). \tag{7.20}$$

Proof It suffices to prove that (7.15) and (7.20) are equivalent. Assume (7.15) is true, so that $X_1, X_2, \cdots, X_n$ are mutually independent conditioning on Y. Since for all i, X_i is independent of $X_j, j \neq i$ conditioning on Y,

$$H(X_i|Y) = H(X_i|Y, X_j, j \neq i) \tag{7.21}$$

Therefore, (7.15) implies (7.20).

Now assume that (7.20) is true. Consider

$$\begin{aligned}
& H(X_1, X_2, \cdots, X_n|Y) \\
&= \sum_{i=1}^{n} H(X_i|Y, X_1, \cdots, X_{i-1}) && (7.22) \\
&= \sum_{i=1}^{n} [H(X_i|Y, X_j, j \neq i) + I(X_i; X_{i+1}, \cdots, X_n|Y, X_1, \cdots, X_{i-1})] && (7.23) \\
&= \sum_{i=1}^{n} H(X_i|Y, X_j, j \neq i) + \sum_{i=1}^{n} I(X_i; X_{i+1}, \cdots, X_n|Y, X_1, \cdots, X_{i-1}). && (7.24)
\end{aligned}$$

Then (7.20) implies

$$\sum_{i=1}^{n} I(X_i; X_{i+1}, \cdots, X_n|Y, X_1, \cdots, X_{i-1}) = 0. \tag{7.25}$$

Since all the terms in the above summation are nonnegative, they must all be equal to 0. In particular, for $i = 1$, we have

$$I(X_1; X_2, \cdots, X_n|Y) = 0. \tag{7.26}$$

By symmetry, it can be shown that

$$I(X_i; X_j, j \neq i|Y) = 0 \tag{7.27}$$

for all $1 \leq i \leq n$. Then this implies (7.15) by the last theorem, completing the proof. □

THEOREM 7.5 *Let C and Q_i be disjoint index sets and W_i be a subset of Q_i for $1 \leq i \leq k$, where $k \geq 2$. Assume that there exist at least two i such that $W_i \neq \emptyset$. Let $X_{Q_i} = (X_l, l \in Q_i), 1 \leq i \leq k$, and $X_C = (X_l, l \in C)$ be collections of random variables. If $X_{Q_i}, 1 \leq i \leq k$, are mutually independent conditioning on X_C, then X_{W_i} such that $W_i \neq \emptyset$ are mutually independent conditioning on $(X_C, X_{Q_i - W_i}, 1 \leq i \leq k)$.*

We first give an example before we prove the theorem.

EXAMPLE 7.6 *Suppose $X_1, (X_2, X_3, X_4)$, and (X_5, X_6) are mutually independent conditioning on X_7. By Theorem 7.5, X_1, X_2, and (X_5, X_6) are mutually independent conditioning on (X_3, X_4, X_7).*

Proof of Theorem 7.5 Assume $X_{Q_i}, 1 \leq i \leq k$, are mutually independent conditioning on X_C, i.e.,

$$H(X_{Q_i}, 1 \leq i \leq k|X_C) = \sum_{i=1}^{k} H(X_{Q_i}|X_C). \tag{7.28}$$

Consider

$$\begin{aligned}
&H(X_{W_i}, 1 \leq i \leq k|X_C, X_{Q_i - W_i}, 1 \leq i \leq k) \\
&= H(X_{Q_i}, 1 \leq i \leq k|X_C) - H(X_{Q_i - W_i}, 1 \leq i \leq k|X_C) && (7.29) \\
&= \sum_{i=1}^{k} H(X_{Q_i}|X_C) \\
&\quad - \sum_{i=1}^{k} H(X_{Q_i - W_i}|X_C, X_{Q_j - W_j}, 1 \leq j \leq i-1) && (7.30) \\
&\geq \sum_{i=1}^{k} H(X_{Q_i}|X_C, X_{Q_j - W_j}, 1 \leq j \leq i-1) \\
&\quad - \sum_{i=1}^{k} H(X_{Q_i - W_i}|X_C, X_{Q_j - W_j}, 1 \leq j \leq i-1) && (7.31) \\
&= \sum_{i=1}^{k} H(X_{W_i}|X_C, X_{Q_j - W_j}, 1 \leq j \leq i) && (7.32) \\
&\geq \sum_{i=1}^{k} H(X_{W_i}|X_C, X_{Q_j - W_j}, 1 \leq j \leq k). && (7.33)
\end{aligned}$$

In the second step we have used (7.28), and the two inequalities follow because conditioning does not increase entropy. On the other hand, by the chain rule for entropy, we have

$$H(X_{W_i}, 1 \le i \le k | X_C, X_{Q_i - W_i}, 1 \le i \le k)$$
$$= \sum_{i=1}^{k} H(X_{W_i} | X_C, (X_{Q_j - W_j}, 1 \le j \le k), (X_{W_l}, 1 \le l \le i-1)). \tag{7.34}$$

Therefore, it follows from (7.33) that

$$\sum_{i=1}^{k} H(X_{W_i} | X_C, X_{Q_j - W_j}, 1 \le j \le k) \tag{7.35}$$
$$\le H(X_{W_i}, 1 \le i \le k | X_C, X_{Q_i - W_i}, 1 \le i \le k) \tag{7.36}$$
$$= \sum_{i=1}^{k} H(X_{W_i} | X_C, (X_{Q_j - W_j}, 1 \le j \le k), (X_{W_l}, 1 \le l \le i-1)). \tag{7.37}$$

However, since conditioning does not increase entropy, the ith term in the summation in (7.35) is lower bounded by the ith term in the summation in (7.37). Thus we conclude that the inequality in (7.36) is an equality. Hence, the conditional entropy in (7.36) is equal to the summation in (7.35), i.e.,

$$H(X_{W_i}, 1 \le i \le k | X_C, X_{Q_i - W_i}, 1 \le i \le k) \tag{7.38}$$
$$= \sum_{i=1}^{k} H(X_{W_i} | X_C, X_{Q_j - W_j}, 1 \le j \le k). \tag{7.39}$$

The theorem is proved. □

Theorem 7.5 specifies a set of conditional mutual independencies (CMI's) which is implied by a CMI. This theorem is crucial for understanding the effect of a CMI on the structure of the I-Measure μ^*, which we discuss next.

LEMMA 7.7 *Let $(Z_{i1}, \cdots, Z_{it_i}), 1 \le i \le r$, be r collections of random variables, where $r \ge 2$, and let Y be a random variable, such that $(Z_{i1}, \cdots, Z_{it_i})$, $1 \le i \le r$, are mutually independent conditioning on Y. Then*

$$\mu^* \left(\bigcap_{i=1}^{r} \bigcap_{j=1}^{t_i} \tilde{Z}_{ij} - \tilde{Y} \right) = 0. \tag{7.40}$$

We first prove the following set identity which will be used in proving this lemma.

LEMMA 7.8 *Let S and T be disjoint index sets, and A_i and B be sets. Let μ be a set-additive function. Then*

$$\mu\left(\left(\bigcap_{i\in S} A_i\right)\cap\left(\bigcap_{j\in T} A_j\right)-B\right)$$

$$= \sum_{S'\subset S}\sum_{T'\subset T}(-1)^{|S'|+|T'|}\left(\mu(A_{S'}-B)+\mu(A_{T'}-B)-\mu(A_{S'\cup T'}-B)\right), \tag{7.41}$$

where $A_{S'}$ denotes $\cup_{i\in S'} A_i$.

Proof The right hand side of (7.41) is equal to

$$\sum_{S'\subset S}\sum_{T'\subset T}(-1)^{|S'|+|T'|}\mu(A_{S'}-B)+\sum_{S'\subset S}\sum_{T'\subset T}(-1)^{|S'|+|T'|}\mu(A_{T'}-B)$$
$$-\sum_{S'\subset S}\sum_{T'\subset T}(-1)^{|S'|+|T'|}\mu(A_{S'\cup T'}-B). \tag{7.42}$$

Now

$$\sum_{S'\subset S}\sum_{T'\subset T}(-1)^{|S'|+|T'|}\mu(A_{S'}-B)=\sum_{S'\subset S}(-1)^{|S'|}\mu(A_{S'}-B)\sum_{T'\subset T}(-1)^{|T'|}. \tag{7.43}$$

Since

$$\sum_{T'\subset T}(-1)^{|T'|}=\sum_{k=0}^{|T|}\binom{|T|}{k}(-1)^k=0 \tag{7.44}$$

by the binomial formula[1], we conclude that

$$\sum_{S'\subset S}\sum_{T'\subset T}(-1)^{|S'|+|T'|}\mu(A_{S'}-B)=0. \tag{7.45}$$

Similarly,

$$\sum_{S'\subset S}\sum_{T'\subset T}(-1)^{|S'|+|T'|}\mu(A_{T'}-B)=0. \tag{7.46}$$

---

[1]This can be obtained by letting $a = 1$ and $b = -1$ in the binomial formula

$$(a+b)^{|T|}=\sum_{k=0}^{|T|}\binom{|T|}{k}a^k b^{|T|-k}.$$

Therefore, (7.41) is equivalent to

$$\mu\left(\left(\bigcap_{i\in S} A_i\right) \cap \left(\bigcap_{j\in T} A_j\right) - B\right) = \sum_{S'\subset S}\sum_{T'\subset T}(-1)^{|S'|+|T'|+1}\mu(A_{S'\cup T'}-B) \tag{7.47}$$

which can readily be obtained from Theorem 6.A.1. Hence, the lemma is proved. □

Proof of Lemma 7.7 We first prove the lemma for $r = 2$. By Lemma 7.8,

$$\mu^*\left(\bigcap_{i=1}^{2}\bigcap_{j=1}^{t_i}\tilde{Z}_{ij} - \tilde{Y}\right) =$$

$$\sum_{S'\subset\{1,\cdots,t_1\}}\sum_{T'\subset\{1,\cdots,t_2\}}(-1)^{|S'|+|T'|}\left[\mu^*\left(\bigcup_{j\in S'}\tilde{Z}_{1j} - \tilde{Y}\right)\right.$$

$$\left. +\mu^*\left(\bigcup_{k\in T'}\tilde{Z}_{2k} - \tilde{Y}\right) - \mu^*\left(\left(\bigcup_{j\in S'}\tilde{Z}_{1j}\right)\cup\left(\bigcup_{k\in T'}\tilde{Z}_{2k}\right) - \tilde{Y}\right)\right]. \tag{7.48}$$

The expression in the square bracket is equal to

$$H(Z_{1j}, j\in S'|Y) + H(Z_{2k}, k\in T'|Y)$$
$$-H((Z_{1j}, j\in S'), (Z_{2k}, k\in T')|Y), \tag{7.49}$$

which is equal to 0 because $(Z_{1j}, j \in S')$ and $(Z_{2k}, k \in T')$ are independent conditioning on Y. Therefore the lemma is proved for $r = 2$.

For $r > 2$, we write

$$\mu^*\left(\bigcap_{i=1}^{r}\bigcap_{j=1}^{t_i}\tilde{Z}_{ij} - \tilde{Y}\right) = \mu^*\left(\left(\bigcap_{i=1}^{r-1}\bigcap_{j=1}^{t_i}\tilde{Z}_{ij}\right)\cap\left(\bigcap_{j=1}^{t_r}\tilde{Z}_{rj}\right) - \tilde{Y}\right). \tag{7.50}$$

Since $((Z_{i1}, \cdots, Z_{it_i}), 1 \le i \le r-1)$ and $(Z_{r1}, \cdots, Z_{rt_r})$ are independent conditioning on Y, upon applying the lemma for $r = 2$, we see that

$$\mu^*\left(\bigcap_{i=1}^{r}\bigcap_{j=1}^{t_i}\tilde{Z}_{ij} - \tilde{Y}\right) = 0. \tag{7.51}$$

The lemma is proved. □

THEOREM 7.9 *Let T and $Q_i, 1 \le i \le k$, be disjoint index sets, where $k \ge 2$, and let $X_{Q_i} = (X_l, l \in Q_i), 1 \le i \le k$, and $X_T = (X_l, l \in T)$ be collections of random variables. Then $X_{Q_i}, 1 \le i \le k$, are mutually independent* *conditioning on X_T if and only if for any $W_1, W_2, \cdots, W_k$, where $W_i \subset Q_i$, $1 \leq i \leq k$, if there exist at least two i such that $W_i \neq \emptyset$, then*

$$\mu^* \left(\left(\bigcap_{i=1}^{k} \bigcap_{j \in W_i} \tilde{X}_j \right) - \tilde{X}_{T \cup (\cup_{i=1}^{k} (Q_i - W_i))} \right) = 0. \tag{7.52}$$

We first give an example before proving this fundamental result. The reader should compare this example with Example 7.6.

EXAMPLE 7.10 *Suppose $X_1, (X_2, X_3, X_4)$, and (X_5, X_6) are mutually independent conditioning on X_7. By Theorem 7.9,*

$$\mu^*(\tilde{X}_1 \cap \tilde{X}_2 \cap \tilde{X}_5 \cap \tilde{X}_6 - (\tilde{X}_3 \cup \tilde{X}_4 \cup \tilde{X}_7)) = 0. \tag{7.53}$$

However, the theorem does not say, for instance, that

$$\mu^*(\tilde{X}_2 \cap \tilde{X}_4 - (\tilde{X}_1 \cup \tilde{X}_3 \cup \tilde{X}_5 \cup \tilde{X}_6 \cup \tilde{X}_7)) \tag{7.54}$$

is equal to 0.

Proof of Theorem 7.9 We first prove the 'if' part. Assume that for any $W_1, W_2, \cdots, W_k$, where $W_i \subset Q_i$, $1 \leq i \leq k$, if there exist at least two i such that $W_i \neq \emptyset$, then (7.52) holds. Consider

$$H(X_{Q_i}, 1 \leq i \leq k | X_T) = \mu^* \left(\tilde{X}_{\cup_{i=1}^{k} Q_i} - \tilde{X}_T \right) \tag{7.55}$$

$$= \sum_{B \in S} \mu^*(B), \tag{7.56}$$

where S consists of sets of the form

$$\left(\bigcap_{i=1}^{k} \bigcap_{j \in W_i} \tilde{X}_j \right) - \tilde{X}_{T \cup (\cup_{i=1}^{k} (Q_i - W_i))} \tag{7.57}$$

with $W_i \subset Q_i$ for $1 \leq i \leq k$ and there exists at least one i such that $W_i \neq \emptyset$. By our assumption, if $B \in S$ is such that there exist at least two i for which $W_i \neq \emptyset$, then $\mu^*(B) = 0$. Therefore, if $\mu^*(B)$ is possibly nonzero, then B must be such that there exists a unique i for which $W_i \neq \emptyset$. Now for $1 \leq i \leq k$, let S_l be the set consisting of sets of the form in (7.57) with $W_i \subset Q_i$, $W_i \neq \emptyset$, and $W_l = \emptyset$ for $l \neq i$. In other words, S_i consists of atoms of the form

$$\left(\bigcap_{j \in W_i} \tilde{X}_j \right) - \tilde{X}_{T \cup (\cup_{l \neq i} Q_l) \cup (Q_i - W_i)} \tag{7.58}$$

with $W_i \subset Q_i$ and $W_i \neq \emptyset$. Then

$$\sum_{B \in S} \mu^*(B) = \sum_{i=1}^{k} \sum_{B \in S_i} \mu^*(B). \tag{7.59}$$

Now

$$\begin{aligned} & \tilde{X}_{Q_i} - \tilde{X}_{T \cup (\cup_{l \neq i} Q_l)} \\ &= \bigcup_{\substack{W_i \subset Q_i \\ W_i \neq \emptyset}} \left[\left(\bigcap_{j \in W_i} \tilde{X}_j \right) - \tilde{X}_{T \cup (\cup_{l \neq i} Q_l) \cup (Q_i - W_i)} \right] & (7.60) \\ &= \bigcup_{B \in S_i} B. & (7.61) \end{aligned}$$

Since μ^* is set-additive, we have

$$\mu^* \left(\tilde{X}_{Q_i} - \tilde{X}_{T \cup (\cup_{l \neq i} Q_l)} \right) = \sum_{B \in S_i} \mu^*(B). \tag{7.62}$$

Hence, from (7.56) and (7.59), we have

$$\begin{aligned} & H(X_{Q_i}, 1 \leq i \leq k | X_T) \\ &= \sum_{i=1}^{k} \sum_{B \in S_i} \mu^*(B) & (7.63) \\ &= \sum_{i=1}^{k} \mu^* \left(\tilde{X}_{Q_i} - \tilde{X}_{T \cup (\cup_{l \neq i} Q_l)} \right) & (7.64) \\ &= \sum_{i=1}^{k} H(X_{Q_i} | X_T, X_{Q_l}, l \neq i), & (7.65) \end{aligned}$$

where (7.64) follows from (7.62). By Theorem 7.4, $X_{Q_i}, 1 \leq i \leq k$, are mutually independent conditioning on X_T.

We now prove the 'only if' part. Assume $X_{Q_i}, 1 \leq i \leq k$, are mutually independent conditioning on X_T. For any collection of sets $W_1, W_2, \cdots, W_k$, where $W_i \subset Q_i$, $1 \leq i \leq k$, if there exist at least two i such that $W_i \neq \emptyset$, by Theorem 7.5, $X_{W_i}, 1 \leq i \leq k$, are mutually independent conditioning on $(X_T, X_{Q_i - W_i}, 1 \leq i \leq k)$. By Lemma 7.7, we obtain (7.52). The theorem is proved. □

7.2 FULL CONDITIONAL MUTUAL INDEPENDENCE

DEFINITION 7.11 *A conditional mutual independency on $X_1, X_2, \cdots, X_n$ is full if all $X_1, X_2, \cdots, X_n$ are involved. Such a conditional mutual independency is called a full conditional mutual independency (FCMI).*

EXAMPLE 7.12 *For $n = 5$,*

X_1, X_2, X_4, and X_5 are mutually independent conditioning on X_3

is an FCMI. However,

X_1, X_2, and X_5 are mutually independent conditioning on X_3

is not an FCMI because X_4 is not involved.

As in the previous chapters, we let

$$\mathcal{N}_n = \{1, 2, \cdots, n\}. \tag{7.66}$$

In Theorem 7.9, if

$$T \cup \left(\bigcup_{i=1}^{k} Q_i \right) = \mathcal{N}_n, \tag{7.67}$$

then the tuple $(T, Q_i, 1 \le i \le k)$ defines the following FCMI on $X_1, X_2, \cdots, X_n$:

K : $X_{Q_1}, X_{Q_2}, \cdots, X_{Q_k}$ are mutually independent conditioning on X_T.

We will denote K by $(T, Q_i, 1 \le i \le k)$.

DEFINITION 7.13 *Let $K = (T, Q_i, 1 \le i \le k)$ be an FCMI on $X_1, X_2, \cdots, X_n$. The image of K, denoted by $Im(K)$, is the set of all atoms of $\mathcal{F}_n$ which has the form of the set in (7.57), where $W_i \subset Q_i$, $1 \le i \le k$, and there exist at least two i such that $W_i \neq \emptyset$.*

PROPOSITION 7.14 *Let $K = (T, Q_1, Q_2)$ be an FCI (full conditional independency) on $X_1, X_2, \cdots, X_n$. Then*

$$Im(K) = \{A \in \mathcal{A} : A \subset (\tilde{X}_{Q_1} \cap \tilde{X}_{Q_2} - \tilde{X}_T)\}. \tag{7.68}$$

PROPOSITION 7.15 *Let $K = (T, Q_i, 1 \le i \le k)$ be an FCMI on $X_1, X_2, \cdots, X_n$. Then*

$$Im(K) = \left\{ A \in \mathcal{A} : A \subset \bigcup_{1 \le i < j \le k} (\tilde{X}_{Q_i} \cap \tilde{X}_{Q_j} - \tilde{X}_T) \right\}. \tag{7.69}$$

These two propositions greatly simplify the description of $Im(K)$. Their proofs are elementary and they are left as exercises. We first illustrate these two propositions in the following example.

EXAMPLE 7.16 *Consider $n = 4$ and FCMI's $K_1 = (\{3\}, \{1\}, \{2,4\})$ and $K_2 = (\emptyset, \{1\}, \{2,3\}, \{4\})$. Then*

$$Im(K_1) = \{A \in \mathcal{A} : A \subset (\tilde{X}_1 \cap \tilde{X}_{\{2,4\}} - \tilde{X}_3)\} \qquad (7.70)$$

and

$$Im(K_2) = \{A \in \mathcal{A} : A \subset (\tilde{X}_1 \cap \tilde{X}_{\{2,3\}}) \cup (\tilde{X}_{\{2,3\}} \cap \tilde{X}_4) \cup (\tilde{X}_1 \cap \tilde{X}_4)\}. \qquad (7.71)$$

THEOREM 7.17 *Let K be an FCMI on $X_1, X_2, \cdots, X_n$. Then K holds if and only if $\mu^*(A) = 0$ for all $A \in Im(K)$.*

Proof First, (7.67) is true if K is an FCMI. Then the set in (7.57) can be written as

$$\left(\bigcap_{j \in \cup_{i=1}^k W_i} \tilde{X}_j \right) - \tilde{X}_{\mathcal{N}_n - \cup_{i=1}^k W_i}, \qquad (7.72)$$

which is seen to be an atom of $\mathcal{F}_n$. The theorem can then be proved by a direct application of Theorem 7.9 to the FCMI K. □

Let $A = \cap_{i=1}^n \tilde{Y}_i$ be a nonempty atom of $\mathcal{F}_n$. Define the set

$$U_A = \{i \in \mathcal{N}_n : \tilde{Y}_i = \tilde{X}_i^c\}. \qquad (7.73)$$

Note that A is uniquely specified by U_A because

$$A = \left(\bigcap_{i \in \mathcal{N}_n - U_A} \tilde{X}_i \right) \cap \left(\bigcap_{i \in U_A} \tilde{X}_i^c \right) = \left(\bigcap_{i \in \mathcal{N}_n - U_A} \tilde{X}_i \right) - \tilde{X}_{U_A}. \qquad (7.74)$$

Define $w(A) = n - |U_A|$ as the *weight* of the atom A, the number of $\tilde{X}_i$ in A which are not complemented. We now show that an FCMI $K = (T, Q_i, 1 \leq i \leq k)$ is uniquely specified by $Im(K)$. First, by letting $W_i = Q_i$ for $1 \leq i \leq k$ in Definition 7.13, we see that the atom

$$\left(\bigcap_{j \in \cup_{i=1}^k Q_i} \tilde{X}_j \right) - \tilde{X}_T \qquad (7.75)$$

is in $Im(K)$, and it is the unique atom in $Im(K)$ with the largest weight. From this atom, T can be determined. To determine $Q_i, 1 \leq i \leq k$, we define a relation q on $T^c = \mathcal{N}_n \backslash T$ as follows. For $l, l' \in T^c$, (l, l') is in q if and only if

i) $l = l'$; or

ii) there exists an atom of the form

$$\tilde{X}_l \cap \tilde{X}_{l'} \cap \bigcap_{\substack{1\le j\le n \\ j\ne l,l'}} \tilde{Y}_j \tag{7.76}$$

in $\mathcal{A} - Im(K)$, where $\tilde{Y}_j = \tilde{X}_j$ or $\tilde{X}_j^c$.

Recall that $\mathcal{A}$ is the set of nonempty atoms of $\mathcal{F}_n$. The idea of ii) is that (l, l') is in q if and only if $l, l' \in Q_i$ for some $1 \le i \le k$. Then q is reflexive and symmetric by construction, and is transitive by virtue of the structure of $Im(K)$. In other words, q is an *equivalence relation* which partitions T^c into $\{Q_i, 1 \le i \le k\}$. Therefore, K and $Im(K)$ uniquely specify each other.

The image of an FCMI K completely characterizes the effect of K on the I-Measure for $X_1, X_2, \cdots, X_n$. The joint effect of more than one FCMI can easily be described in terms of the images of the individual FCMI's. Let

$$\Pi = \{K_l, 1 \le l \le m\} \tag{7.77}$$

be a set of FCMI's. By Theorem 7.9, K_l holds if and only if μ^* vanishes on the atoms in $Im(K_l)$. Then $K_l, 1 \le l \le m$ hold simultaneously if and only if μ^* vanishes on the atoms in $\cup_{l=1}^{k} Im(K_l)$. This is summarized as follows.

DEFINITION 7.18 *The image of a set of FCMI's* $\Pi = \{K_l, 1 \le l \le m\}$ *is defined as*

$$Im(\Pi) = \bigcup_{l=1}^{k} Im(K_l). \tag{7.78}$$

THEOREM 7.19 *Let* Π *be a set of FCMI's for* $X_1, X_2, \cdots, X_n$. *Then* Π *holds if and only if* $\mu^*(A) = 0$ *for all* $A \in Im(\Pi)$.

In probability problems, we are often given a set of conditional independencies and we need to see whether another given conditional independency is logically implied. This is called the *implication problem* which will be discussed in detail in Section 12.5. The next theorem gives a solution to this problem if only FCMI's are involved.

THEOREM 7.20 *Let* Π_1 *and* Π_2 *be two sets of FCMI's. Then* Π_1 *implies* Π_2 *if and only if* $Im(\Pi_2) \subset Im(\Pi_1)$.

Proof We first prove that if $Im(\Pi_2) \subset Im(\Pi_1)$, then Π_1 implies Π_2. Assume $Im(\Pi_2) \subset Im(\Pi_1)$ and Π_1 holds. Then by Theorem 7.19, $\mu^*(A) = 0$ for all $A \in Im(\Pi_1)$. Since $Im(\Pi_2) \subset Im(\Pi_1)$, this implies that $\mu^*(A) = 0$ for all $A \in Im(\Pi_2)$. Again by Theorem 7.19, this implies Π_2 also holds. Therefore, if $Im(\Pi_2) \subset Im(\Pi_1)$, then Π_1 implies Π_2.

We now prove that if Π_1 implies Π_2, then $Im(\Pi_2) \subset Im(\Pi_1)$. To prove this, we assume that Π_1 implies Π_2 but $Im(\Pi_2) \not\subset Im(\Pi_1)$, and we will show that this leads to a contradiction. Fix a nonempty atom $A \in Im(\Pi_2) - Im(\Pi_1)$. By Theorem 6.11, we can construct random variables $X_1, X_2, \cdots, X_n$ such that μ^* vanishes on all the atoms of $\mathcal{F}_n$ except for A. Then μ^* vanishes on all the atoms in $Im(\Pi_1)$ but not on all the atoms in $Im(\Pi_2)$. By Theorem 7.19, this implies that for $X_1, X_2, \cdots, X_n$ so constructed, Π_1 holds but Π_2 does not hold. Therefore, Π_1 does not imply Π_2, which is a contradiction. The theorem is proved. □

Remark In the course of proving this theorem and all its preliminaries, we have used nothing more than the basic inequalities. Therefore, we have shown that the basic inequalities are a sufficient set of tools to solve the implication problem if only FCMI's are involved.

COROLLARY 7.21 *Two sets of FCMI's are equivalent if and only if their images are identical.*

Proof Two set of FCMI's Π_1 and Π_2 are equivalent if and only if

$$\Pi_1 \Rightarrow \Pi_2 \quad \text{and} \quad \Pi_2 \Rightarrow \Pi_1. \tag{7.79}$$

Then by the last theorem, this is equivalent to $Im(\Pi_2) \subset Im(\Pi_1)$ and $Im(\Pi_1) \subset Im(\Pi_2)$, i.e., $Im(\Pi_2) = Im(\Pi_1)$. The corollary is proved. □

Thus a set of FCMI's is completely characterized by its image. A set of FCMI's is a set of probabilistic constraints, but the characterization by its image is purely set-theoretic! This characterization offers an intuitive set-theoretic interpretation of the joint effect of FCMI's on the I-Measure for $X_1, X_2, \cdots, X_n$. For example, $Im(K_1) \cap Im(K_2)$ is interpreted as the effect commonly due to K_1 and K_2, $Im(K_1) - Im(K_2)$ is interpreted as the effect due to K_1 but not K_2, etc. We end this section with an example.

EXAMPLE 7.22 *Consider $n = 4$. Let*

$$K_1 = (\emptyset, \{1,2,3\}, \{4\}), \quad K_2 = (\emptyset, \{1,2,4\}, \{3\}) \tag{7.80}$$

$$K_3 = (\emptyset, \{1,2\}, \{3,4\}), \quad K_4 = (\emptyset, \{1,3\}, \{2,4\}) \tag{7.81}$$

and let $\Pi_1 = \{K_1, K_2\}$ and $\Pi_2 = \{K_3, K_4\}$. Then

$$Im(\Pi_1) = Im(K_1) \cup Im(K_2) \tag{7.82}$$

and

$$Im(\Pi_2) = Im(K_3) \cup Im(K_4), \tag{7.83}$$

where

$$Im(K_1) = \{A \in \mathcal{A} : A \subset (\tilde{X}_{\{1,2,3\}} \cap \tilde{X}_4)\} \quad (7.84)$$

$$Im(K_2) = \{A \in \mathcal{A} : A \subset (\tilde{X}_{\{1,2,4\}} \cap \tilde{X}_3)\} \quad (7.85)$$

$$Im(K_3) = \{A \in \mathcal{A} : A \subset (\tilde{X}_{\{1,2\}} \cap \tilde{X}_{\{3,4\}})\} \quad (7.86)$$

$$Im(K_4) = \{A \in \mathcal{A} : A \subset (\tilde{X}_{\{1,3\}} \cap \tilde{X}_{\{2,4\}})\}. \quad (7.87)$$

It can readily be seen by using an information diagram that $Im(\Pi_1) \subset Im(\Pi_2)$*. Therefore,* Π_2 *implies* Π_1*. Note that no probabilistic argument is involved in this proof.*

7.3 MARKOV RANDOM FIELD

A *Markov random field* is a generalization of a discrete time Markov chain in the sense that the time index for the latter, regarded as a chain, is replaced by a general graph for the former. Historically, the study of Markov random field stems from statistical physics. The classical Ising model, which is defined on a rectangular lattice, was used to explain certain empirically observed facts about ferromagnetic materials. In this section, we explore the structure of the I-Measure for a Markov random field.

We refer the reader to textbooks on graph theory (e.g. [34]) for formal definitions of the graph-theoretic terminologies to be used in the rest of the chapter. Let $G = (V, E)$ be an undirected graph, where V is the set of vertices and E is the set of edges. We assume that there is no *loop* in G, i.e., there is no edge in G which connects a vertex to itself. For any (possibly empty) subset U of V, denote by $G\backslash U$ the graph obtained from G by eliminating all the vertices in U and all the edges joining a vertex in U.

The connectivity of a graph partitions the graph into subgraphs called *components*, i.e., two vertices are in the same component if and only if they are connected. Let $s(U)$ be the number of distinct components in $G\backslash U$. Denote the sets of vertices of these components by $V_1(U), V_2(U), \cdots, V_{s(U)}(U)$. If $s(U) > 1$, we say that U is a cutset in G.

DEFINITION 7.23 (MARKOV RANDOM FIELD) *Let* $G = (V, E)$ *be an undirected graph with* $V = \mathcal{N}_n = \{1, 2, \cdots, n\}$*, and let* X_i *be a random variable corresponding to vertex* i*. Then* $X_1, X_2, \cdots, X_n$ *form a Markov random field represented by* G *if for all cutsets* U *in* G*, the sets of random variables* $X_{V_1(U)}, X_{V_2(U)}, \cdots, X_{V_{s(U)}(U)}$ *are mutually independent conditioning on* X_U*.*

This definition of a Markov random field is referred to as the *global Markov property* in the literature. If $X_1, X_2, \cdots, X_n$ form a Markov random field represented by a graph G, we also say that $X_1, X_2, \cdots, X_n$ form a *Markov graph* G. When G is a chain, we say that $X_1, X_2, \cdots, X_n$ form a Markov chain.

In the definition of a Markov random field, each cutset U in G specifies an FCMI on $X_1, X_2, \cdots, X_n$, denoted by $[U]$. Formally,

$$[U] : X_{V_1(U)}, \cdots, X_{V_{s(U)}(U)} \text{ are mutually independent conditioning on } X_U.$$

For a collection of cutsets $U_1, U_2, \cdots, U_k$ in G, we introduce the notation

$$[U_1, U_2, \cdots, U_k] = [U_1] \wedge [U_2] \wedge \cdots \wedge [U_k] \tag{7.88}$$

where '$\wedge$' denotes 'logical AND.' Using this notation, $X_1, X_2, \cdots, X_n$ form a Markov graph G if and only if

$$[U \subset V : U \neq V \text{ and } s(U) > 1] \tag{7.89}$$

holds. Therefore, a Markov random field is simply a collection of FCMI's induced by a graph.

We now define two types of nonempty atoms of $\mathcal{F}_n$ with respect to a graph G. Recall the definition of the set U_A for a nonempty atom A of $\mathcal{F}_n$ in (7.73).

DEFINITION 7.24 *For a nonempty atom A of $\mathcal{F}_n$, if $s(U_A) = 1$, i.e., $G\backslash U_A$ is connected, then A is a Type I atom, otherwise A is a Type II atom. The sets of all Type I and Type II atoms of $\mathcal{F}_n$ are denoted by T_1 and T_2, respectively.*

THEOREM 7.25 *$X_1, X_2, \cdots, X_n$ form a Markov graph G if and only if μ^* vanishes on all Type II atoms.*

Before we prove this theorem, we first state the following proposition which is the graph-theoretic analog of Theorem 7.5. The proof is trivial and is omitted. This proposition and Theorem 7.5 together establish an analogy between the structure of conditional mutual independence and the connectivity of a graph. This analogy will play a key role in proving Theorem 7.25.

PROPOSITION 7.26 *Let C and Q_i be disjoint subsets of the vertex set V of a graph G and W_i be a subset of Q_i for $1 \leq i \leq k$, where $k \geq 2$. Assume that there exist at least two i such that $W_i \neq \emptyset$. If $Q_i, 1 \leq i \leq k$, are disconnected in $G\backslash C$, then those W_i which are nonempty are disconnected in $G\backslash(C \cup \bigcup_{i=1}^{k}(Q_i - W_i))$.*

EXAMPLE 7.27 *In the graph G in Figure 7.4, $\{1\}$, $\{2, 3, 4\}$, and $\{5, 6\}$ are disjoint in $G\backslash\{7\}$. Then Proposition 7.26 says that $\{1\}$, $\{2\}$, and $\{5, 6\}$ are disjoint in $G\backslash\{3, 4, 7\}$.*

Proof of Theorem 7.25 Recall the definition of the set U_A for a nonempty atom A in (7.73). We note that $\{U_A, A \in \mathcal{A}\}$ contains precisely all the proper subsets of $\mathcal{N}_n$. Thus the set of FCMI's specified by the graph G can be written as

$$[U_A : A \in \mathcal{A} \text{ and } s(U_A) > 1] \tag{7.90}$$

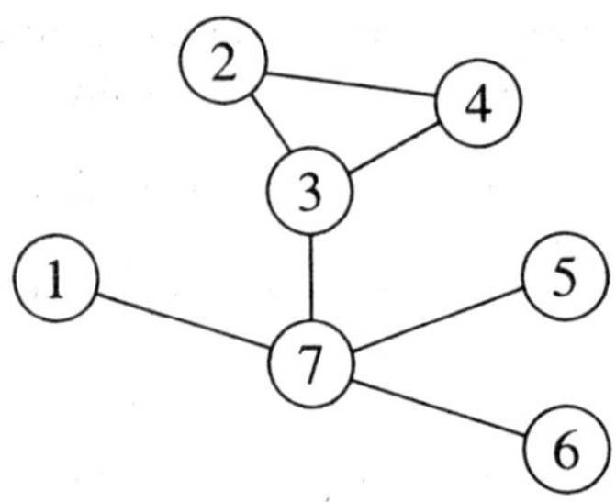

Figure 7.4. The graph G in Example 7.27.

(cf. (7.89)). By Theorem 7.19, it suffices to prove that

$$Im([U_A : A \in \mathcal{A} \text{ and } s(U_A) > 1]) = T_2, \tag{7.91}$$

where T_2 was defined in Definition 7.24.

We first prove that

$$T_2 \subset Im([U_A : A \in \mathcal{A} \text{ and } s(U_A) > 1]). \tag{7.92}$$

Consider an atom $A \in T_2$ so that $s(U_A) > 1$. In Definition 7.13, let $T = U_A$, $k = s(U_A)$, and $Q_i = V_i(U_A)$ for $1 \le i \le s(U_A)$. By considering $W_i = V_i(U_A)$ for $1 \le i \le s(U_A)$, we see that $A \in Im([U_A])$. Therefore,

$$\begin{aligned} T_2 &= \{A \in \mathcal{A} : s(U_A) > 1\} && (7.93)\\ &\subset \bigcup_{A \in \mathcal{A}: s(U_A) > 1} Im([U_A]) && (7.94)\\ &= Im([U_A : A \in \mathcal{A} \text{ and } s(U_A) > 1]). && (7.95) \end{aligned}$$

We now prove that

$$Im([U_A : A \in \mathcal{A} \text{ and } s(U_A) > 1]) \subset T_2. \tag{7.96}$$

Consider $A \in Im([U_A : A \in \mathcal{A} \text{ and } s(U_A) > 1])$. Then there exists $A^* \in \mathcal{A}$ with $s(U_{A^*}) > 1$ such that $A \in Im([U_{A^*}])$. From Definition 7.13,

$$A = \left(\bigcap_{j \in \cup_{i=1}^{s(U_{A^*})} W_i} \tilde{X}_j \right) - \tilde{X}_{U_{A^*} \cup \left(\cup_{i=1}^{s(U_{A^*})} (V_i(U_{A^*}) - W_i) \right)}, \tag{7.97}$$

where $W_i \subset V_i(U_{A^*})$, $1 \le i \le s(U_{A^*})$, and there exist at least two i such that $W_i \ne \emptyset$. It follows from (7.97) and the definition of U_A that

$$U_A = U_{A^*} \cup \bigcup_{i=1}^{s(U_{A^*})} (V_i(U_{A^*}) - W_i). \tag{7.98}$$

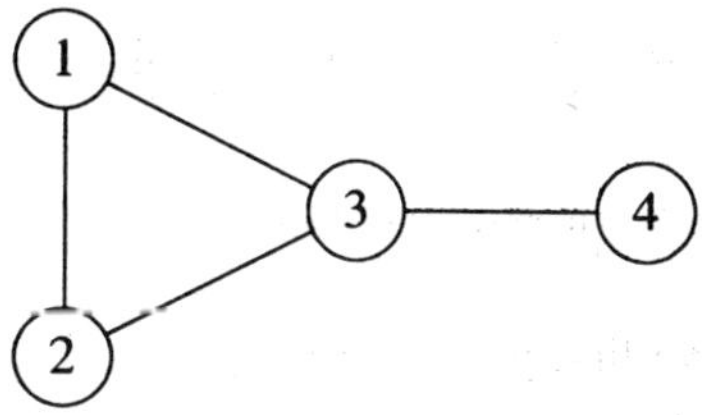

Figure 7.5. The graph G in Example 7.28.

With U_{A^*} playing the role of C and $V_i(U_{A^*})$ playing the role of Q_i in Proposition 7.26, we see by applying the proposition that those (at least two) W_i which are nonempty are disjoint in

$$G \backslash \left(U_{A^*} \cup \left(\bigcup_{i=1}^{s(U_{A^*})} (V_i(U_{A^*}) - W_i) \right) \right) = G \backslash U_{A^*}. \tag{7.99}$$

This implies $s(U_A) > 1$, i.e., $A \in T_2$. Therefore, we have proved (7.96), and hence the theorem is proved. □

EXAMPLE 7.28 *With respect to the graph G in Figure 7.5, the Type II atoms are*

$$\tilde{X}_1 \cap \tilde{X}_2 \cap \tilde{X}_3^c \cap \tilde{X}_4, \ \tilde{X}_1^c \cap \tilde{X}_2 \cap \tilde{X}_3^c \cap \tilde{X}_4, \ \tilde{X}_1 \cap \tilde{X}_2^c \cap \tilde{X}_3^c \cap \tilde{X}_4, \tag{7.100}$$

while the other twelve nonempty atoms of $\mathcal{F}_4$ are Type I atoms. The random variables X_1, X_2, X_3, and X_4 form a Markov graph G if and only if $\mu^*(A) = 0$ for all Type II atoms A.

7.4 MARKOV CHAIN

When the graph G representing a Markov random field is a chain, the Markov random field becomes a Markov chain. In this section, we will show that the information diagram for a Markov chain can be displayed in two dimensions. We will also show that the I-Measure μ^* for a Markov chain is always nonnegative. This characteristic of μ^* facilitates the use of the information diagram because if B is seen to be a subset of B' in the information diagram, then

$$\mu^*(B') = \mu^*(B) + \mu^*(B' - B) \geq \mu^*(B). \tag{7.101}$$

These two properties are not possessed by a general Markov random field.

Without loss of generality, we assume that the Markov chain is represented by the graph G in Figure 7.6. This corresponds to the Markov chain $X_1 \rightarrow$

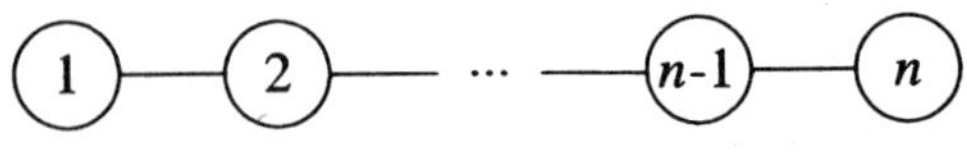

Figure 7.6. The graph G representing the Markov chain $X_1 \to X_2 \to \cdots \to X_n$.

$X_2 \to \cdots \to X_n$. We first prove the following characterization of a Type I atom for a Markov chain.

LEMMA 7.29 *For the Markov chain represented by the graph G in Figure 7.6, a nonempty atom A of $\mathcal{F}_n$ is a Type I atom if and only if*

$$\mathcal{N}_n \backslash U_A = \{l, l+1, \cdots, u\}, \tag{7.102}$$

where $1 \leq l \leq u \leq n$, i.e., the indices of the set variables in A which are not complemented are consecutive.

Proof It is easy to see that for a nonempty atom A, if (7.102) is satisfied, then $G\backslash U_A$ is connected, i.e., $s(U_A) = 1$. Therefore, A is a Type I atom of $\mathcal{F}_n$. On the other hand, if (7.102) is not satisfied, then $G\backslash U_A$ is not connected, i.e., $s(U_A) > 1$, or A is a Type II atom of $\mathcal{F}_n$. The lemma is proved. □

We now show how the information diagram for a Markov chain with any length $n \geq 3$ can be constructed in two dimensions. Since μ^* vanishes on all the Type II atoms of $\mathcal{F}_n$, it is not necessary to display these atoms in the information diagram. In constructing the information diagram, the regions representing the random variables $X_1, X_2, \cdots, X_n$ should overlap with each other such that the regions corresponding to all the Type II atoms are empty, while the regions corresponding to all the Type I atoms are nonempty. Figure 7.7 shows such a construction. Note that this information diagram includes Figures 6.7 and 6.9 as special cases, which are information diagrams for Markov chains with lengths 3 and 4, respectively.

We have already shown that μ^* is nonnegative for a Markov chain with length 3 or 4. Toward proving that this is true for any length $n \geq 3$, it suffices

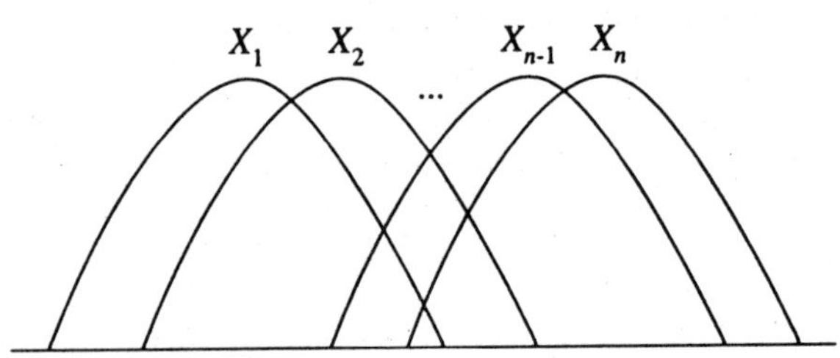

Figure 7.7. The information diagram for the Markov chain $X_1 \to X_2 \to \cdots \to X_n$.

to show that $\mu^*(A) \geq 0$ for all Type I atoms A of $\mathcal{F}_n$ because $\mu^*(A) = 0$ for all Type II atoms A of $\mathcal{F}_n$. We have seen in Lemma 7.29 that for a Type I atom A of $\mathcal{F}_n$, U_A has the form as prescribed in (7.102). Consider any such atom A. Then an inspection of the information diagram in Figure 7.7 reveals that

$$\begin{aligned} \mu^*(A) &= \mu^*(\tilde{X}_l \cap \tilde{X}_{l+1} \cap \cdots \cap \tilde{X}_u - \tilde{X}_{U_A}) && (7.103) \\ &= I(X_l; X_u | X_{U_A}) && (7.104) \\ &\geq 0. && (7.105) \end{aligned}$$

This shows that μ^* is always nonnegative. However, since Figure 7.7 involves an indefinite number of random variables, we give a formal proof of this result in the following theorem.

THEOREM 7.30 *For a Markov chain $X_1 \to X_2 \to \cdots \to X_n$, μ^* is nonnegative.*

Proof Since $\mu^*(A) = 0$ for all Type II atoms A of $\mathcal{F}_n$, it suffices to show that $\mu^*(A) \geq 0$ for all Type I atoms A of $\mathcal{F}_n$. We have seen in Lemma 7.29 that for a Type I atom A of $\mathcal{F}_n$, U_A has the form as prescribed in (7.102). Consider any such atom A and define the set

$$W = \{l+1, \cdots, u-1\}. \qquad (7.106)$$

Then

$$\begin{aligned} & I(X_l; X_u | X_{U_A}) \\ &= \mu^*(\tilde{X}_l \cap \tilde{X}_u - \tilde{X}_{U_A}) && (7.107) \\ &= \mu^*\left(\bigcup_{S \subset W}\left(\tilde{X}_l \cap \left(\bigcap_{t \in S} \tilde{X}_t\right) \cap \tilde{X}_u - \tilde{X}_{U_{A \cup (W \setminus S)}}\right)\right) && (7.108) \\ &= \sum_{S \subset W} \mu^*\left(\tilde{X}_l \cap \left(\bigcap_{t \in S} \tilde{X}_t\right) \cap \tilde{X}_u - \tilde{X}_{U_{A \cup (W \setminus S)}}\right). && (7.109) \end{aligned}$$

In the above summation, except for the atom corresponding to $S = W$, namely $(\tilde{X}_l \cap \tilde{X}_{l+1} \cap \cdots \cap \tilde{X}_u - \tilde{X}_{U_A})$, all the atoms are Type II atoms. Therefore,

$$I(X_l; X_u | X_{U_A}) = \mu^*(\tilde{X}_l \cap \tilde{X}_{l+1} \cap \cdots \cap \tilde{X}_u - \tilde{X}_{U_A}). \qquad (7.110)$$

Hence,

$$\begin{aligned} \mu^*(A) &= \mu^*(\tilde{X}_l \cap \tilde{X}_{l+1} \cap \cdots \cap \tilde{X}_u - \tilde{X}_{U_A}) && (7.111) \\ &= I(X_l; X_u | X_{U_A}) && (7.112) \\ &\geq 0. && (7.113) \end{aligned}$$

The theorem is proved. □

PROBLEMS

1. Prove Proposition 7.14 and Proposition 7.15.

2. In Example 7.22, it was shown that Π_2 implies Π_1. Show that Π_1 does not imply Π_2. Hint: Use an information diagram to determine $Im(\Pi_2)\backslash Im(\Pi_1)$.

3. *Alternative definition of the global Markov property:* For any partition $\{U, V_1, V_2\}$ of V such that the sets of vertices V_1 and V_2 are disconnected in $G\backslash U$, the sets of random variables X_{V_1} and X_{V_2} are independent conditioning on X_U.

   Show that this definition is equivalent to the global Markov property in Definition 7.23.

4. *The local Markov property:* For $1 \leq i \leq n$, X_i and X_{V-N_i-i} are independent conditioning on X_{N_i}, where N_i is the set of neighbors[2] of vertex i in G.

   a) Show that the global Markov property implies the local Markov property.

   b) Show that the local Markov property does not imply the global Markov property by giving a counterexample. Hint: Consider a joint distribution which is not strictly positive.

5. Construct a Markov random field whose I-Measure μ^* can take negative values. Hint: Consider a Markov "star."

6. a) Show that X_1, X_2, X_3, and X_4 are mutually independent if and only if

   $$X_1 \perp (X_2, X_3, X_4),\ X_2 \perp (X_3, X_4)|X_1,\ X_3 \perp X_4|(X_1, X_2).$$

   Hint: Use an information diagram.

   b) Generalize the result in a) to n random variables.

7. Determine the Markov random field with four random variables X_1, X_2, X_3, and X_4 which is characterized by the following conditional independencies:

   $$(X_1, X_2, X_5) \perp X_4|X_3$$
   $$X_2 \perp (X_4, X_5)|(X_1, X_3)$$
   $$X_1 \perp (X_3, X_4)|(X_2, X_5).$$

   What are the other conditional independencies pertaining to this Markov random field?

[2] Vertices i and j in an undirected graph are neighbors if i and j are connected by an edge.

HISTORICAL NOTES

A Markov random field can be regarded as a generalization of a discrete-time Markov chain. Historically, the study of Markov random field stems from statistical physics. The classical Ising model, which is defined on a rectangular lattice, was used to explain certain empirically observed facts about ferromagnetic materials. The foundation of the theory of Markov random fields can be found in Preston [166] or Spitzer [199].

The structure of the I-Measure for a Markov chain was first investigated in the unpublished work of Kawabata [114]. Essentially the same result was independently obtained by Yeung eleven years later in the context of the I-Measure, and the result was eventually published in Kawabata and Yeung [115]. Full conditional independencies were shown to be axiomatizable by Malvestuto [137]. The results in this chapter are due to Yeung *et al.* [233], where they obtained a set-theoretic characterization of full conditional independencies and investigated the structure of the I-Measure for a Markov random field. In this paper, they also obtained a hypergraph characterization of a Markov random field based on the I-Measure characterization in Theorem 7.25. Ge and Ye [82] have applied these results to characterize a class of graphical models for conditional independence of random variables.

Chapter 8

CHANNEL CAPACITY

In all practical communication systems, when a signal is transmitted from one point to another point, the signal is inevitably contaminated by random noise, i.e., the signal received is correlated with but possibly different from the signal transmitted. We use a *noisy channel* to model such a situation. A noisy channel is a "system" which has one input terminal and one output terminal[1], with the input connected to the transmission point and the output connected to the receiving point. When the signal is transmitted through the channel, it is distorted in a random way which depends on the channel characteristics. As such, the signal received may be different from the signal transmitted.

In communication engineering, we are interested in conveying messages reliably through a noisy channel at the maximum possible rate. We first look at a simple channel called the *binary symmetric channel* (BSC), which is represented by the transition diagram in Figure 8.1. In this channel both the input X and the output Y take values in the set $\{0, 1\}$. There is a certain probability, denoted by ϵ, that the output is not equal to the input. That is, if the input is 0, then the output is 0 with probability $1 - \epsilon$, and is 1 with probability ϵ. Likewise, if the input is 1, then the output is 1 with probability $1 - \epsilon$, and is 0 with probability ϵ. The parameter ϵ is called the *crossover probability* of the BSC.

Let $\{A, B\}$ be the message set which contains two possible messages to be conveyed through a BSC with $0 \leq \epsilon < 0.5$. We further assume that the two messages A and B are equally likely. If the message is A, we map it to the codeword 0, and if the message is B, we map it to the codeword 1. This is the simplest example of a *channel code*. The codeword is then transmitted

[1]The discussion on noisy channels here is confined to point-to-point channels.

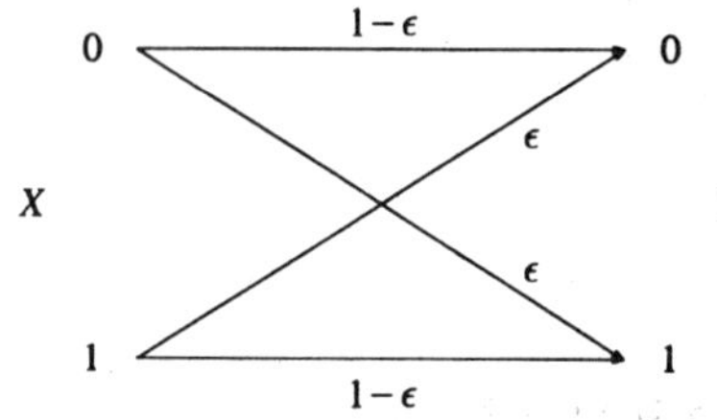

Figure 8.1. A binary symmetric channel.

through the channel. Our task is to decode the message based on the output of the channel, and an error is said to occur if the message is decoded incorrectly.

Consider

$$\Pr\{A|Y=0\} = \Pr\{X=0|Y=0\} \tag{8.1}$$

$$= \frac{\Pr\{X=0\}\Pr\{Y=0|X=0\}}{\Pr\{Y=0\}} \tag{8.2}$$

$$= \frac{0.5(1-\epsilon)}{\Pr\{Y=0\}}. \tag{8.3}$$

Since

$$\Pr\{Y=0\} = \Pr\{Y=1\} = 0.5 \tag{8.4}$$

by symmetry[2], it follows that

$$\Pr\{A|Y=0\} = 1-\epsilon \tag{8.5}$$

and

$$\Pr\{B|Y=0\} = 1 - \Pr\{A|Y=0\} = \epsilon. \tag{8.6}$$

Since $\epsilon < 0.5$,

$$\Pr\{B|Y=0\} < \Pr\{A|Y=0\}. \tag{8.7}$$

Therefore, in order to minimize the probability of error, we decode a received 0 to the message A. By symmetry, we decode a received 1 to the message B.

[2]More explicitly,

$$\begin{aligned}\Pr\{Y=0\} &= \Pr\{A\}\Pr\{Y=0|A\} + \Pr\{B\}\Pr\{Y=0|B\} \\ &= 0.5\Pr\{Y=0|X=0\} + 0.5\Pr\{Y=0|X=1\} \\ &= 0.5(1-\epsilon) + 0.5\epsilon \\ &= 0.5.\end{aligned}$$

An error occurs if a 0 is received and the message is B, or if a 1 is received and the message is A. Therefore, the probability of error, denoted by P_e, is given by

$$\begin{aligned} P_e &= \Pr\{Y=0\}\Pr\{B|Y=0\} + \Pr\{Y=1\}\Pr\{A|Y=1\} && (8.8) \\ &= 0.5\epsilon + 0.5\epsilon && (8.9) \\ &= \epsilon, && (8.10) \end{aligned}$$

where (8.9) follows from (8.6) because

$$\Pr\{A|Y=1\} = \Pr\{B|Y=0\} = \epsilon \tag{8.11}$$

by symmetry.

Let us assume that $\epsilon \neq 0$. Then the above scheme obviously does not provide perfectly reliable communication. If we are allowed to use the channel only once, then this is already the best we can do. However, if we are allowed to use the same channel repeatedly, then we can improve the reliability by generalizing the above scheme.

We now consider the following channel code which we refer to as the binary *repetition code*. Let $n \geq 1$ be an odd positive integer which is called the *block length* of the code. In this code, the message A is mapped to the sequence of n 0's, and the message B is mapped to the sequence of n 1's. The codeword, which consists of a sequence of either n 0's or n 1's, is transmitted through the channel in n uses. Upon receiving a sequence of n bits at the output of the channel, we use the majority vote to decode the message, i.e., if there are more 0's than 1's in the sequence, we decode the sequence to the message A, otherwise we decode the sequence to the message B. Note that the block length is chosen to be odd so that there cannot be a tie. When $n = 1$, this scheme reduces to the previous scheme.

For this more general scheme, we continue to denote the probability of error by P_e. Let N_0 and N_1 be the number of 0's and 1's in the received sequence, respectively. Clearly,

$$N_0 + N_1 = n. \tag{8.12}$$

For large n, if the message is A, the number of 0's received is approximately equal to

$$E[N_0|A] = n(1-\epsilon) \tag{8.13}$$

and the number of 1's received is approximately equal to

$$E[N_1|A] = n\epsilon \tag{8.14}$$

with high probability by the weak law of large numbers. This implies that the probability of an error, namely the event $\{N_0 < N_1\}$, is small because

$$n(1-\epsilon) > n\epsilon \tag{8.15}$$

with the assumption that $\epsilon < 0.5$. Specifically,

$$\begin{aligned} \Pr\{\text{error}|A\} &= \Pr\{N_0 < N_1|A\} &(8.16)\\ &= \Pr\{n - N_1 < N_1|A\} &(8.17)\\ &= \Pr\{N_1 > 0.5n|A\} &(8.18)\\ &\leq \Pr\{N_1 > (\epsilon + \phi)n|A\}, &(8.19) \end{aligned}$$

where

$$0 < \phi < 0.5 - \epsilon, \tag{8.20}$$

so that ϕ is positive and

$$\epsilon + \phi < 0.5. \tag{8.21}$$

Note that such a ϕ exists because $\epsilon < 0.5$. Then by the weak law of large numbers, the upper bound in (8.19) tends to 0 as $n \to \infty$. By symmetry, $\Pr\{\text{error}|B\}$ also tends to 0 as $n \to \infty$. Therefore,

$$P_e = \Pr\{A\}\Pr\{\text{error}|A\} + \Pr\{B\}\Pr\{\text{error}|B\} \tag{8.22}$$

tends to 0 as $n \to \infty$. In other words, by using a long enough repetition code, we can make P_e arbitrarily small. In this sense, we say that reliable communication is achieved asymptotically.

We point out that for a BSC with $\epsilon > 0$, for any given transmitted sequence of length n, the probability of receiving any given sequence of length n is nonzero. It follows that for any two distinct input sequences, there is always a nonzero probability that the same output sequence is produced so that the two input sequences become indistinguishable. Therefore, except for very special channels (e.g., the BSC with $\epsilon = 0$), no matter how the encoding/decoding scheme is devised, a nonzero probability of error is inevitable, and asymptotically reliable communication is the best we can hope for.

Though a rather naive approach, asymptotically reliable communication can be achieved by using the repetition code. The repetition code, however, is not without catch. For a channel code, the *rate* of the code in bit(s) per use, is defined as the ratio of the logarithm of the size of the message set in the base 2 to the block length of the code. Roughly speaking, the rate of a channel code is the average number of bits the channel code attempts to convey through the channel per use of the channel. For a binary repetition code with block length n, the rate is $\frac{1}{n}\log 2 = \frac{1}{n}$, which tends to 0 as $n \to \infty$. Thus in order to achieve asymptotic reliability by using the repetition code, we cannot communicate through the noisy channel at any positive rate!

In this chapter, we characterize the maximum rate at which information can be communicated through a *discrete memoryless channel* (DMC) with an arbitrarily small probability of error. This maximum rate, which is generally positive, is known as the *channel capacity*. Then we discuss the use of feedback in communicating through a channel, and show that feedback does not increase the capacity. At the end of the chapter, we discuss transmitting an information source through a DMC, and we show that asymptotic optimality can be achieved by separating source coding and channel coding.

8.1 DISCRETE MEMORYLESS CHANNELS

DEFINITION 8.1 *Let $\mathcal{X}$ and $\mathcal{Y}$ be discrete alphabets, and $p(y|x)$ be a transition matrix from $\mathcal{X}$ to $\mathcal{Y}$. A discrete channel $p(y|x)$ is a single-input single-output system with input random variable X taking values in $\mathcal{X}$ and output random variable Y taking values in $\mathcal{Y}$ such that*

$$\Pr\{X = x, Y = y\} = \Pr\{X = x\}p(y|x) \tag{8.23}$$

for all $(x, y) \in \mathcal{X} \times \mathcal{Y}$.

Remark From (8.23), we see that if $\Pr\{X = x\} > 0$, then

$$p(y|x) = \frac{\Pr\{X = x, Y = y\}}{\Pr\{X = x\}} = \Pr\{Y = y|X = x\}. \tag{8.24}$$

However, $\Pr\{Y = y|X = x\}$ is not defined if $\Pr\{X = x\} = 0$. Nevertheless, (8.23) is valid for both cases.

DEFINITION 8.2 *A discrete memoryless channel (DMC) is a sequence of replicates of a generic discrete channel. These discrete channels are indexed by a discrete-time index i, where $i \geq 1$, with the ith channel being available for transmission at time i. Transmission through a channel is assumed to be instantaneous.*

The DMC is the simplest nontrivial model for a communication channel. For simplicity, a DMC will be specified by $p(y|x)$, the transition matrix of the generic discrete channel.

Let X_i and Y_i be respectively the input and the output of a DMC at time i, where $i \geq 1$. Figure 8.2 is an illustration of a discrete memoryless channel. In the figure, the memoryless attribute of the channel manifests itself by the fact that for all i, the ith discrete channel only has X_i as its input.

For each $i \geq 1$,

$$\Pr\{X_i = x, Y_i = y\} = \Pr\{X_i = x\}p(y|x) \tag{8.25}$$

since the ith discrete channel is a replicate of the generic discrete channel $p(y|x)$. However, the dependency of the output random variables $\{Y_i\}$ on the input random variables $\{X_i\}$ cannot be further described without specifying how the DMC is being used. In this chapter we will study two representative

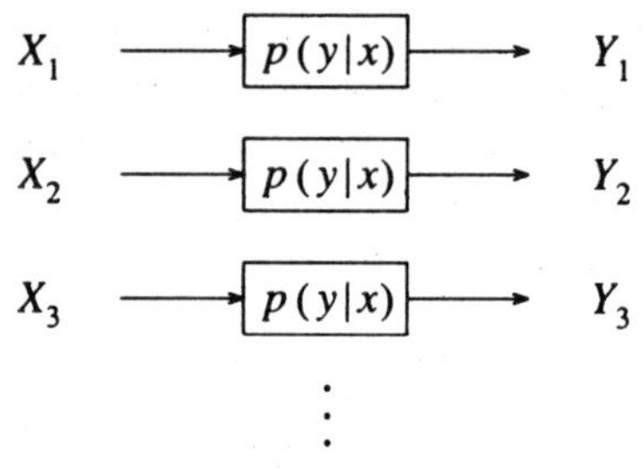

Figure 8.2. A discrete memoryless channel.

models. In the first model the channel is used without feedback. This will be discussed in Section 8.2 through Section 8.5. In the second model, the channel is used with complete feedback. This will be discussed in Section 8.6. To keep our discussion simple, we will assume that the alphabets $\mathcal{X}$ and $\mathcal{Y}$ are finite.

DEFINITION 8.3 *The capacity of a discrete memoryless channel* $p(y|x)$ *is defined as*

$$C = \max_{p(x)} I(X;Y), \tag{8.26}$$

where X *and* Y *are respectively the input and the output of the generic discrete channel, and the maximum is taken over all input distributions* $p(x)$.

From the above definition, we see that

$$C \geq 0 \tag{8.27}$$

because

$$I(X;Y) \geq 0 \tag{8.28}$$

for all input distributions $p(x)$. By Theorem 2.43, we have

$$C = \max_{p(x)} I(X;Y) \leq \max_{p(x)} H(X) = \log |\mathcal{X}|. \tag{8.29}$$

Likewise, we have

$$C \leq \log |\mathcal{Y}|. \tag{8.30}$$

Therefore,

$$C \leq \min(\log |\mathcal{X}|, \log |\mathcal{Y}|). \tag{8.31}$$

Since $I(X;Y)$ is a continuous functional of $p(x)$ and the set of all $p(x)$ is a compact set (i.e., closed and bounded) in $\Re^{|\mathcal{X}|}$, the maximum value of $I(X;Y)$ can be attained[3]. This justifies taking the maximum rather than the supremum in the definition of channel capacity in (8.26).

[3]The assumption that $\mathcal{X}$ is finite is essential in this argument.

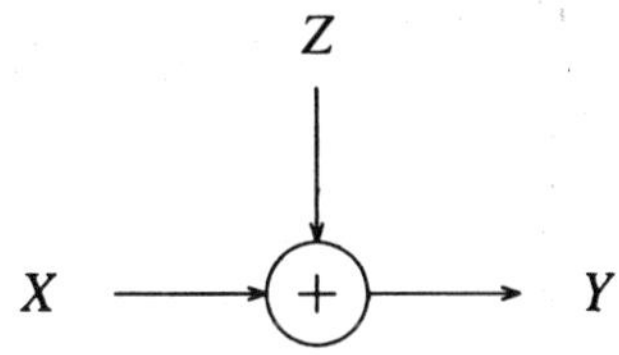

Figure 8.3. An alternative representation for a binary symmetric channel.

We will prove subsequently that C is in fact the maximum rate at which information can be communicated reliably through a DMC. We first give some examples of DMC's for which the capacities can be obtained in closed form. In the following, X and Y denote, respectively, the input and the output of the generic discrete channel, and all logarithms are in the base 2.

EXAMPLE 8.4 (BINARY SYMMETRIC CHANNEL) *The binary symmetric channel (BSC) has been shown in Figure 8.1. Alternatively, a BSC can be represented by Figure 8.3. Here, Z is a binary random variable representing the noise of the channel, with*

$$\Pr\{Z=0\} = 1-\epsilon \quad \textit{and} \quad \Pr\{Z=1\} = \epsilon, \tag{8.32}$$

and Z is independent of X. Then

$$Y = X + Z \bmod 2. \tag{8.33}$$

See Problem 1 for a generalization of this representation.

In order to determine the capacity of the BSC, we first bound $I(X;Y)$ as follows.

$$\begin{aligned} I(X;Y) &= H(Y) - H(Y|X) & (8.34)\\ &= H(Y) - \sum_x p(x) H(Y|X=x) & (8.35)\\ &= H(Y) - \sum_x p(x) h_b(\epsilon) & (8.36)\\ &= H(Y) - h_b(\epsilon) & (8.37)\\ &\le 1 - h_b(\epsilon), & (8.38) \end{aligned}$$

where we have used h_b to denote the binary entropy function in the base 2. In order to achieve this upper bound, we have to make $H(Y)=1$, i.e., the output distribution of the BSC is uniform. This can be done by letting $p(x)$ be the uniform distribution on $\{0,1\}$. Therefore, the upper bound on $I(X;Y)$ can be achieved, and we conclude that

$$C = 1 - h_b(\epsilon) \text{ bit per use.} \tag{8.39}$$

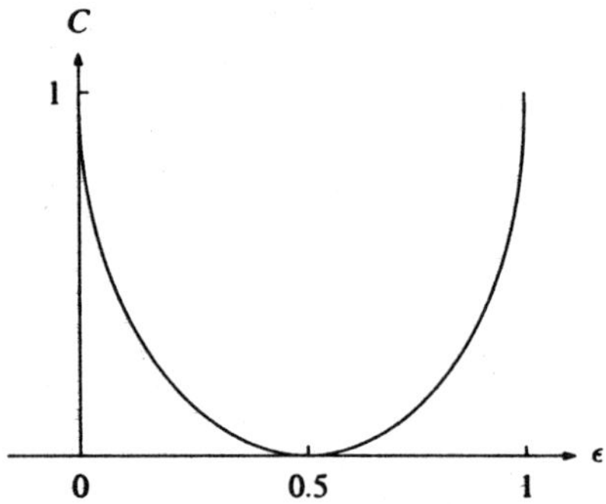

Figure 8.4. The capacity of a binary symmetric channel.

Figure 8.4 is a plot of the capacity C versus the crossover probability ϵ. We see from the plot that C attains the maximum value 1 when $\epsilon = 0$ or $\epsilon = 1$, and attains the minimum value 0 when $\epsilon = 0.5$. When $\epsilon = 0$, it is easy to see that $C = 1$ is the maximum rate at which information can be communicated through the channel reliably. This can be achieved simply by transmitting unencoded bits through the channel, and no decoding is necessary because all the bits are received unchanged. When $\epsilon = 1$, the same can be achieved with the additional decoding step which complements all the received bits. By doing so, the bits transmitted through the channel can be recovered without error. Thus from the communication point of view, for binary channels, a channel which never makes error and a channel which always makes errors are equally good. When $\epsilon = 0.5$, the channel output is independent of the channel input. Therefore, no information can possibly be communicated through the channel.

EXAMPLE 8.5 (BINARY ERASURE CHANNEL) *The binary erasure channel is shown in Figure 8.5. In this channel, the input alphabet is $\{0, 1\}$, while the output alphabet is $\{0, 1, e\}$. With probability γ, the erasure symbol e is produced at the output, which means that the input bit is lost; otherwise the input bit is reproduced at the output without error. The parameter γ is called the erasure probability.*

To determine the capacity of this channel, we first consider

$$C = \max_{p(x)} I(X;Y) \tag{8.40}$$

$$= \max_{p(x)} (H(Y) - H(Y|X)) \tag{8.41}$$

$$= \max_{p(x)} H(Y) - h_b(\gamma). \tag{8.42}$$

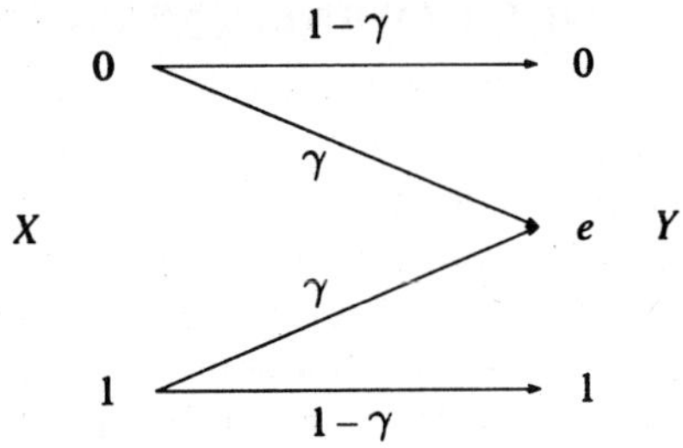

Figure 8.5. A binary erasure channel.

Thus we only have to maximize $H(Y)$. To this end, let

$$\Pr\{X = 0\} = a \tag{8.43}$$

and define a binary random variable E by

$$E = \begin{cases} 0 & \textit{if } Y \neq e \\ 1 & \textit{if } Y = e. \end{cases} \tag{8.44}$$

The random variable E indicates whether an erasure has occurred, and it is a function of Y. Then

$$\begin{aligned} H(Y) &= H(Y, E) & (8.45)\\ &= H(E) + H(Y|E) & (8.46)\\ &= h_b(\gamma) + (1-\gamma)h_b(a). & (8.47) \end{aligned}$$

Hence,

$$\begin{aligned} C &= \max_{p(x)} H(Y) - h_b(\gamma) & (8.48)\\ &= \max_{a}[h_b(\gamma) + (1-\gamma)h_b(a)] - h_b(\gamma) & (8.49)\\ &= (1-\gamma)\max_{a} h_b(a) & (8.50)\\ &= (1-\gamma), & (8.51) \end{aligned}$$

where capacity is achieved by letting $a = 0.5$, i.e., the input distribution is uniform.

It is in general not possible to obtain the capacity of a DMC in closed form, and we have to resort to numerical computation. In Chapter 10 we will discuss the Blahut-Arimoto algorithm for computing channel capacity.

8.2 THE CHANNEL CODING THEOREM

We will justify the definition of the capacity of a DMC by the proving the *channel coding theorem*. This theorem, which consists of two parts, will be formally stated at the end of the section. The direct part of the theorem asserts that information can be communicated through a DMC with an arbitrarily small probability of error at any rate less than the channel capacity. Here it is assumed that the decoder knows the transition matrix of the DMC. The converse part of the theorem asserts that if information is communicated through a DMC at a rate higher than the capacity, then the probability of error is bounded away from zero. For better appreciation of the definition of channel capacity, we will first prove the converse part in Section 8.3 and then prove the direct part in Section 8.4.

DEFINITION 8.6 *An (n, M) code for a discrete memoryless channel with input alphabet $\mathcal{X}$ and output alphabet $\mathcal{Y}$ is defined by an encoding function*

$$f : \{1, 2, \cdots, M\} \to \mathcal{X}^n \tag{8.52}$$

and a decoding function

$$g : \mathcal{Y}^n \to \{1, 2, \cdots, M\}. \tag{8.53}$$

The set $\{1, 2, \cdots, M\}$, denoted by $\mathcal{W}$, is called the message set. The sequences $f(1), f(2), \cdots, f(M)$ in $\mathcal{X}^n$ are called codewords, and the set of codewords is called the codebook.

In order to distinguish a channel code as defined above from a channel code with feedback which will be discussed in Section 8.6, we will refer to the former as a channel code without feedback.

We assume that a message W is randomly chosen from the message set $\mathcal{W}$ according to the uniform distribution. Therefore,

$$H(W) = \log M. \tag{8.54}$$

With respect to a channel code for a DMC $p(y|x)$, we let

$$\mathbf{X} = (X_1, X_2, \cdots, X_n) \tag{8.55}$$

and

$$\mathbf{Y} = (Y_1, Y_2, \cdots, Y_n) \tag{8.56}$$

be the input sequence and the output sequence of the channel, respectively. Evidently,

$$\mathbf{X} = f(W). \tag{8.57}$$

We also let

$$\hat{W} = g(\mathbf{Y}). \tag{8.58}$$

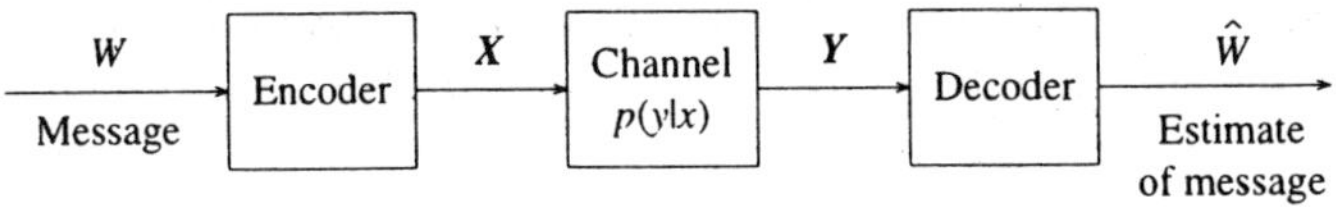

Figure 8.6. A channel code with block length n.

Figure 8.6 is the block diagram for a channel code.

DEFINITION 8.7 *For all $1 \leq w \leq M$, let*

$$\lambda_w = \Pr\{\hat{W} \neq w | W = w\} = \sum_{\mathbf{y} \in \mathcal{Y}^n : g(\mathbf{y}) \neq w} \Pr\{\mathbf{Y} = \mathbf{y} | \mathbf{X} = f(w)\} \quad (8.59)$$

be the conditional probability of error given that the message is w.

We now define two performance measures for a channel code.

DEFINITION 8.8 *The maximal probability of error of an (n, M) code is defined as*

$$\lambda_{max} = \max_w \lambda_w. \quad (8.60)$$

DEFINITION 8.9 *The average probability of error of an (n, M) code is defined as*

$$P_e = \Pr\{\hat{W} \neq W\}. \quad (8.61)$$

From the definition of P_e, we have

$$\begin{aligned} P_e &= \Pr\{\hat{W} \neq W\} & (8.62)\\ &= \sum_w \Pr\{W = w\} \Pr\{\hat{W} \neq W | W = w\} & (8.63)\\ &= \sum_w \frac{1}{M} \Pr\{\hat{W} \neq w | W = w\} & (8.64)\\ &= \frac{1}{M} \sum_w \lambda_w, & (8.65) \end{aligned}$$

i.e., P_e is the arithmetic mean of λ_w, $1 \leq w \leq M$. It then follows that

$$P_e \leq \lambda_{max}. \quad (8.66)$$

In fact, it can be readily seen that this inequality remains valid even without the assumption that W distributes uniformly on the message set $\mathcal{W}$.

DEFINITION 8.10 *The rate of an (n, M) channel code is $n^{-1} \log M$ in bits per use.*

DEFINITION 8.11 *A rate R is asymptotically achievable for a discrete memoryless channel $p(y|x)$ if for any $\epsilon > 0$, there exists for sufficiently large n an (n, M) code such that*

$$\frac{1}{n} \log M > R - \epsilon \tag{8.67}$$

and

$$\lambda_{max} < \epsilon. \tag{8.68}$$

For brevity, an asymptotically achievable rate will be referred to as an achievable rate.

In other words, a rate R is achievable if there exists a sequence of codes whose rates approach R and whose probabilities of error approach zero. We end this section by stating the channel coding theorem, which gives a characterization of all achievable rates. This theorem will be proved in the next two sections.

THEOREM 8.12 (CHANNEL CODING THEOREM) *A rate R is achievable for a discrete memoryless channel $p(y|x)$ if and only if $R \leq C$, the capacity of the channel.*

8.3 THE CONVERSE

Let us consider a channel code with block length n. The random variables involved in this code are W, X_i and Y_i for $1 \leq i \leq n$, and $\hat{W}$. We see from the definition of a channel code in Definition 8.6 that all the random variables are generated sequentially according to some deterministic or probabilistic rules. Specifically, the random variables are generated in the order $W, X_1, Y_1, X_2, Y_2, \cdots, X_n, Y_n, \hat{W}$. The generation of these random variables can be represented by the dependency graph[4] in Figure 8.7. In this graph, a node represents a random variable. If there is a (directed) edge from node X to node Y, then node X is called a *parent* of node Y. We further distinguish a *solid* edge and a *dotted* edge: a solid edge represents functional (deterministic) dependency, while a dotted edge represents the probabilistic dependency induced by the transition matrix $p(y|x)$ of the generic discrete channel. For a node X, its parent nodes represent all the random variables on which random variable X depends when it is generated. We will use q to denote the joint distribution of these random variables as well as all the marginals, and let x_i denote the ith component of a sequence $\mathbf{x}$. From the dependency graph, we see that for all $(w, \mathbf{x}, \mathbf{y}, \hat{w}) \in \mathcal{W} \times \mathcal{X}^n \times \mathcal{Y}^n \times \hat{\mathcal{W}}$ such that $q(\mathbf{x}) > 0$ and

[4]A dependency graph can be regarded as a Bayesian network [160].

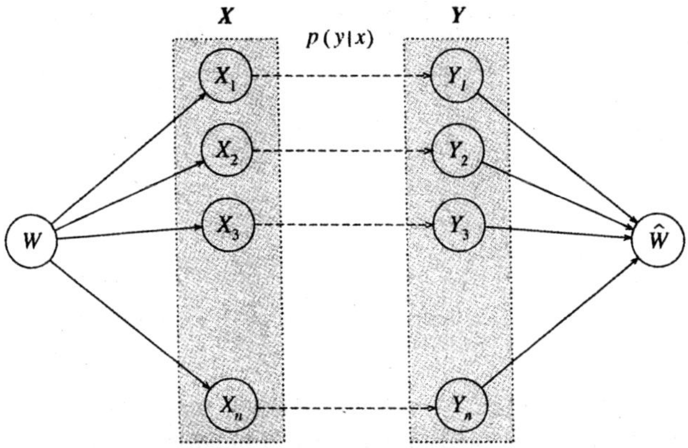

Figure 8.7. The dependency graph for a channel code without feedback.

$q(\mathbf{y}) > 0$,

$$q(w, \mathbf{x}, \mathbf{y}\, \hat{w}) = q(w) \left(\prod_{i=1}^{n} q(x_i|w) \right) \left(\prod_{i=1}^{n} p(y_i|x_i) \right) q(\hat{w}|\mathbf{y}). \tag{8.69}$$

Note that $q(w) > 0$ for all w so that $q(x_i|w)$ are well-defined, and $q(x_i|w)$ and $q(\hat{w}|\mathbf{y})$ are both deterministic. Denote the set of nodes $X_1, X_2, \cdots, X_n$ by $\mathbf{X}$ and the set of nodes $Y_1, Y_2, \cdots, Y_n$ by $\mathbf{Y}$. We notice the following structure in the dependency graph: all the edges from W end in $\mathbf{X}$, all the edges from $\mathbf{X}$ end in $\mathbf{Y}$, and all the edges from $\mathbf{Y}$ end in $\hat{W}$. this suggests that the random variables $W, \mathbf{X}, \mathbf{Y}$, and $\hat{W}$ form the Markov chain

$$W \to \mathbf{X} \to \mathbf{Y} \to \hat{W}. \tag{8.70}$$

The validity of this Markov chain can be formally justified by applying Proposition 2.9 to (8.69), so that for all $(w, \mathbf{x}, \mathbf{y}, \hat{w}) \in \mathcal{W} \times \mathcal{X}^n \times \mathcal{Y}^n \times \hat{\mathcal{W}}$ such that $q(\mathbf{x}) > 0$ and $q(\mathbf{y}) > 0$, we can write

$$q(w, \mathbf{x}, \mathbf{y}, \hat{w}) = q(w) q(\mathbf{x}|w) q(\mathbf{y}|\mathbf{x}) q(\hat{w}|\mathbf{y}). \tag{8.71}$$

Now $q(\mathbf{x}, \mathbf{y})$ is obtained by summing over all w and $\hat{w}$ in (8.69), and $q(\mathbf{x})$ is obtained by further summing over all $\mathbf{y}$. After some straightforward algebra and using

$$q(\mathbf{y}|\mathbf{x}) = \frac{q(\mathbf{x}, \mathbf{y})}{q(\mathbf{x})} \tag{8.72}$$

for all $\mathbf{x}$ such that $q(\mathbf{x}) > 0$, we obtain

$$q(\mathbf{y}|\mathbf{x}) = \prod_{i=1}^{n} p(y_i|x_i). \tag{8.73}$$

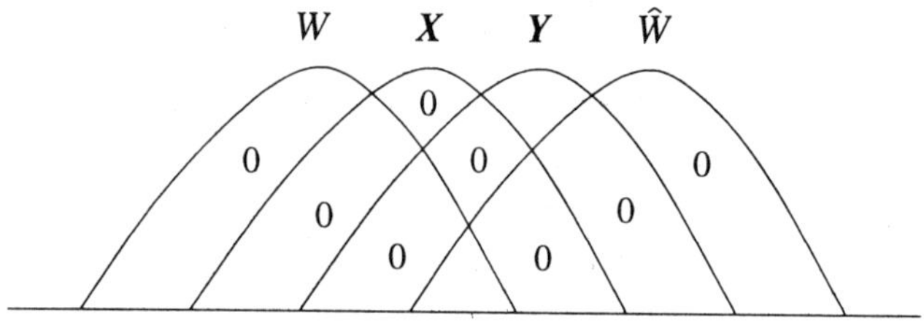

Figure 8.8. The information diagram for $W \to \mathbf{X} \to \mathbf{Y} \to \hat{W}$.

The Markov chain in (8.70) and the relation in (8.73) are apparent from the setup of the problem, and the above justification may seem superfluous. However, the methodology developed here is necessary for handling the more delicate situation which arises when the channel is used with feedback. This will be discussed in Section 8.6.

Consider a channel code whose probability of error is arbitrarily small. Since $W, \mathbf{X}, \mathbf{Y}$, and $\hat{W}$ form the Markov chain in (8.70), the information diagram for these four random variables is as shown in Figure 8.8. Moreover, $\mathbf{X}$ is a function of W, and $\hat{W}$ is a function of $\mathbf{Y}$. These two relations are equivalent to

$$H(\mathbf{X}|W) = 0, \tag{8.74}$$

and

$$H(\hat{W}|\mathbf{Y}) = 0, \tag{8.75}$$

respectively. Since the probability of error is arbitrarily small, W and $\hat{W}$ are essentially identical. To gain insight into the problem, we assume for the time being that W and $\hat{W}$ are equivalent, so that

$$H(\hat{W}|W) = H(W|\hat{W}) = 0. \tag{8.76}$$

Since the I-Measure μ^* for a Markov chain is nonnegative, the constraints in (8.74) to (8.76) imply that μ^* vanishes on all the atoms in Figure 8.8 marked with a '0.' Immediately, we see that

$$H(W) = I(\mathbf{X}; \mathbf{Y}). \tag{8.77}$$

That is, the amount of information conveyed through the channel is essentially the mutual information between the input sequence and the output sequence of the channel.

For a single transmission, we see from the definition of channel capacity that the mutual information between the input and the output cannot exceed the capacity of the channel, i.e., for all $1 \le i \le n$,

$$I(X_i; Y_i) \le C. \tag{8.78}$$

Summing i from 1 to n, we have

$$\sum_{i=1}^{n} I(X_i;Y_i) \leq nC. \tag{8.79}$$

Upon establishing in the next lemma that

$$I(\mathbf{X};\mathbf{Y}) \leq \sum_{i=1}^{n} I(X_i;Y_i), \tag{8.80}$$

the converse of the channel coding theorem then follows from

$$\begin{aligned} \frac{1}{n}\log M &= \frac{1}{n}H(W) && (8.81)\\ &= \frac{1}{n}I(\mathbf{X};\mathbf{Y}) && (8.82)\\ &\leq \frac{1}{n}\sum_{i=1}^{n} I(X_i;Y_i) && (8.83)\\ &\leq C. && (8.84)\end{aligned}$$

LEMMA 8.13 *For a discrete memoryless channel used with a channel code without feedback, for any $n \geq 1$,*

$$I(\mathbf{X};\mathbf{Y}) \leq \sum_{i=1}^{n} I(X_i;Y_i), \tag{8.85}$$

where X_i and Y_i are, respectively, the input and the output of the channel at time i.

Proof For any $(\mathbf{x},\mathbf{y}) \in \mathcal{X}^n \times \mathcal{Y}^n$, if $q(\mathbf{x},\mathbf{y}) > 0$, then $q(\mathbf{x}) > 0$ and (8.73) holds. Therefore,

$$q(\mathbf{Y}|\mathbf{X}) = \prod_{i=1}^{n} p(Y_i|X_i) \tag{8.86}$$

holds for all $(\mathbf{x},\mathbf{y})$ in the support of $q(\mathbf{x},\mathbf{y})$. Then

$$-E\log q(\mathbf{Y}|\mathbf{X}) = -E\log\prod_{i=1}^{n} p(Y_i|X_i) = -\sum_{i=1}^{n} E\log p(Y_i|X_i), \tag{8.87}$$

or

$$H(\mathbf{Y}|\mathbf{X}) = \sum_{i=1}^{n} H(Y_i|X_i). \tag{8.88}$$

Hence,

$$
\begin{aligned}
I(\mathbf{X};\mathbf{Y}) &= H(\mathbf{Y}) - H(\mathbf{Y}|\mathbf{X}) && (8.89)\\
&\le \sum_{i=1}^{n} H(Y_i) - \sum_{i=1}^{n} H(Y_i|X_i) && (8.90)\\
&= \sum_{i=1}^{n} I(X_i;Y_i). && (8.91)
\end{aligned}
$$

The lemma is proved. □

We now formally prove the converse of the channel coding theorem. Let R be an achievable rate, i.e., for any $\epsilon > 0$, there exists for sufficiently large n an (n, M) code such that

$$\frac{1}{n}\log M > R - \epsilon \tag{8.92}$$

and

$$\lambda_{max} < \epsilon. \tag{8.93}$$

Consider

$$
\begin{aligned}
\log M &\stackrel{a)}{=} H(W) && (8.94)\\
&= H(W|\hat{W}) + I(W;\hat{W}) && (8.95)\\
&\stackrel{b)}{\le} H(W|\hat{W}) + I(\mathbf{X};\mathbf{Y}) && (8.96)\\
&\stackrel{c)}{\le} H(W|\hat{W}) + \sum_{i=1}^{n} I(X_i;Y_i) && (8.97)\\
&\stackrel{d)}{\le} H(W|\hat{W}) + nC, && (8.98)
\end{aligned}
$$

where

a) follows from (8.54);

b) follows from the data processing theorem since $W \to \mathbf{X} \to \mathbf{Y} \to \hat{W}$;

c) follows from Lemma 8.13;

d) follows from (8.79).

From (8.61) and Fano's inequality (cf. Corollary 2.48), we have

$$H(W|\hat{W}) < 1 + P_e \log M. \tag{8.99}$$

Therefore, from (8.98),

$$\begin{aligned} \log M &< 1 + P_e \log M + nC && (8.100) \\ &\leq 1 + \lambda_{max} \log M + nC && (8.101) \\ &< 1 + \epsilon \log M + nC, && (8.102) \end{aligned}$$

where we have used (8.66) and (8.93), respectively, to obtain the last two inequalities. Dividing by n and rearranging the terms, we have

$$\frac{1}{n} \log M < \frac{\frac{1}{n} + C}{1 - \epsilon}, \tag{8.103}$$

and from (8.92), we obtain

$$R - \epsilon < \frac{\frac{1}{n} + C}{1 - \epsilon}. \tag{8.104}$$

For any $\epsilon > 0$, the above inequality holds for all sufficiently large n. Letting $n \to \infty$ and then $\epsilon \to 0$, we conclude that

$$R \leq C. \tag{8.105}$$

This completes the proof for the converse of the channel coding theorem.

From the above proof, we can obtain an asymptotic bound on P_e when the rate of the code $\frac{1}{n} \log M$ is greater than C. Consider (8.100) and obtain

$$P_e \geq 1 - \frac{1 + nC}{\log M} = 1 - \frac{\frac{1}{n} + C}{\frac{1}{n} \log M}. \tag{8.106}$$

Then

$$P_e \geq 1 - \frac{\frac{1}{n} + C}{\frac{1}{n} \log M} \approx 1 - \frac{C}{\frac{1}{n} \log M} \tag{8.107}$$

when n is large. This asymptotic bound on P_e, which is strictly positive if $\frac{1}{n} \log M > C$, is illustrated in Figure 8.9.

In fact, the lower bound in (8.106) implies that $P_e > 0$ for all n if $\frac{1}{n} \log M > C$ because if $P_e^{(n_0)} = 0$ for some n_0, then for all $k \geq 1$, by concatenating k copies of the code, we obtain a code with the same rate and block length equal to kn_0 such that $P_e^{(kn_0)} = 0$, which is a contradiction to our conclusion that $P_e > 0$ when n is large. Therefore, if we use a code whose rate is greater than the channel capacity, the probability of error is non-zero for all block lengths.

The converse of the channel coding theorem we have proved is called the *weak converse*. A stronger version of this result called the *strong converse* can be proved, which says that $P_e \to 1$ as $n \to \infty$ if there exists an $\epsilon > 0$ such that $\frac{1}{n} \log M \geq C + \epsilon$ for all n.

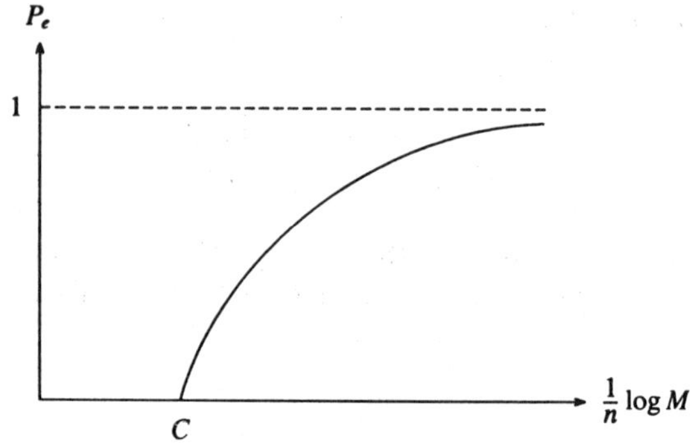

Figure 8.9. An asymptotic upper bound on P_e.

8.4 ACHIEVABILITY OF THE CHANNEL CAPACITY

We have shown in the last section that the channel capacity C is an upper bound on all achievable rates for a DMC. In this section, we show that the rate C is achievable, which implies that any rate $R \leq C$ is achievable.

Consider a DMC $p(y|x)$, and denote the input and the output of the generic discrete channel by X and Y, respectively. For every input distribution $p(x)$, we will prove that the rate $I(X;Y)$ is achievable by showing for large n the existence of a channel code such that

1. the rate of the code is arbitrarily close to $I(X;Y)$;
2. the maximal probability of error λ_{max} is arbitrarily small.

Then by choosing the input distribution $p(x)$ to be one which achieves the channel capacity, i.e., $I(X;Y) = C$, we conclude that the rate C is achievable.

Fix any $\epsilon > 0$ and let δ be a small positive quantity to be specified later. Toward proving the existence of a desired code, we fix an input distribution $p(x)$ for the generic discrete channel $p(y|x)$, and let M be an *even* integer satisfying

$$I(X;Y) - \frac{\epsilon}{2} < \frac{1}{n}\log M < I(X;Y) - \frac{\epsilon}{4}, \tag{8.108}$$

where n is sufficiently large. We now describe a *random coding scheme* in the following steps:

1. Construct the codebook $\mathcal{C}$ of an (n, M) code randomly by generating M codewords in $\mathcal{X}^n$ independently and identically according to $p(x)^n$. Denote these codewords by $\mathbf{X}(1), \mathbf{X}(2), \cdots, \mathbf{X}(M)$.
2. Reveal the codebook $\mathcal{C}$ to both the encoder and the decoder.

3. A message W is chosen from $\mathcal{W}$ according to the uniform distribution.

4. The message W is encoded into the codeword $\mathbf{X}(W)$, which is then transmitted through the channel.

5. The channel outputs a sequence $\mathbf{Y}$ according to

$$\Pr\{\mathbf{Y} = \mathbf{y}|\mathbf{X}(W) = \mathbf{x}\} = \prod_{i=1}^{n} p(y_i|x_i) \tag{8.109}$$

(cf. (8.73) and Remark 1 in Section 8.6).

6. The sequence $\mathbf{Y}$ is decoded to the message w if $(\mathbf{X}(w), \mathbf{Y}) \in T^n_{[XY]\delta}$ and there does not exists $w' \neq w$ such that $(\mathbf{X}(w'), \mathbf{Y}) \in T^n_{[XY]\delta}$. Otherwise, $\mathbf{Y}$ is decoded to a constant message in $\mathcal{W}$. Denote by $\hat{W}$ the message to which $\mathbf{Y}$ is decoded.

Remark 1 There are a total of $|\mathcal{X}|^{Mn}$ possible codebooks which can be constructed by the random procedure in Step 2, where we regard two codebooks whose sets of codewords are permutations of each other as two different codebooks.

Remark 2 Strong typicality is used in defining the decoding function in Step 6. This is made possible by the assumption that the alphabets $\mathcal{X}$ and $\mathcal{Y}$ are finite.

We now analyze the performance of this random coding scheme. Let

$$Err = \{\hat{W} \neq W\} \tag{8.110}$$

be the event of a decoding error. In the following, we analyze $\Pr\{Err\}$, the probability of a decoding error for the random code constructed above. For all $1 \le w \le M$, define the event

$$E_w = \{(\mathbf{X}(w), \mathbf{Y}) \in T^n_{[XY]\delta}\}. \tag{8.111}$$

Now

$$\Pr\{Err\} = \sum_{w=1}^{M} \Pr\{Err|W = w\}\Pr\{W = w\}. \tag{8.112}$$

Since $\Pr\{Err|W = w\}$ are identical for all w by symmetry in the code construction, we have

$$\begin{aligned} \Pr\{Err\} &= \Pr\{Err|W = 1\} \sum_{w=1}^{M} \Pr\{W = w\} && (8.113) \\ &= \Pr\{Err|W = 1\}, && (8.114) \end{aligned}$$

i.e., we can assume without loss of generality that the message 1 is chosen. Then decoding is correct if the received sequence $\mathbf{Y}$ is decoded to the message 1. This is the case if E_1 occurs but E_w does not occur for all $2 \le w \le M$. It follows that[5]

$$\Pr\{Err^c|W = 1\} \ge \Pr\{E_1 \cap E_2^c \cap E_3^c \cap \cdots \cap E_M^c|W = 1\}, \quad (8.115)$$

which implies

$$\begin{aligned} &\Pr\{Err|W = 1\} \\ &= 1 - \Pr\{Err^c|W = 1\} & (8.116) \\ &\le 1 - \Pr\{E_1 \cap E_2^c \cap E_3^c \cap \cdots \cap E_M^c|W = 1\} & (8.117) \\ &= \Pr\{(E_1 \cap E_2^c \cap E_3^c \cap \cdots \cap E_M^c)^c|W = 1\} & (8.118) \\ &= \Pr\{E_1^c \cup E_2 \cup E_3 \cup \cdots \cup E_M|W = 1\}. & (8.119) \end{aligned}$$

By the union bound, we have

$$\Pr\{Err|W = 1\} \le \Pr\{E_1^c|W = 1\} + \sum_{w=2}^{M} \Pr\{E_w|W = 1\}. \quad (8.120)$$

First, if $W = 1$, then $(\mathbf{X}(1), \mathbf{Y})$ are n i.i.d. copies of the pair of generic random variables (X, Y). By the strong JAEP (Theorem 5.8), for any $\nu > 0$,

$$\Pr\{E_1^c|W = 1\} = \Pr\{(\mathbf{X}(1), \mathbf{Y}) \notin T^n_{[XY]\delta}|W = 1\} < \nu \quad (8.121)$$

for sufficiently large n. Second, if $W = 1$, then for $2 \le w \le M$, $(\mathbf{X}(w), \mathbf{Y})$ are n i.i.d. copies of the pair of generic random variables (X', Y'), where X' and Y' have the same marginal distributions as X and Y, respectively, and X' and Y' are independent because $\mathbf{X}(1)$ and $\mathbf{X}(w)$ are independent and $\mathbf{Y}$ depends only on $\mathbf{X}(1)$. Therefore,

$$\begin{aligned} &\Pr\{E_w|W = 1\} \\ &= \Pr\{(\mathbf{X}(w), \mathbf{Y}) \in T^n_{[XY]\delta}|W = 1\} & (8.122) \\ &= \sum_{(\mathbf{x},\mathbf{y}) \in T^n_{[XY]\delta}} p(\mathbf{x})p(\mathbf{y}). & (8.123) \end{aligned}$$

By the consistency of strong typicality, for $(\mathbf{x}, \mathbf{y}) \in T^n_{[XY]\delta}$, $\mathbf{x} \in T^n_{[X]\delta}$ and $\mathbf{y} \in T^n_{[Y]\delta}$. By the strong AEP, all $p(\mathbf{x})$ and $p(\mathbf{y})$ in the above summation satisfy

$$p(\mathbf{x}) \le 2^{-n(H(X)-\eta)} \quad (8.124)$$

[5]If E_1 does not occur or E_w occurs for some $1 \le w \le M$, the received sequence $\mathbf{Y}$ is decoded to the constant message, which may happen to be the message 1. Therefore, the inequality in (8.115) is not an equality in general.

and

$$p(\mathbf{y}) \leq 2^{-n(H(Y)-\zeta)}, \tag{8.125}$$

where $\eta, \zeta \to 0$ as $\delta \to 0$. Again by the strong JAEP,

$$|T^n_{[XY]\delta}| \leq 2^{n(H(X,Y)+\xi)}, \tag{8.126}$$

where $\xi \to 0$ as $\delta \to 0$. Then from (8.123), we have

$$\begin{aligned}
&\Pr\{E_w|W=1\} \\
&\leq 2^{n(H(X,Y)+\xi)} \cdot 2^{-n(H(X)-\eta)} \cdot 2^{-n(H(Y)-\zeta)} && (8.127)\\
&= 2^{-n(H(X)+H(Y)-H(X,Y)-\xi-\eta-\zeta)} && (8.128)\\
&= 2^{-n(I(X;Y)-\xi-\eta-\zeta)} && (8.129)\\
&= 2^{-n(I(X;Y)-\tau)}, && (8.130)
\end{aligned}$$

where

$$\tau = \xi + \eta + \zeta \to 0 \tag{8.131}$$

as $\delta \to 0$.

From the upper bound in (8.108), we have

$$M < 2^{n(I(X;Y)-\frac{\epsilon}{4})}. \tag{8.132}$$

Using (8.121), (8.130), and the above upper bound on M, it follows from (8.114) and (8.120) that

$$\begin{aligned}
\Pr\{Err\} &< \nu + 2^{n(I(X;Y)-\frac{\epsilon}{4})} \cdot 2^{-n(I(X;Y)-\tau)} && (8.133)\\
&= \nu + 2^{-n(\frac{\epsilon}{4}-\tau)}. && (8.134)
\end{aligned}$$

Since $\tau \to 0$ as $\delta \to 0$, for sufficiently small δ, we have

$$\frac{\epsilon}{4} - \tau > 0 \tag{8.135}$$

for any $\epsilon > 0$, so that $2^{-n(\frac{\epsilon}{4}-\tau)} \to 0$ as $n \to \infty$. Then by letting $\nu < \frac{\epsilon}{3}$, it follows from (8.134) that

$$\Pr\{Err\} < \frac{\epsilon}{2} \tag{8.136}$$

for sufficiently large n.

The main idea of the above analysis of $\Pr\{Err\}$ is the following. In constructing the codebook, we randomly generate M codewords in $\mathcal{X}^n$ according to $p(x)^n$, and one of the codewords is sent through the channel $p(y|x)$. When n is large, with high probability, the received sequence is jointly typical with the codeword sent with respect to $p(x,y)$. If the number of codewords M grows with n at a rate less than $I(X;Y)$, then the probability that the received sequence is jointly typical with a codeword other than the one sent through the channel is negligible. Accordingly, the message can be decoded correctly with probability arbitrarily close to 1.

In constructing the codebook by the random procedure in Step 2, we choose a codebook $\mathcal{C}$ with a certain probability $\Pr\{\mathcal{C}\}$ from the ensemble of all possible codebooks. By conditioning on the codebook chosen, we have

$$\Pr\{Err\} = \sum_{\mathcal{C}} \Pr\{\mathcal{C}\}\Pr\{Err|\mathcal{C}\}, \tag{8.137}$$

i.e., $\Pr\{Err\}$ is a weighted average of $\Pr\{Err|\mathcal{C}\}$ over all $\mathcal{C}$ in the ensemble of all possible codebooks, where $\Pr\{Err|\mathcal{C}\}$ is the average probability of error of the code, i.e., P_e, when the codebook $\mathcal{C}$ is chosen (cf. Definition 8.9). The reader should compare the two different expansions of $\Pr\{Err\}$ in (8.137) and (8.112).

Therefore, there exists at least one codebook $\mathcal{C}^*$ such that

$$\Pr\{Err|\mathcal{C}^*\} \le \Pr\{Err\} < \frac{\epsilon}{2}. \tag{8.138}$$

Thus we have shown that for any $\epsilon > 0$, there exists for sufficiently large n an (n, M) code such that

$$\frac{1}{n}\log M > I(X;Y) - \frac{\epsilon}{2} \tag{8.139}$$

(cf. (8.108)) and

$$P_e < \frac{\epsilon}{2}. \tag{8.140}$$

We are still one step away from proving that the rate $I(X;Y)$ is achievable because we require that λ_{max} instead of P_e is arbitrarily small. Toward this end, we write (8.140) as

$$\frac{1}{M}\sum_{w=1}^{M} \lambda_w < \frac{\epsilon}{2}, \tag{8.141}$$

or

$$\sum_{w=1}^{M} \lambda_w < \left(\frac{M}{2}\right)\epsilon. \tag{8.142}$$

Upon ordering the codewords according to their conditional probabilities of error, we observe that the conditional probabilities of error of the better half of the M codewords are less than ϵ, otherwise the conditional probabilities of error of the worse half of the codewords are at least ϵ, and they contribute at least $(\frac{M}{2})\epsilon$ to the summation in (8.142), which is a contradiction.

Thus by discarding the worse half of the codewords in C^*, for the resulting codebook, the maximal probability of error λ_{max} is less than ϵ. Using (8.139) and considering

$$\begin{aligned} \frac{1}{n}\log\frac{M}{2} &= \frac{1}{n}\log M - \frac{1}{n} && (8.143)\\ &> \left(I(X;Y) - \frac{\epsilon}{2}\right) - \frac{1}{n} && (8.144)\\ &> I(X;Y) - \epsilon && (8.145) \end{aligned}$$

when n is sufficiently large, we see that the rate of the resulting code is greater than $I(X;Y) - \epsilon$. Hence, we conclude that the rate $I(X;Y)$ is achievable.

Finally, upon letting the input distribution $p(x)$ be one which achieves the channel capacity, i.e., $I(X;Y) = C$, we have proved that the rate C is achievable. This completes the proof of the direct part of the channel coding theorem.

8.5 A DISCUSSION

In the last two sections, we have proved the channel coding theorem which asserts that reliable communication through a DMC at rate R is possible if and only if $R < C$, the channel capacity. By reliable communication at rate R, we mean that the size of the message set grows exponentially with n at rate R, while the message can be decoded correctly with probability arbitrarily close to 1 as $n \to \infty$. Therefore, the capacity C is a fundamental characterization of a DMC.

The capacity of a noisy channel is analogous to the capacity of a water pipe in the following way. For a water pipe, if we pump water through the pipe at a rate higher than its capacity, the pipe would burst and water would be lost. For a communication channel, if we communicate through the channel at a rate higher than the capacity, the probability of error is bounded away from zero, i.e., information is lost.

In proving the direct part of the channel coding theorem, we showed that there exists a channel code whose rate is arbitrarily close to C and whose probability of error is arbitrarily close to zero. Moreover, the existence of such a code is guaranteed only when the block length n is large. However, the proof does not indicate how we can find such a codebook. For this reason, the proof we gave is called an existence proof (as oppose to a constructive proof).

For a fixed block length n, we in principle can search through the ensemble of all possible codebooks for a good one, but this is quite prohibitive even for small n because the number of all possible codebooks grows doubly exponentially with n. Specifically, the total number of all possible (n, M) codebooks is equal to $|\mathcal{X}|^{Mn}$. When the rate of the code is close to C, M is approximately equal to 2^{nC}. Therefore, the number of codebooks we need to search through is about $|\mathcal{X}|^{n2^{nC}}$

Nevertheless, the proof of the direct part of the channel coding theorem does indicate that if we generate a codebook randomly as prescribed, the codebook is most likely to be good. More precisely, we now show that the probability of choosing a code $\mathcal{C}$ such that $\Pr\{Err|\mathcal{C}\}$ is greater than any prescribed $\psi > 0$ is arbitrarily small when n is sufficiently large. Consider

$$\begin{aligned}
\Pr\{Err\} &= \sum_{\mathcal{C}} \Pr\{\mathcal{C}\}\Pr\{Err|\mathcal{C}\} &(8.146)\\
&= \sum_{\mathcal{C}:\Pr\{Err|\mathcal{C}\}\le\psi} \Pr\{\mathcal{C}\}\Pr\{Err|\mathcal{C}\} \\
&\quad + \sum_{\mathcal{C}:\Pr\{Err|\mathcal{C}\}>\psi} \Pr\{\mathcal{C}\}\Pr\{Err|\mathcal{C}\} &(8.147)\\
&\ge \sum_{\mathcal{C}:\Pr\{Err|\mathcal{C}\}>\psi} \Pr\{\mathcal{C}\}\Pr\{Err|\mathcal{C}\} &(8.148)\\
&> \psi \sum_{\mathcal{C}:\Pr\{Err|\mathcal{C}\}>\psi} \Pr\{\mathcal{C}\}, &(8.149)
\end{aligned}$$

which implies

$$\sum_{\mathcal{C}:\Pr\{Err|\mathcal{C}\}>\psi} \Pr\{\mathcal{C}\} < \frac{\Pr\{Err\}}{\psi}. \tag{8.150}$$

From (8.138), we have

$$\Pr\{Err\} < \frac{\epsilon}{2} \tag{8.151}$$

for any $\epsilon > 0$ when n is sufficiently large. Then

$$\sum_{\mathcal{C}:\Pr\{Err|\mathcal{C}\}>\psi} \Pr\{\mathcal{C}\} < \frac{\epsilon}{2\psi}. \tag{8.152}$$

Since ψ is fixed, this upper bound can be made arbitrarily small by choosing a sufficiently small ϵ.

Although the proof of the direct part of the channel coding theorem does not provide an explicit construction of a good code, it does give much insight into what a good code is like. Figure 8.10 is an illustration of a channel code which achieves the channel capacity. Here we assume that the input distribution $p(x)$ is one which achieves the channel capacity, i.e., $I(X;Y) = C$. The idea is that most of the codewords are typical sequences in $\mathcal{X}^n$ with respect to $p(x)$. (For this reason, the repetition code is not a good code.) When such a codeword is transmitted through the channel, the received sequence is likely to be one of about $2^{nH(Y|X)}$ sequences in $\mathcal{Y}^n$ which are jointly typical with the transmitted codeword with respect to $p(x,y)$. The association between a codeword and the about $2^{nH(Y|X)}$ corresponding sequences in $\mathcal{Y}^n$ is shown as a cone in the figure. As we require that the probability of decoding error is small, the cones

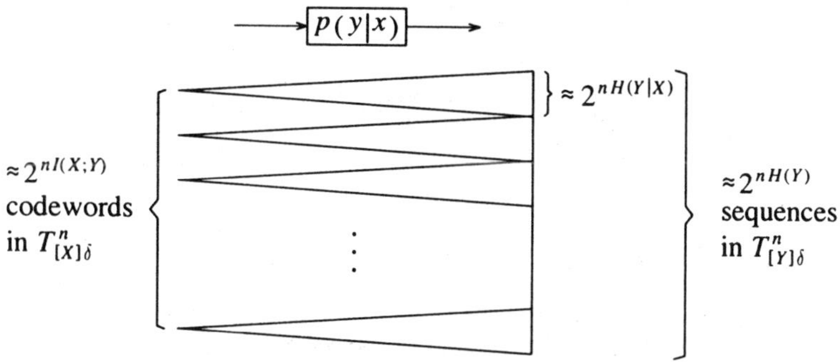

Figure 8.10. A channel code which achieves capacity.

essentially do not overlap with each other. Since the number of typical sequences with respect to $p(y)$ is about $2^{nH(Y)}$, the number of codewords cannot exceed about

$$\frac{2^{nH(Y)}}{2^{nH(Y|X)}} = 2^{nI(X;Y)} = 2^{nC} \tag{8.153}$$

This is consistent with the converse of the channel coding theorem. The direct part of the channel coding theorem says that when n is large, as long as the number of codewords generated randomly is not more than about $2^{n(C-\epsilon)}$, the overlap among the cones is negligible with high probability.

Therefore, instead of searching through the ensemble of all possible codebooks for a good one, we can generate a codebook randomly, and it is likely to be good. However, such a code is difficult to use due to the following implementation issues.

A codebook with block length n and rate R consists of $n2^{nR}$ symbols from the input alphabet $\mathcal{X}$. This means that the size of the codebook, i.e., the amount of storage required to store the codebook, grows exponentially with n. This also makes the encoding process inefficient.

Another issue is regarding the computation required for decoding. Based on the sequence received at the output of the channel, the decoder needs to decide which of the about 2^{nR} codewords was the one transmitted. This requires an exponential amount of computation.

In practice, we are satisfied with the reliability of communication once it exceeds a certain level. Therefore, the above implementation issues may eventually be resolved with the advancement of microelectronics. But before then, we still have to deal with these issues. For this reason, the entire field of *coding theory* has been developed since the 1950's. Researchers in this field are devoted to searching for good codes and devising efficient decoding algorithms.

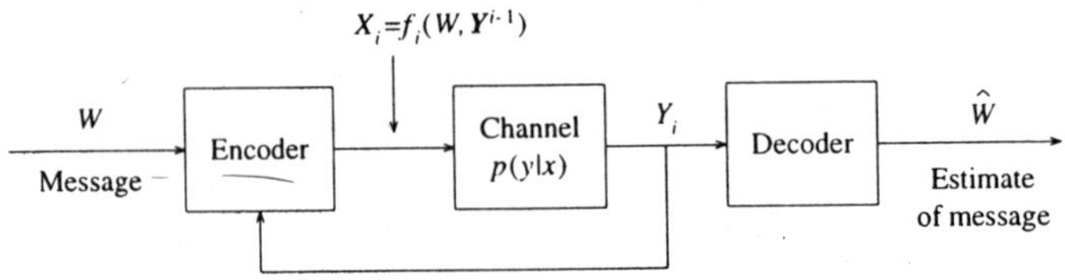

Figure 8.11. A channel code with feedback.

In fact, almost all the codes studied in coding theory are *linear codes*. By taking advantage of the linear structures of these codes, efficient encoding and decoding can be achieved. In particular, Berrou *et al.* [25] have proposed a linear code called the *turbo code*[6] in 1993, which is now generally believed to be the practical way to achieve the channel capacity.

Today, channel coding has been widely used in home entertainment systems (e.g., audio CD and DVD), computer storage systems (e.g., CD-ROM, hard disk, floppy disk, and magnetic tape), computer communication, wireless communication, and deep space communication. The most popular channel codes used in existing systems include the Hamming code, the Reed-Solomon code[7], the BCH code, and convolutional codes. We refer the interested reader to textbooks on coding theory [30] [133] [217] for discussions of this subject.

8.6 FEEDBACK CAPACITY

Feedback is common in practical communication systems for correcting possible errors which occur during transmission. As an example, during a telephone conversation, we often have to request the speaker to repeat due to poor voice quality of the telephone line. As another example, in data communication the receiver may request a packet to be retransmitted if the *parity check* bits received are incorrect. In general, when feedback from the receiver is available at the transmitter, the transmitter can at any time decide what to transmit next based on the feedback so far, and can potentially transmit information through the channel reliably at a higher rate.

In this section, we study a model in which a DMC is used with complete feedback. The block diagram for the model is shown in Figure 8.11. In this model, the symbol Y_i received at the output of the channel at time i is available instantaneously at the encoder without error. Then depending on the message W and all the previous feedback $Y_1, Y_2, \cdots, Y_i$, the encoder decides the value

[6]The turbo code is a special case of the class of *Low-density parity-check* (LDPC) codes proposed by Gallager [78] in 1962 (see MacKay [136]). However, the performance of such codes was not known at that time due to lack of high speed computers for simulation.

[7]The Reed-Solomon code was independently discovered by Arimoto [13].

of X_{i+1}, the next symbol to be transmitted. Such a channel code is formally defined below.

DEFINITION 8.14 *An (n, M) code with complete feedback for a discrete memoryless channel with input alphabet $\mathcal{X}$ and output alphabet $\mathcal{Y}$ is defined by encoding functions*

$$f_i : \{1, 2, \cdots, M\} \times \mathcal{Y}^{i-1} \to \mathcal{X} \tag{8.154}$$

for $1 \leq i \leq n$ and a decoding function

$$g : \mathcal{Y}^n \to \{1, 2, \cdots, M\}. \tag{8.155}$$

We will use $\mathbf{Y}^i$ to denote $(Y_1, Y_2, \cdots, Y_i)$ and X_i to denote $f_i(W, \mathbf{Y}^{i-1})$. We note that a channel code without feedback is a special case of a channel code with complete feedback because for the latter, the encoder can ignore the feedback.

DEFINITION 8.15 *A rate R is achievable with complete feedback for a discrete memoryless channel $p(y|x)$ if for any $\epsilon > 0$, there exists for sufficiently large n an (n, M) code with complete feedback such that*

$$\frac{1}{n} \log M > R - \epsilon \tag{8.156}$$

and

$$\lambda_{max} < \epsilon. \tag{8.157}$$

DEFINITION 8.16 *The feedback capacity, C_{FB}, of a discrete memoryless channel is the supremum of all the rates achievable by codes with complete feedback.*

PROPOSITION 8.17 *The supremum in the definition of C_{FB} in Definition 8.16 is the maximum.*

Proof Consider rates $R^{(k)}$ which are achievable with complete feedback such that

$$\lim_{k \to \infty} R^{(k)} = R. \tag{8.158}$$

Then for any $\epsilon > 0$, for all k, there exists for sufficiently large n an $(n, M^{(k)})$ code with complete feedback such that

$$\frac{1}{n} \log M^{(k)} > R^{(k)} - \epsilon \tag{8.159}$$

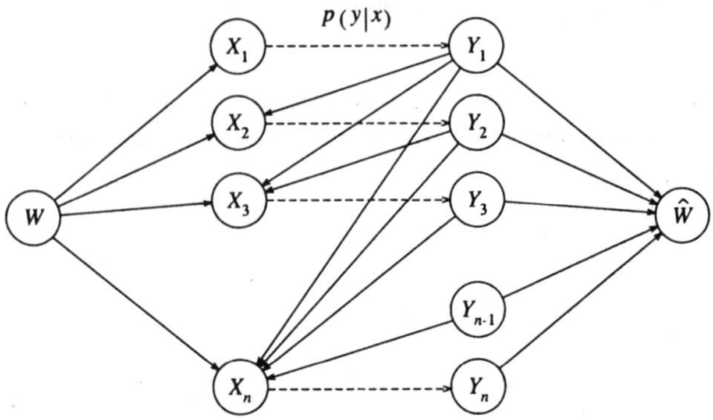

Figure 8.12. The dependency graph for a channel code with feedback.

and

$$\lambda_{max}^{(k)} < \epsilon. \tag{8.160}$$

By virtue of (8.158), let $k(\epsilon)$ be an integer such that for all $k > k(\epsilon)$,

$$|R - R^{(k)}| < \epsilon, \tag{8.161}$$

which implies

$$R^{(k)} > R - \epsilon. \tag{8.162}$$

Then for all $k > k(\epsilon)$,

$$\frac{1}{n} \log M^{(k)} > R^{(k)} - \epsilon > R - 2\epsilon. \tag{8.163}$$

Therefore, it follows from (8.163) and (8.160) that R is achievable with complete feedback. This implies that the supremum in Definition 8.16, which can be achieved, is in fact the maximum. □

Since a channel code without feedback is a special case of a channel code with complete feedback, any rate R achievable by the former is also achievable by the latter. Therefore,

$$C_{FB} \geq C. \tag{8.164}$$

The interesting question is whether C_{FB} is greater than C. The answer surprisingly turns out to be no for a DMC, as we now show. From the description of a channel code with complete feedback, we obtain the dependency graph for the random variables $W, \mathbf{X}, \mathbf{Y}, \hat{W}$ in Figure 8.12. From this dependency graph, we see that

$$q(w, \mathbf{x}, \mathbf{y}, \hat{w}) = q(w) \left(\prod_{i=1}^{n} q(x_i|w, \mathbf{y}^{i-1}) \right) \left(\prod_{i=1}^{n} p(y_i|x_i) \right) q(\hat{w}|\mathbf{y}) \quad (8.165)$$

for all $(w, \mathbf{x}, \mathbf{y}, \hat{w}) \in \mathcal{W} \times \mathcal{X}^n \times \mathcal{Y}^n \times \mathcal{W}$ such that $q(w, \mathbf{y}^{i-1}), q(x_i) > 0$ for $1 \leq i \leq n$ and $q(\mathbf{y}) > 0$, where $\mathbf{y}^i = (y_1, y_2, \cdots, y_i)$. Note that $q(x_i|w, \mathbf{y}^{i-1})$ and $q(\hat{w}|\mathbf{y})$ are deterministic.

LEMMA 8.18 *For all* $1 \leq i \leq n$,

$$(W, \mathbf{Y}^{i-1}) \to X_i \to Y_i \quad (8.166)$$

forms a Markov chain.

Proof The dependency graph for the random variables $W, \mathbf{X}^i$, and $\mathbf{Y}^i$ is shown in Figure 8.13. Denote the set of nodes $W, \mathbf{X}^{i-1}$, and $\mathbf{Y}^{i-1}$ by Z. Then we see that all the edges from Z end at X_i, and the only edge from X_i ends at Y_i. This means that Y_i depends on $(W, \mathbf{X}^{i-1}, \mathbf{Y}^{i-1})$ only through X_i, i.e.,

$$(W, \mathbf{X}^{i-1}, \mathbf{Y}^{i-1}) \to X_i \to Y_i \quad (8.167)$$

forms a Markov chain, or

$$I(W, \mathbf{X}^{i-1}, \mathbf{Y}^{i-1}; Y_i|X_i) = 0. \quad (8.168)$$

This can be formally justified by Proposition 2.9, and the details are omitted here. Since

$$\begin{aligned} 0 &= I(W, \mathbf{X}^{i-1}, \mathbf{Y}^{i-1}; Y_i|X_i) && (8.169) \\ &= I(W, \mathbf{Y}^{i-1}; Y_i|X_i) + I(\mathbf{X}^{i-1}; Y_i|W, X_i, \mathbf{Y}^{i-1}) && (8.170) \end{aligned}$$

and mutual information is nonnegative, we obtain

$$I(W, \mathbf{Y}^{i-1}; Y_i|X_i) = 0, \quad (8.171)$$

or

$$(W, \mathbf{Y}^{i-1}) \to X_i \to Y_i \quad (8.172)$$

forms a Markov chain. The lemma is proved. □

From the definition of C_{FB} and by virtue of Proposition 8.17, if $R \leq C_{FB}$, then R is a rate achievable with complete feedback. We will show that if a rate R is achievable with complete feedback, then $R \leq C$. If so, then $R \leq C_{FB}$ implies $R \leq C$, which can be true if and only if $C_{FB} \leq C$. Then from (8.164), we can conclude that $C_{FB} = C$.

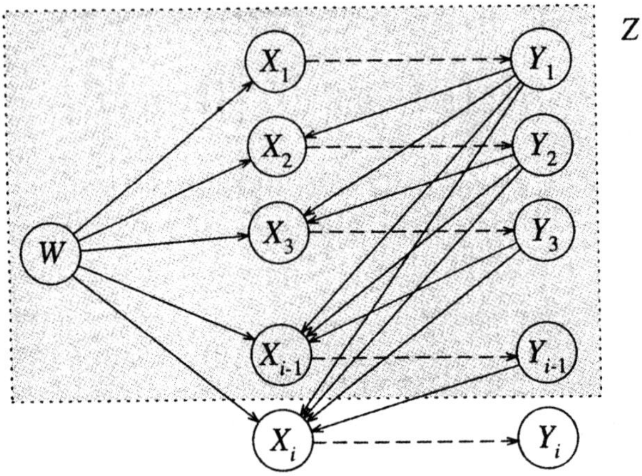

Figure 8.13. The dependency graph for $W, \mathbf{X}^i$, and $\mathbf{Y}^i$.

Let R be a rate achievable with complete feedback, i.e., for any $\epsilon > 0$, there exists for sufficiently large n an (n, M) code with complete feedback such that

$$n^{-1} \log M > R - \epsilon \tag{8.173}$$

and

$$\lambda_{max} < \epsilon. \tag{8.174}$$

Consider

$$\log M = H(W) = I(W;\mathbf{Y}) + H(W|\mathbf{Y}) \tag{8.175}$$

and bound $I(W;\mathbf{Y})$ and $H(W|\mathbf{Y})$ as follows. First,

$$\begin{aligned}
I(W;\mathbf{Y}) &= H(\mathbf{Y}) - H(\mathbf{Y}|W) && (8.176)\\
&= H(\mathbf{Y}) - \sum_{i=1}^{n} H(Y_i|\mathbf{Y}^{i-1}, W) && (8.177)\\
&\overset{a)}{=} H(\mathbf{Y}) - \sum_{i=1}^{n} H(Y_i|\mathbf{Y}^{i-1}, W, X_i) && (8.178)\\
&\overset{b)}{=} H(\mathbf{Y}) - \sum_{i=1}^{n} H(Y_i|X_i) && (8.179)\\
&\leq \sum_{i=1}^{n} H(Y_i) - \sum_{i=1}^{n} H(Y_i|X_i) && (8.180)\\
&= \sum_{i=1}^{n} I(X_i;Y_i) && (8.181)\\
&\leq nC, && (8.182)
\end{aligned}$$

where a) follows because X_i is a function of W and $\mathbf{Y}^{i-1}$ and b) follows from Lemma 8.18. Second,

$$H(W|\mathbf{Y}) = H(W|\mathbf{Y}, \hat{W}) \leq H(W|\hat{W}). \tag{8.183}$$

Thus

$$\log M \leq H(W|\hat{W}) + nC, \tag{8.184}$$

which is the same as (8.98). Then by (8.173) and an application of Fano's inequality, we conclude as in the proof for the converse of the channel coding theorem that

$$R \leq C. \tag{8.185}$$

Hence, we have proved that $C_{FB} = C$.

Remark 1 The proof for the converse of the channel coding theorem in Section 8.3 depends critically on the Markov chain

$$W \rightarrow \mathbf{X} \rightarrow \mathbf{Y} \rightarrow \hat{W} \tag{8.186}$$

and the relation in (8.73) (the latter implies Lemma 8.13). Both of them do not hold in general in the presence of feedback.

Remark 2 The proof for $C_{FB} = C$ in this section is also a proof for the converse of the channel coding theorem, so we actually do not need the proof in Section 8.3. However, the proof here and the proof in Section 8.3 have different spirits. Without comparing the two proofs, one cannot possibly understand the subtlety of the result that feedback does not increase the capacity of a DMC.

Remark 3 Although feedback does not increase the capacity of a DMC, the availability of feedback often makes coding much simpler. For some channels, communication through the channel with zero probability of error can be achieved in the presence of feedback by using a *variable-length* channel code. These are discussed in the next example.

EXAMPLE 8.19 *Consider the binary erasure channel in Example 8.5 whose capacity is* $1-\gamma$, *where* γ *is the erasure probability. In the presence of complete feedback, for every information bit to be transmitted, the encoder can transmit the same information bit through the channel until an erasure does not occur, i.e., the information bit is received correctly. Then the number of uses of the channel it takes to transmit an information bit through the channel correctly has a truncated geometrical distribution whose mean is* $(1-\gamma)^{-1}$. *Therefore, the effective rate at which information can be transmitted through the channel is* $1-\gamma$. *In other words, the channel capacity is achieved by using a very*

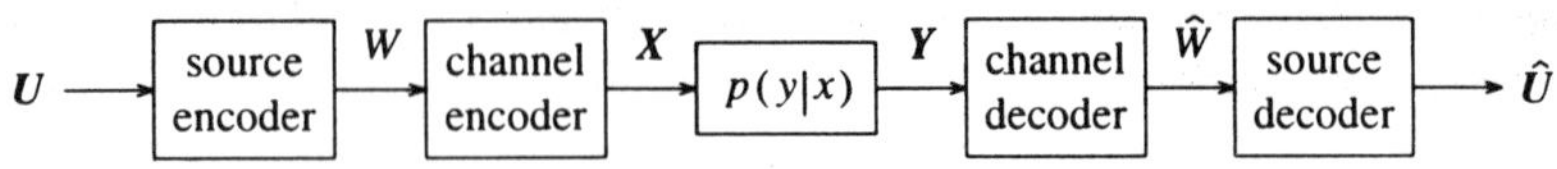

Figure 8.14. Separation of source coding and channel coding.

simple variable-length code. Moreover, the channel capacity is achieved with zero probability of error.

In the absence of feedback, the rate $1-\gamma$ *can also be achieved, but with an arbitrarily small probability of error and a much more complicated code.*

8.7 SEPARATION OF SOURCE AND CHANNEL CODING

We have so far considered the situation in which we want to convey a message through a DMC, where the message is randomly selected from a finite set according to the uniform distribution. However, in most situations, we want to convey an information source through a DMC. Let $\{U_k, k > -n\}$ be an ergodic stationary information source with entropy rate H. Denote the common alphabet by $\mathcal{U}$ and assume that $\mathcal{U}$ is finite. To convey $\{U_k\}$ through the channel, we can employ a source code with rate R_s and a channel code with rate R_c as shown in Figure 8.14 such that $R_s < R_c$.

Let f^s and g^s be respectively the encoding function and the decoding function of the source code, and f^c and g^c be respectively the encoding function and the decoding function of the channel code. The block of n information symbols $\mathbf{U} = (U_{-(n-1)}, U_{-(n-2)}, \cdots, U_0)$ is first encoded by the source encoder into an index

$$W = f^s(\mathbf{U}), \tag{8.187}$$

called the source codeword. Then W is mapped by the channel encoder to a distinct channel codeword

$$\mathbf{X} = f^c(W), \tag{8.188}$$

where $\mathbf{X} = (X_1, X_2, \cdots, X_n)$. This is possible because there are about 2^{nR_s} source codewords and about 2^{nR_c} channel codewords, and we assume that $R_s < R_c$. Then $\mathbf{X}$ is transmitted through the DMC $p(y|x)$, and the sequence $\mathbf{Y} = (Y_1, Y_2, \cdots, Y_n)$ is received. Based on $\mathbf{Y}$, the channel decoder first estimates W as

$$\hat{W} = g^c(\mathbf{Y}). \tag{8.189}$$

Finally, the source decoder decodes $\hat{W}$ to

$$\hat{\mathbf{U}} = g^s(\hat{W}). \tag{8.190}$$

For this scheme, an error occurs if $\mathbf{U} \neq \hat{\mathbf{U}}$, and we denote the probability of error by P_e.

We now show that if $H < C$, the capacity of the DMC $p(y|x)$, then it is possible to convey $\mathbf{U}$ through the channel with an arbitrarily small probability of error. First, we choose R_s and R_c such that

$$H < R_s < R_c < C. \tag{8.191}$$

Observe that if $\hat{W} = W$ and $g^s(W) = \mathbf{U}$, then from (8.190),

$$\hat{\mathbf{U}} = g^s(\hat{W}) = g^s(W) = \mathbf{U}, \tag{8.192}$$

i.e., an error does not occur. In other words, if an error occurs, either $\hat{W} \neq W$ or $g^s(W) \neq \mathbf{U}$. Then by the union bound, we have

$$P_e \leq \Pr\{\hat{W} \neq W\} + \Pr\{g^s(W) \neq \mathbf{U}\}. \tag{8.193}$$

For any $\epsilon > 0$ and sufficiently large n, by the Shannon-McMillan-Breiman theorem, there exists a source code such that

$$\Pr\{g^s(W) \neq \mathbf{U}\} \leq \epsilon. \tag{8.194}$$

By the channel coding theorem, there exists a channel code such that $\lambda_{max} \leq \epsilon$, where λ_{max} is the maximal probability of error. This implies

$$\begin{aligned}
\Pr\{\hat{W} \neq W\} &= \sum_w \Pr\{\hat{W} \neq W | W = w\}\Pr\{W = w\} && (8.195)\\
&\leq \lambda_{max} \sum_w \Pr\{W = w\} && (8.196)\\
&= \lambda_{max} && (8.197)\\
&\leq \epsilon. && (8.198)
\end{aligned}$$

Combining (8.194) and (8.198), we have

$$P_e \leq 2\epsilon. \tag{8.199}$$

Therefore, we conclude that as long as $H < C$, it is possible to convey $\{U_k\}$ through the DMC reliably.

In the scheme we have discussed, source coding and channel coding are separated. In general, source coding and channel coding can be combined. This technique is called *joint source-channel coding*. It is then natural to ask whether it is possible to convey information through the channel reliably at a higher rate by using joint source-channel coding. In the rest of the section, we show that the answer to this question is no to the extent that for asymptotic reliability, we must have $H \leq C$. However, whether asymptotical reliability

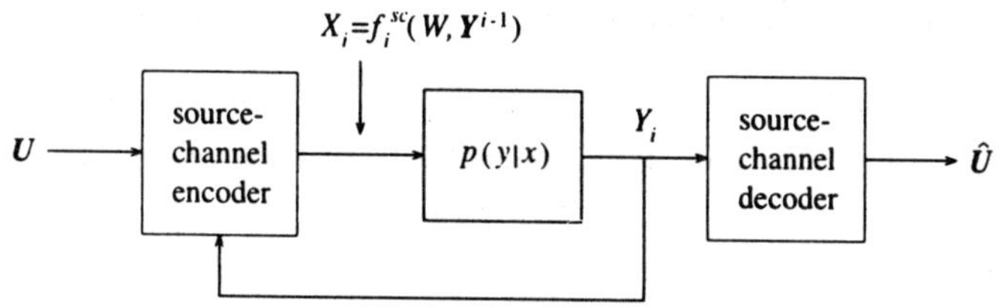

Figure 8.15. Joint source-channel coding.

can be achieved for $H = C$ depends on the specific information source and channel.

We base our discussion on the general assumption that complete feedback is available at the encoder as shown in Figure 8.15. Let f_i^{sc}, $1 \leq i \leq n$, be the encoding functions and g^{sc} be the decoding function of the source-channel code. Then

$$X_i = f_i^{sc}(\mathbf{U}, \mathbf{Y}^{i-1}) \tag{8.200}$$

for $1 \leq i \leq n$, where $\mathbf{Y}^{i-1} = (Y_1, Y_2, \cdots, Y_{i-1})$, and

$$\hat{\mathbf{U}} = g^{sc}(\mathbf{Y}), \tag{8.201}$$

where $\hat{\mathbf{U}} = (\hat{U}_1, \hat{U}_2, \cdots, \hat{U}_n)$. In exactly the same way as we proved (8.182) in the last section, we can prove that

$$I(\mathbf{U}; \mathbf{Y}) \leq nC. \tag{8.202}$$

Since $\hat{\mathbf{U}}$ is a function of $\mathbf{Y}$,

$$\begin{aligned} I(\mathbf{U}; \hat{\mathbf{U}}) &\leq I(\mathbf{U}; \hat{\mathbf{U}}, \mathbf{Y}) & (8.203)\\ &= I(\mathbf{U}; \mathbf{Y}) & (8.204)\\ &\leq nC. & (8.205) \end{aligned}$$

For any $\epsilon > 0$,

$$H(\mathbf{U}) \geq n(H - \epsilon) \tag{8.206}$$

for sufficiently large n. Then

$$n(H - \epsilon) \leq H(\mathbf{U}) = H(\mathbf{U}|\hat{\mathbf{U}}) + I(\mathbf{U}; \hat{\mathbf{U}}) \leq H(\mathbf{U}|\hat{\mathbf{U}}) + nC. \tag{8.207}$$

Applying Fano's inequality (Corollary 2.48), we obtain

$$n(H - \epsilon) \leq 1 + nP_e \log |\mathcal{U}| + nC, \tag{8.208}$$

or

$$H - \epsilon \leq \frac{1}{n} + P_e \log |\mathcal{U}| + C. \tag{8.209}$$

for all i. We assume that $\{X_i\}$ and $\{Z_i\}$ are independent, but we make no assumption that Z_i are i.i.d. so that the channel may have memory.

a) Prove that

$$I(\mathbf{X};\mathbf{Y}) \leq n - h_b(\epsilon).$$

b) Show that the upper bound in a) can be achieved by letting X_i be i.i.d. bits taking the values 0 and 1 with equal probability and $Z_1 = Z_2 = \cdots = Z_n$.

c) Show that with the assumptions in b), $I(\mathbf{X};\mathbf{Y}) > nC$, where $C = 1 - h_b(\epsilon)$ is the capacity of the BSC if it is memoryless.

Show that the capacity of a channel can be increased by feedback if the channel has memory.

In Remark 1 toward the end of Section 8.6, it was mentioned that in the presence of feedback, both the Markov chain $W \rightarrow \mathbf{X} \rightarrow \mathbf{Y} \rightarrow \hat{W}$ and Lemma 8.13 do not hold in general. Give examples to substantiate this remark.

Prove that when a DMC is used with complete feedback,

$$\Pr\{Y_i = y_i | \mathbf{X}^i = \mathbf{x}^i, \mathbf{Y}^{i-1} = \mathbf{y}^{i-1}\} = \Pr\{Y_i = y_i | X_i = x_i\}$$

for all $i \geq 1$. This relation, which is a consequence of the causality of the code, says that given the current input, the current output does not depend on all the past inputs and outputs of the DMC.

Let

$$P(\epsilon) = \begin{bmatrix} 1-\epsilon & \epsilon \\ \epsilon & 1-\epsilon \end{bmatrix}$$

be the transition matrix for a BSC with crossover probability ϵ. Define $a * b = (1-a)b + a(1-b)$ for $0 \leq a, b \leq 1$.

a) Prove that a DMC with transition matrix $P(\epsilon_1)P(\epsilon_2)$ is equivalent to a BSC with crossover probability $\epsilon_1 * \epsilon_2$. Such a channel is the cascade of two BSC's with crossover probabilities ϵ_1 and ϵ_2, respectively.

b) Repeat a) for a DMC with transition matrix $P(\epsilon_2)P(\epsilon_1)$.

c) Prove that

$$1 - h_b(\epsilon_1 * \epsilon_2) \leq \min(1 - h_b(\epsilon_1), 1 - h_b(\epsilon_2)).$$

This means that the capacity of the cascade of two BSC's is upper bounded by the capacity of either of the two BSC's.

For asymptotic reliability, $P_e \to 0$ as $n \to \infty$. Therefore, b and then $\epsilon \to 0$, we conclude that

$$H \leq C.$$

This result, sometimes called the *separation theorem for so coding*, says that asymptotic optimality can be achieved by coding and channel coding. This theorem has significant eng tion because the source code and the channel code can be de without losing asymptotic optimality. Specifically, we only best source code for the information source and design the for the channel. Moreover, separation of source coding and c cilitates the transmission of different information sources on because we need only change the source code for different inf Likewise, separation of source coding and channel coding transmission of an information source on different channels only change the channel code for different channels.

We remark that although asymptotic optimality can be ac ing source coding and channel coding, for finite block leng of error generally can be reduced by using joint source-char

PROBLEMS

In the following, $\mathbf{X} = (X_1, X_2, \cdots, X_n)$, $\mathbf{x} = (x_1, x_2, \cdots,$

1. *Alternative representation of a discrete channel* Let X tively the input and the output of a discrete channel $p(y|$ exists a random variable Z and a function f such that Y is independent of X. This is a generalization of the alte tion of a BSC in Example 8.4.

   Let X_i and Y_i be respectively the input and the output o Then Y_i depends causally on X_i via $Y_i = f(X_i, Z_i)$, wl random variable at time i which has the same distribut independent of all the random variables which have alre in the system.

2. Show that the capacity of a DMC with complete feed creased by using probabilistic encoding and/or decodin

3. *Memory increases capacity* Consider a BSC with cr $0 < \epsilon < 1$ represented by $X_i = Y_i + Z_i \bmod 2$, where respectively the input, the output, and the noise randon Then

$$\Pr\{Z_i = 0\} = 1 - \epsilon \quad \text{and} \quad \Pr\{Z_i = 1\}$$

d) Prove that a DMC with transition matrix $P(\epsilon)^n$ is equivalent to a BSC with crossover probabilities $\frac{1}{2}(1-(1-2\epsilon)^n)$.

8. *Symmetric channel* A DMC is *symmetric* if the rows of the transition matrix $p(y|x)$ are permutations of each other and so are the columns. Determine the capacity of such a channel.

   See Section 4.5 in Gallager [80] for a more general discussion.

9. Let C_1 and C_2 be the capacities of two DMC's with transition matrices P_1 and P_2, respectively, and let C be the capacity of the DMC with transition matrix P_1P_2. Prove that $C \leq \min(C_1, C_2)$.

10. *Two independent parallel channels* Let C_1 and C_2 be the capacities of two DMC's $p_1(y_1|x_1)$ and $p_2(y_2|x_2)$, respectively. Determine the capacity of the DMC

$$p(y_1, y_2|x_1, x_2) = p_1(y_1|x_1)p_2(y_2|x_2).$$

   Hint: Prove that

$$I(X_1, X_2; Y_1, Y_2) \leq I(X_1; Y_1) + I(X_2; Y_2)$$

   if $p(y_1, y_2|x_1, x_2) = p_1(y_1|x_1)p_2(y_2|x_2)$.

11. *Maximum likelihood decoding* In maximum likelihood decoding for a given channel and a given codebook, if a received sequence $\mathbf{y}$ is decoded to a codeword $\mathbf{x}$, then $\mathbf{x}$ maximizes $\Pr\{\mathbf{y}|\mathbf{x}'\}$ among all codewords $\mathbf{x}'$ in the codebook.

    a) Prove that maximum likelihood decoding minimizes the average probability of error.

    b) Does maximum likelihood decoding also minimize the maximal probability of error? Give an example if your answer is no.

12. *Minimum distance decoding* The Hamming distance between two binary sequences $\mathbf{x}$ and $\mathbf{y}$, denoted by $d(\mathbf{x}, \mathbf{y})$, is the number of places where $\mathbf{x}$ and $\mathbf{y}$ differ. In minimum distance decoding for a memoryless BSC, if a received sequence $\mathbf{y}$ is decoded to a codeword $\mathbf{x}$, then $\mathbf{x}$ minimizes $d(\mathbf{x}', \mathbf{y})$ over all codewords $\mathbf{x}'$ in the codebook. Prove that minimum distance decoding is equivalent to maximum likelihood decoding if the crossover probability of the BSC is less than 0.5.

13. The following figure shows a communication system with two DMC's with complete feedback. The capacities of the two channels are respectively C_1 and C_2.

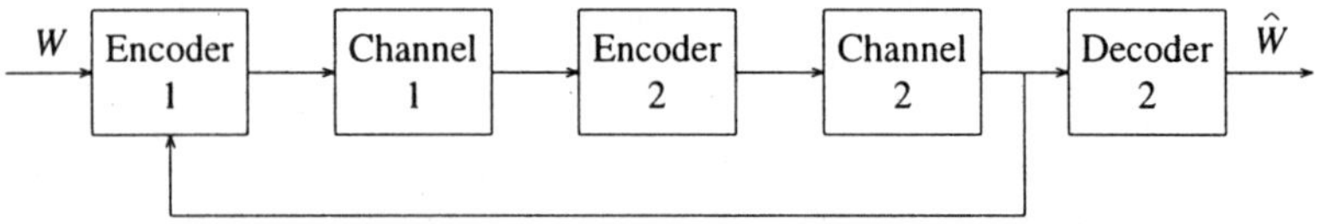

a) Give the dependency graph for all the random variables involved in the coding scheme.

b) Prove that the capacity of the system is $\min(C_1, C_2)$.

14. *Binary arbitrarily varying channel* Consider a memoryless BSC whose crossover probability is time-varying. Specifically, the crossover probability $\epsilon(i)$ at time i is an arbitrary value in $[\epsilon_1, \epsilon_2]$, where $0 \leq \epsilon_1 < \epsilon_2 < 0.5$. Prove that the capacity of this channel is $1 - h_b(\epsilon_2)$. (Ahlswede and Wolfowitz [9].)

15. Consider a BSC with crossover probability $\epsilon \in [\epsilon_1, \epsilon_2]$, where $0 < \epsilon_1 < \epsilon_2 < 0.5$, but the exact value of ϵ is unknown. Prove that the capacity of this channel is $1 - h_b(\epsilon_2)$.

HISTORICAL NOTES

The concept of channel capacity was introduced in Shannon's original paper [184], where he stated the channel coding theorem and outlined a proof. The first rigorous proof was due to Feinstein [68]. The random coding error exponent was developed by Gallager [79] in a simplified proof.

The converse of the channel coding theorem was proved by Fano [66], where he used an inequality now bearing his name. The strong converse was first proved by Wolfowitz [220]. An iterative algorithm for calculating the channel capacity developed independently by Arimoto [14] and Blahut [28] will be discussed in Chapter 10.

It was proved by Shannon [187] that the capacity of a discrete memoryless channel cannot be increased by feedback. The proof here based on dependency graphs is inspired by Bayesian networks.

Chapter 9

RATE-DISTORTION THEORY

Let H be the entropy rate of an information source. By the source coding theorem, it is possible to design a source code with rate R which reconstructs the source sequence $\mathbf{X} = (X_1, X_2, \cdots, X_n)$ with an arbitrarily small probability of error provided $R > H$ and the block length n is sufficiently large. However, there are situations in which we want to convey an information source by a source code with rate less than H. Then we are motivated to ask: what is the best we can do when $R < H$?

A natural approach is to design a source code such that for part of the time the source sequence is reconstructed correctly, while for the other part of the time the source sequence is reconstructed incorrectly, i.e., an error occurs. In designing such a code, we try to minimize the probability of error. However, this approach is not viable asymptotically because the converse of the source coding theorem says that if $R < H$, then the probability of error inevitably tends to 1 as $n \to \infty$.

Therefore, if $R < H$, no matter how the source code is designed, the source sequence is almost always reconstructed incorrectly when n is large. An alternative approach is to design a source code called a *rate-distortion code* which reproduces the source sequence with distortion. In order to formulate the problem properly, we need a *distortion measure* between each source sequence and each reproduction sequence. Then we try to design a rate-distortion code which with high probability reproduces the source sequence with a distortion within a tolerance level.

Clearly, a smaller distortion can potentially be achieved if we are allowed to use a higher coding rate. *Rate-distortion theory*, the subject matter of this chapter, gives a characterization of the asymptotic optimal tradeoff between the coding rate of a rate-distortion code for a given information source and the allowed distortion in the reproduction sequence with respect to a distortion measure.

9.1 SINGLE-LETTER DISTORTION MEASURES

Let $\{X_k, k \geq 1\}$ be an i.i.d. information source with generic random variable X. We assume that the source alphabet $\mathcal{X}$ is finite. Let $p(x)$ be the probability distribution of X, and we assume without loss of generality that the support of X is equal to $\mathcal{X}$. Consider a source sequence

$$\mathbf{x} = (x_1, x_2, \cdots, x_n) \tag{9.1}$$

and a reproduction sequence

$$\hat{\mathbf{x}} = (\hat{x}_1, \hat{x}_2, \cdots, \hat{x}_n). \tag{9.2}$$

The components of $\hat{\mathbf{x}}$ can take values in $\mathcal{X}$, but more generally, they can take values in any finite set $\hat{\mathcal{X}}$ which may be different from $\mathcal{X}$. The set $\hat{\mathcal{X}}$, which is also assumed to be finite, is called the reproduction alphabet. To measure the distortion between $\mathbf{x}$ and $\hat{\mathbf{x}}$, we introduce the single-letter distortion measure and the average distortion measure.

DEFINITION 9.1 *A single-letter distortion measure is a mapping*

$$d : \mathcal{X} \times \hat{\mathcal{X}} \to \Re^+, \tag{9.3}$$

where $\Re^+$ is the set of nonnegative real numbers[1]*. The value $d(x, \hat{x})$ denotes the distortion incurred when a source symbol x is reproduced as $\hat{x}$.*

DEFINITION 9.2 *The average distortion between a source sequence $\mathbf{x} \in \mathcal{X}^n$ and a reproduction sequence $\hat{\mathbf{x}} \in \hat{\mathcal{X}}^n$ induced by a single-letter distortion measure d is defined by*

$$d(\mathbf{x}, \hat{\mathbf{x}}) = \frac{1}{n} \sum_{k=1}^{n} d(x_k, \hat{x}_k). \tag{9.4}$$

In Definition 9.2, we have used d to denote both the single-letter distortion measure and the average distortion measure, but this abuse of notation should cause no ambiguity. Henceforth, we will refer to a single-letter distortion measure simply as a distortion measure.

Very often, the source sequence $\mathbf{x}$ represents quantized samples of a continuous signal, and the user attempts to recognize certain objects and derive

[1]Note that $d(x, \hat{x})$ is finite for all $(x, \hat{x}) \in \mathcal{X} \times \hat{\mathcal{X}}$.

meaning from the reproduction sequence $\hat{\mathbf{x}}$. For example, $\mathbf{x}$ may represent a video signal, an audio signal, or an image. The ultimate purpose of a distortion measure is to reflect the distortion between $\mathbf{x}$ and $\hat{\mathbf{x}}$ as *perceived* by the user. This goal is difficult to achieve in general because measurements of the distortion between $\mathbf{x}$ and $\hat{\mathbf{x}}$ must be made within context unless the symbols in $\mathcal{X}$ carry no physical meaning. Specifically, when the user derives meaning from $\hat{\mathbf{x}}$, the distortion in $\hat{\mathbf{x}}$ as perceived by the user depends on the context. For example, the perceived distortion is small for a portrait contaminated by a fairly large noise, while the perceived distortion is large for the image of a book page contaminated by the same noise. Hence, a good distortion measure should be context dependent.

Although the average distortion is not necessarily the best way to measure the distortion between a source sequence and a reproduction sequence, it has the merit of being simple and easy to use. Moreover, rate-distortion theory, which is based on the average distortion measure, provides a framework for data compression when distortion is inevitable.

EXAMPLE 9.3 *When the symbols in $\mathcal{X}$ and $\hat{\mathcal{X}}$ represent real values, a popular distortion measure is the square-error distortion measure which is defined by*

$$d(x, \hat{x}) = (x - \hat{x})^2. \tag{9.5}$$

The average distortion measure so induced is often referred to as the mean-square error.

EXAMPLE 9.4 *When $\mathcal{X}$ and $\hat{\mathcal{X}}$ are identical and the symbols in $\mathcal{X}$ do not carry any particular meaning, a frequently used distortion measure is the Hamming distortion measure, which is defined by*

$$d(x, \hat{x}) = \begin{cases} 0 & \text{if } x = \hat{x} \\ 1 & \text{if } x \neq \hat{x}. \end{cases} \tag{9.6}$$

The Hamming distortion measure indicates the occurrence of an error. In particular, for an estimate $\hat{X}$ of X, we have

$$Ed(X, \hat{X}) = \Pr\{X = \hat{X}\} \cdot 0 + \Pr\{X \neq \hat{X}\} \cdot 1 = \Pr\{X \neq \hat{X}\}, \tag{9.7}$$

i.e., the expectation of the Hamming distortion measure between X and $\hat{X}$ is the probability of error.

For $\mathbf{x} \in \mathcal{X}^n$ and $\hat{\mathbf{x}} \in \hat{\mathcal{X}}^n$, the average distortion $d(\mathbf{x}, \hat{\mathbf{x}})$ induced by the Hamming distortion measure gives the frequency of error in the reproduction sequence $\hat{\mathbf{x}}$.

DEFINITION 9.5 *For a distortion measure d, for each $x \in \mathcal{X}$, let $\hat{x}^*(x) \in \hat{\mathcal{X}}$ minimize $d(x, \hat{x})$ over all $\hat{x} \in \hat{\mathcal{X}}$. A distortion measure d is said to be normal if*

$$c_x \stackrel{\text{def}}{=} d(x, \hat{x}^*(x)) = 0 \tag{9.8}$$

for all $x \in \mathcal{X}$.

The square-error distortion measure and the Hamming distortion measure are examples of normal distortion measures. Basically, a normal distortion measure is one which allows X to be reproduced with zero distortion. Although a distortion measure d is not normal in general, a normalization of d can always be obtained by defining the distortion measure

$$\tilde{d}(x, \hat{x}) = d(x, \hat{x}) - c_x \tag{9.9}$$

for all $(x, \hat{x}) \in \mathcal{X} \times \hat{\mathcal{X}}$. Evidently, $\tilde{d}$ is a normal distortion measure, and it is referred to as the normalization of d.

EXAMPLE 9.6 *Let d be a distortion measure defined by*

$d(x, \hat{x})$	a	b	c
1	2	7	5
2	4	3	8

Then $\tilde{d}$, the normalization of d, is given by

$\tilde{d}(x, \hat{x})$	a	b	c
1	0	5	3
2	1	0	5

Note that for every $x \in \mathcal{X}$, there exists an $\hat{x} \in \hat{\mathcal{X}}$ such that $\tilde{d}(x, \hat{x}) = 0$.

Let $\hat{X}$ be any estimate of X which takes values in $\hat{\mathcal{X}}$, and denote the joint distribution for X and $\hat{X}$ by $p(x, \hat{x})$. Then

$$\begin{aligned}
Ed(X, \hat{X}) &= \sum_x \sum_{\hat{x}} p(x, \hat{x}) d(x, \hat{x}) && (9.10)\\
&= \sum_x \sum_{\hat{x}} p(x, \hat{x}) \left[\tilde{d}(x, \hat{x}) + c_x\right] && (9.11)\\
&= E\tilde{d}(X, \hat{X}) + \sum_x p(x) \sum_{\hat{x}} p(\hat{x}|x) c_x && (9.12)\\
&= E\tilde{d}(X, \hat{X}) + \sum_x p(x) c_x \left(\sum_{\hat{x}} p(\hat{x}|x)\right) && (9.13)\\
&= E\tilde{d}(X, \hat{X}) + \sum_x p(x) c_x && (9.14)\\
&= E\tilde{d}(X, \hat{X}) + \Delta, && (9.15)
\end{aligned}$$

where

$$\Delta = \sum_x p(x)c_x \tag{9.16}$$

is a constant which depends only on $p(x)$ and d but not on the conditional distribution $p(\hat{x}|x)$. In other words, for a given X and a distortion measure d, the expected distortion between X and an estimate $\hat{X}$ of X is always reduced by a constant upon using $\tilde{d}$ instead of d as the distortion measure. For reasons which will be explained in Section 9.3, it is sufficient for us to assume that a distortion measure is normal.

DEFINITION 9.7 *Let $\hat{x}^*$ minimizes $Ed(X, \hat{x})$ over all $\hat{x} \in \hat{\mathcal{X}}$, and define*

$$D_{max} = Ed(X, \hat{x}^*). \tag{9.17}$$

$\hat{x}^*$ is the best estimate of X if we know nothing about X, and D_{max} is the minimum expected distortion between X and a constant estimate of X. The significance of D_{max} can be seen by taking the reproduction sequence $\hat{\mathbf{X}}$ to be $(\hat{x}^*, \hat{x}^*, \cdots, \hat{x}^*)$. Since $d(X_k, \hat{x}^*)$ are i.i.d., by the weak law of large numbers

$$d(\mathbf{X}, \hat{\mathbf{X}}) = \frac{1}{n}\sum_{k=1}^{n} d(X_k, \hat{x}^*) \rightarrow Ed(X, \hat{x}^*) = D_{max} \tag{9.18}$$

in probability, i.e., for any $\epsilon > 0$,

$$\Pr\{d(\mathbf{X}, \hat{\mathbf{X}}) > D_{max} + \epsilon\} \leq \epsilon \tag{9.19}$$

for sufficiently large n. Note that $\hat{\mathbf{X}}$ is a constant sequence which does not depend on $\mathbf{X}$. In other words, even when no description of $\mathbf{X}$ is available, we can still achieve an average distortion no more than $D_{max} + \epsilon$ with probability arbitrarily close to 1 when n is sufficiently large.

The notation D_{max} may seem confusing because the quantity stands for the minimum rather than the maximum expected distortion between X and a constant estimate of X. But we see from the above discussion that this notation is in fact appropriate because D_{max} is the maximum distortion we have to be concerned about. Specifically, it is not meanful to impose a constraint $D \geq D_{max}$ on the reproduction sequence because it can be achieved even without receiving any information about the sequence produced by the source.

9.2 THE RATE-DISTORTION FUNCTION $R(D)$

Throughout this chapter, all the discussions are with respect to an i.i.d. information source $\{X_k, k \geq 1\}$ with generic random variable X and a distortion measure d. All logarithms are in the base 2 unless otherwise specified.

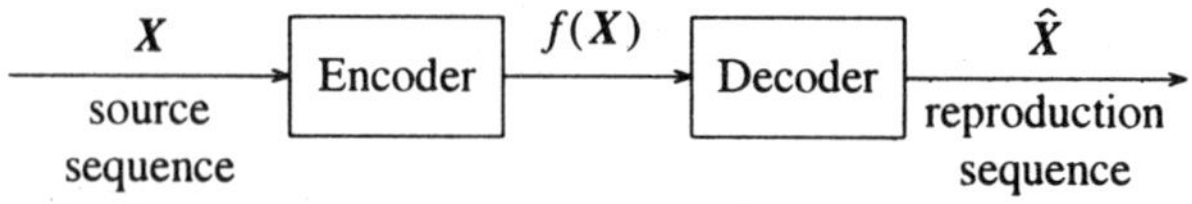

Figure 9.1. A rate-distortion code with block length n.

DEFINITION 9.8 *An (n, M) rate-distortion code is defined by an encoding function*

$$f : \mathcal{X}^n \to \{1, 2, \cdots, M\} \tag{9.20}$$

and a decoding function

$$g : \{1, 2, \cdots, M\} \to \hat{\mathcal{X}}^n. \tag{9.21}$$

The set $\{1, 2, \cdots, M\}$, denoted by $\mathcal{I}$, is called the index set. The reproduction sequences $g(f(1)), g(f(2)), \cdots, g(f(M))$ in $\hat{\mathcal{X}}^n$ are called codewords, and the set of codewords is called the codebook.

Figure 9.1 is an illustration of a rate-distortion code.

DEFINITION 9.9 *The rate of an (n, M) rate-distortion code is $n^{-1} \log M$ in bits per symbol.*

DEFINITION 9.10 *A rate-distortion pair (R, D) is asymptotically achievable if for any $\epsilon > 0$, there exists for sufficiently large n an (n, M) rate-distortion code such that*

$$\frac{1}{n} \log M \le R + \epsilon \tag{9.22}$$

and

$$\Pr\{d(\mathbf{X}, \hat{\mathbf{X}}) > D + \epsilon\} \le \epsilon, \tag{9.23}$$

where $\hat{\mathbf{X}} = g(f(\mathbf{X}))$. For brevity, an asymptotically achievable pair will be referred to as an achievable pair.

Remark It is clear from the definition that if (R, D) is achievable, then (R', D) and (R, D') are also achievable for all $R' \ge R$ and $D' \ge D$.

DEFINITION 9.11 *The rate-distortion region is the subset of $\Re^2$ containing all achievable pairs (R, D).*

THEOREM 9.12 *The rate-distortion region is closed and convex.*

Proof We first show that the rate-distortion region is closed. Consider achievable rate-distortion pairs $(R^{(k)}, D^{(k)})$ such that

$$\lim_{k \to \infty} (R^{(k)}, D^{(k)}) = (R, D). \tag{9.24}$$

Then for any $\epsilon > 0$, for all k, there exists for sufficiently large n an $(n, M^{(k)})$ code such that

$$\frac{1}{n} \log M^{(k)} \leq R^{(k)} + \epsilon \tag{9.25}$$

and

$$\Pr\{d(\mathbf{X}^{(k)}, \hat{\mathbf{X}}^{(k)}) > D^{(k)} + \epsilon\} \leq \epsilon, \tag{9.26}$$

where $f^{(k)}$ and $g^{(k)}$ are respectively the encoding function and the decoding function of the $(n, M^{(k)})$ code, and $\hat{\mathbf{X}}^{(k)} = g^{(k)}(f^{(k)}(\mathbf{X}))$. By virtue of (9.24), let $k(\epsilon)$ be an integer such that for all $k > k(\epsilon)$,

$$|R - R^{(k)}| < \epsilon \tag{9.27}$$

and

$$|D - D^{(k)}| < \epsilon, \tag{9.28}$$

which imply

$$R^{(k)} < R + \epsilon \tag{9.29}$$

and

$$D^{(k)} < D + \epsilon, \tag{9.30}$$

respectively. Then for all $k > k(\epsilon)$,

$$\frac{1}{n} \log M^{(k)} \leq R^{(k)} + \epsilon < R + 2\epsilon \tag{9.31}$$

and

$$\begin{aligned} \Pr\{d(\mathbf{X}^{(k)}, \hat{\mathbf{X}}^{(k)}) > D + 2\epsilon\} &\leq \Pr\{d(\mathbf{X}^{(k)}, \hat{\mathbf{X}}^{(k)}) > D^{(k)} + \epsilon\} && (9.32) \\ &\leq \epsilon. && (9.33) \end{aligned}$$

Note that (9.32) follows because

$$D + 2\epsilon > D^{(k)} + \epsilon \tag{9.34}$$

by (9.30). From (9.31) and (9.33), we see that (R, D) is also achievable. Thus we have proved that the rate-distortion region is closed.

We will prove the convexity of the rate-distortion region by a time-sharing argument whose idea is the following. Roughly speaking, if we can use a code $\mathcal{C}_1$ to achieve $(R^{(1)}, D^{(1)})$ and a code $\mathcal{C}_2$ to achieve $(R^{(2)}, D^{(2)})$, then for any rational number λ between 0 and 1, we can use $\mathcal{C}_1$ for a fraction λ of the time and use $\mathcal{C}_2$ for a fraction $\bar{\lambda}$ of the time to achieve $(R^{(\lambda)}, D^{(\lambda)})$, where

$$\begin{aligned} R^{(\lambda)} &= \lambda R^{(1)} + \bar{\lambda} R^{(2)}, && (9.35) \\ D^{(\lambda)} &= \lambda D^{(1)} + \bar{\lambda} D^{(2)}, && (9.36) \end{aligned}$$

and $\bar{\lambda} = 1 - \lambda$. Since the rate-distortion region is closed as we have proved, λ can be taken as any real number between 0 and 1, and the convexity of the region follows.

We now give a formal proof for the convexity of the rate-distortion region. Let

$$\lambda = \frac{r}{r+s}, \tag{9.37}$$

where r and s are positive integers. Then λ is a rational number between 0 and 1. We now prove that if $(R^{(1)}, D^{(1)})$ and $(R^{(2)}, D^{(2)})$ are achievable, then $(R^{(\lambda)}, D^{(\lambda)})$ is also achievable. Assume $(R^{(1)}, D^{(1)})$ and $(R^{(2)}, D^{(2)})$ are achievable. Then for any $\epsilon > 0$ and sufficiently large n, there exist an $(n, M^{(1)})$ code and an $(n, M^{(2)})$ code such that

$$\frac{1}{n} \log M^{(i)} \leq R^{(i)} + \epsilon \tag{9.38}$$

and

$$\Pr\{d(\mathbf{X}, \hat{\mathbf{X}}^{(i)}) > D^{(i)} + \epsilon\} \leq \epsilon, \tag{9.39}$$

$i = 1, 2$. Let

$$M(\lambda) = (M^{(1)})^r (M^{(2)})^s \tag{9.40}$$

and

$$n(\lambda) = (r+s)n. \tag{9.41}$$

We now construct an $(n(\lambda), M(\lambda))$ code by concatenating r copies of the $(n, M^{(1)})$ code followed by s copies of the $(n, M^{(2)})$ code. We call these $r + s$ codes subcodes of the $(n(\lambda), M(\lambda))$ code. For this code, let

$$\mathbf{Y} = (\mathbf{X}(1), \mathbf{X}(2), \cdots, \mathbf{X}(r+s)) \tag{9.42}$$

and

$$\hat{\mathbf{Y}} = (\hat{\mathbf{X}}(1), \hat{\mathbf{X}}(2), \cdots, \hat{\mathbf{X}}(r+s)), \tag{9.43}$$

where $\mathbf{X}(j)$ and $\hat{\mathbf{X}}(j)$ are the source sequence and the reproduction sequence of the jth subcode, respectively. Then for this $(n(\lambda), M(\lambda))$ code,

$$\begin{aligned}
\frac{1}{n(\lambda)} \log M(\lambda) &= \frac{1}{(r+s)n} \log[(M^{(1)})^r (M^{(2)})^s] && (9.44)\\
&= \frac{1}{(r+s)n} (r \log M^{(1)} + s \log M^{(2)}) && (9.45)\\
&= \lambda \left(\frac{1}{n} \log M^{(1)} \right) + \bar{\lambda} \left(\frac{1}{n} \log M^{(2)} \right) && (9.46)\\
&\leq \lambda (R^{(1)} + \epsilon) + \bar{\lambda} (R^{(2)} + \epsilon) && (9.47)\\
&= (\lambda R^{(1)} + \bar{\lambda} R^{(2)}) + \epsilon && (9.48)\\
&= R^{(\lambda)} + \epsilon, && (9.49)
\end{aligned}$$

where (9.47) follows from (9.38), and

$$
\begin{aligned}
&\Pr\{d(\mathbf{Y},\hat{\mathbf{Y}}) > D^{(\lambda)} + \epsilon\} \\
&= \Pr\left\{\frac{1}{r+s}\sum_{j=1}^{r+s} d(\mathbf{X}(j),\hat{\mathbf{X}}(j)) > D^{(\lambda)} + \epsilon\right\} && (9.50)\\
&\leq \Pr\Big\{d(\mathbf{X}(j),\hat{\mathbf{X}}(j)) > D^{(1)} + \epsilon \text{ for some } 1 \leq j \leq r \text{ or} \\
&\qquad d(\mathbf{X}(j),\hat{\mathbf{X}}(j)) > D^{(2)} + \epsilon \text{ for some } r+1 \leq j \leq r+s\Big\} && (9.51)\\
&\leq \sum_{j=1}^{r} \Pr\{d(\mathbf{X}(j),\hat{\mathbf{X}}(j)) > D^{(1)} + \epsilon\} \\
&\quad + \sum_{j=r+1}^{r+s} \Pr\{d(\mathbf{X}(j),\hat{\mathbf{X}}(j)) > D^{(2)} + \epsilon\} && (9.52)\\
&\leq (r+s)\epsilon, && (9.53)
\end{aligned}
$$

where (9.52) follows from the union bound and (9.53) follows from (9.39). Hence, we conclude that the rate-distortion pair $(R^{(\lambda)}, D^{(\lambda)})$ is achievable. This completes the proof of the theorem. □

DEFINITION 9.13 *The rate-distortion function $R(D)$ is the minimum of all rates R for a given distortion D such that (R, D) is achievable.*

DEFINITION 9.14 *The distortion-rate function $D(R)$ is the minimum of all distortions D for a given rate R such that (R, D) is achievable.*

Both the functions $R(D)$ and $D(R)$ are equivalent descriptions of the boundary of the rate-distortion region. They are sufficient to describe the rate-distortion region because the region is closed. Note that in defining $R(D)$, the minimum instead of the infimum is taken because for a fixed D, the set of all R such that (R, D) is achievable is closed and lower bounded by zero. Similarly, the minimum instead of the infimum is taken in defining $D(R)$. In the subsequent discussions, only $R(D)$ will be used.

THEOREM 9.15 *The following properties hold for the rate-distortion function $R(D)$:*

1. *$R(D)$ is non-increasing in D.*
2. *$R(D)$ is convex.*
3. *$R(D) = 0$ for $D \geq D_{max}$.*
4. *$R(0) \leq H(X)$.*

Proof From the remark following Definition 9.10, since $(R(D), D)$ is achievable, $(R(D), D')$ is also achievable for all $D' \geq D$. Therefore, $R(D) \geq R(D')$ because $R(D')$ is the minimum of all R such that (R, D') is achievable. This proves Property 1.

Property 2 follows immediately from the convexity of the rate-distortion region which was proved in Theorem 9.12. From the discussion toward the end of the last section, we see for any $\epsilon > 0$, it is possible to achieve

$$\Pr\{d(\mathbf{X}, \hat{\mathbf{X}}) > D_{max} + \epsilon\} \leq \epsilon \tag{9.54}$$

for sufficiently large n with no description of $\mathbf{X}$ available. Therefore, $(0, D)$ is achievable for all $D \geq D_{max}$, proving Property 3.

Property 4 is a consequence of the assumption that the distortion measure d is normalized, which can be seen as follows. By the source coding theorem, for any $\epsilon > 0$, by using a rate no more than $H(X) + \epsilon$, we can describe the source sequence $\mathbf{X}$ of length n with probability of error less than ϵ when n is sufficiently large. Since d is normalized, for each $k \geq 1$, let

$$\hat{X}_k = \hat{x}^*(X_k) \tag{9.55}$$

(cf. Definition 9.5), so that whenever an error does not occur,

$$d(X_k, \hat{X}_k) = d(X_k, \hat{x}^*(X_k)) = 0 \tag{9.56}$$

by (9.8) for each k, and

$$d(\mathbf{X}, \hat{\mathbf{X}}) = \frac{1}{n}\sum_{k=1}^{n} d(X_k, \hat{X}_k) = \frac{1}{n}\sum_{k=1}^{n} d(X_k, \hat{x}^*(X_k)) = 0. \tag{9.57}$$

Therefore,

$$\Pr\{d(\mathbf{X}, \hat{\mathbf{X}}) > \epsilon\} \leq \epsilon, \tag{9.58}$$

which shows that the pair $(H(X), 0)$ is achievable. This in turn implies that $R(0) \leq H(X)$ because $R(0)$ is the minimum of all R such that $(R, 0)$ is achievable. □

Figure 9.2 is an illustration of a rate-distortion function $R(D)$. The reader should note the four properties of $R(D)$ in Theorem 9.15. The rate-distortion theorem, which will be stated in the next section, gives a characterization of $R(D)$.

9.3 THE RATE-DISTORTION THEOREM

DEFINITION 9.16 *For $D \geq 0$, the information rate-distortion function is defined by*

$$R_I(D) = \min_{\hat{X}: Ed(X,\hat{X}) \leq D} I(X; \hat{X}). \tag{9.59}$$

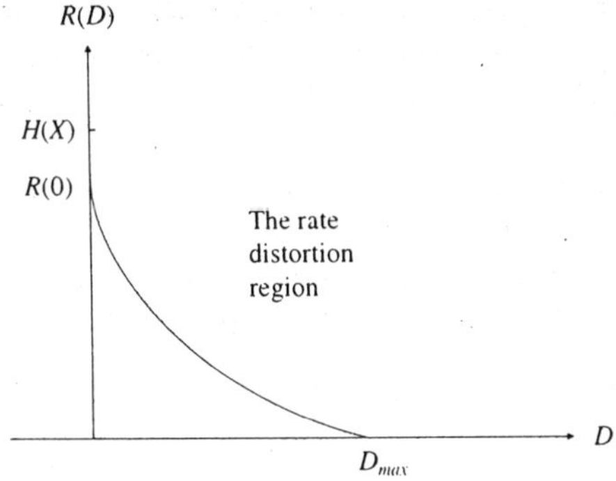

Figure 9.2. A rate-distortion function $R(D)$.

In defining $R_I(D)$, the minimization is taken over all random variables $\hat{X}$ jointly distributed with X such that

$$Ed(X, \hat{X}) \leq D. \tag{9.60}$$

Since $p(x)$ is given, the minimization is taken over the set of all $p(\hat{x}|x)$ such that (9.60) is satisfied, namely the set

$$\left\{ p(\hat{x}|x) : \sum_{x,\hat{x}} p(x)p(\hat{x}|x)d(x,\hat{x}) \leq D \right\}. \tag{9.61}$$

Since this set is compact in $\Re^{|\mathcal{X}||\hat{\mathcal{X}}|}$ and $I(X;\hat{X})$ is a continuous functional of $p(\hat{x}|x)$, the minimum value of $I(X;\hat{X})$ can be attained[2]. This justifies taking the minimum instead of the infimum in the definition of $R_I(D)$.

We have seen in Section 9.1 that we can obtain a normalization $\tilde{d}$ for any distortion measure d with

$$E\tilde{d}(X,\hat{X}) = Ed(X,\hat{X}) - \Delta \tag{9.62}$$

for any $\hat{X}$, where Δ is a constant which depends only on $p(x)$ and d. Thus if d is not normal, we can always replace d by $\tilde{d}$ and D by $D - \Delta$ in the definition of $R_I(D)$ without changing the minimization problem. Therefore, we do not lose any generality by assuming that a distortion measure d is normal.

THEOREM 9.17 (THE RATE-DISTORTION THEOREM) $R(D) = R_I(D)$.

[2]The assumption that both $\mathcal{X}$ and $\hat{\mathcal{X}}$ are finite is essential in this argument.

The rate-distortion theorem, which is the main result in rate-distortion theory, says that the minimum coding rate for achieving a distortion D is $R_I(D)$. This theorem will be proved in the next two sections. In the next section, we will prove the converse of this theorem, i.e., $R(D) \geq R_I(D)$, and in Section 9.5, we will prove the achievability of $R_I(D)$, i.e., $R(D) \leq R_I(D)$.

In order for $R_I(D)$ to be a characterization of $R(D)$, it has to satisfy the same properties as $R(D)$. In particular, the four properties of $R(D)$ in Theorem 9.15 should also be satisfied by $R_I(D)$.

THEOREM 9.18 *The following properties hold for the information rate-distortion function $R_I(D)$:*

1. *$R_I(D)$ is non-increasing in D.*
2. *$R_I(D)$ is convex.*
3. *$R_I(D) = 0$ for $D \geq D_{max}$.*
4. *$R_I(0) \leq H(X)$.*

Proof Referring to the definition of $R_I(D)$ in (9.59), for a larger D, the minimization is taken over a larger set. Therefore, $R_I(D)$ is non-increasing in D, proving Property 1.

To prove Property 2, consider any $D^{(1)}, D^{(2)} \geq 0$ and let λ be any number between 0 and 1. Let $\hat{X}^{(i)}$ achieves $R_I(D^{(i)})$ for $i = 1, 2$, i.e.,

$$R_I(D^{(i)}) = I(X; \hat{X}^{(i)}), \tag{9.63}$$

where

$$Ed(X, \hat{X}^{(i)}) \leq D^{(i)}, \tag{9.64}$$

and let $\hat{X}^{(i)}$ be defined by the transition matrix $p_i(\hat{x}|x)$. Let $\hat{X}^{(\lambda)}$ be jointly distributed with X which is defined by

$$p_\lambda(\hat{x}|x) = \lambda p_1(\hat{x}|x) + \bar{\lambda} p_2(\hat{x}|x), \tag{9.65}$$

where $\bar{\lambda} = 1 - \lambda$. Then

$$\begin{aligned}
&Ed(X, \hat{X}^{(\lambda)}) \\
&= \sum_{x,\hat{x}} p(x) p_\lambda(\hat{x}|x) d(x, \hat{x}) &(9.66)\\
&= \sum_{x,\hat{x}} p(x)(\lambda p_1(\hat{x}|x) + \bar{\lambda} p_2(\hat{x}|x)) d(x, \hat{x}) &(9.67)\\
&= \lambda \left(\sum_{x,\hat{x}} p(x) p_1(\hat{x}|x) d(x, \hat{x}) \right) + \bar{\lambda} \left(\sum_{x,\hat{x}} p(x) p_2(\hat{x}|x) d(x, \hat{x}) \right) &(9.68)
\end{aligned}$$

$$
\begin{aligned}
&= \lambda Ed(X,\hat{X}^{(1)}) + \bar{\lambda} Ed(X,\hat{X}^{(2)}) && (9.69)\\
&\le \lambda D^{(1)} + \bar{\lambda} D^{(2)} && (9.70)\\
&= D^{(\lambda)}, && (9.71)
\end{aligned}
$$

where

$$D^{(\lambda)} = \lambda D^{(1)} + \bar{\lambda} D^{(2)}, \tag{9.72}$$

and (9.70) follows from (9.64). Now consider

$$
\begin{aligned}
\lambda R_I(D^{(1)}) + \bar{\lambda} R_I(D^{(2)}) &= \lambda I(X;\hat{X}^{(1)}) + \bar{\lambda} I(X;\hat{X}^{(2)}) && (9.73)\\
&\ge I(X;\hat{X}^{(\lambda)}) && (9.74)\\
&\ge R_I(D^{(\lambda)}), && (9.75)
\end{aligned}
$$

where the inequality in (9.74) follows from the convexity of mutual information with respect to the transition matrix $p(\hat{x}|x)$ (see Example 6.14), and the inequality in (9.75) follows from (9.71) and the definition of $R_I(D)$. Therefore, we have proved Property 2.

To prove Property 3, let $\hat{X}$ take the value $\hat{x}^*$ as defined in Definition 9.7 with probability 1. Then

$$I(X;\hat{X}) = 0 \tag{9.76}$$

and

$$Ed(X;\hat{X}) = Ed(X;\hat{x}^*) = D_{max}. \tag{9.77}$$

Then for $D \ge D_{max}$,

$$R_I(D) \le I(X;\hat{X}) = 0. \tag{9.78}$$

On the other hand, since $R_I(D)$ is nonnegative, we conclude that

$$R_I(D) = 0. \tag{9.79}$$

This proves Property 3.

Finally, to prove Property 4, we let

$$\hat{X} = \hat{x}^*(X), \tag{9.80}$$

where $\hat{x}^*(x)$ is defined in Definition 9.5. Then

$$
\begin{aligned}
Ed(X,\hat{X}) &= Ed(X,\hat{x}^*(X)) && (9.81)\\
&= \sum_x p(x) d(x,\hat{x}^*(x)) && (9.82)\\
&= 0 && (9.83)
\end{aligned}
$$

by (9.8) since we assume that d is a normal distortion measure. Moreover,

$$R_I(0) \le I(X;\hat{X}) \le H(X). \tag{9.84}$$

Then Property 4 and hence the theorem is proved. □

COROLLARY 9.19 *If $R_I(0) > 0$, then $R_I(D)$ is strictly decreasing for $0 \leq D \leq D_{max}$, and the inequality constraint in Definition 9.16 for $R_I(D)$ can be replaced by an equality constraint.*

Proof Assume that $R_I(0) > 0$. We first show that $R_I(D) > 0$ for $0 \leq D < D_{max}$ by contradiction. Suppose $R_I(D') = 0$ for some $0 \leq D' < D_{max}$, and let $R_I(D')$ be achieved by some $\hat{X}$. Then

$$R_I(D') = I(X; \hat{X}) = 0 \tag{9.85}$$

implies that X and $\hat{X}$ are independent, or

$$p(x, \hat{x}) = p(x)p(\hat{x}) \tag{9.86}$$

for all x and $\hat{x}$. It follows that

$$\begin{aligned}
D' &\geq Ed(X, \hat{X}) && (9.87)\\
&= \sum_x \sum_{\hat{x}} p(x, \hat{x}) d(x, \hat{x}) && (9.88)\\
&= \sum_x \sum_{\hat{x}} p(x)p(\hat{x}) d(x, \hat{x}) && (9.89)\\
&= \sum_{\hat{x}} p(\hat{x}) \sum_x p(x) d(x, \hat{x}) && (9.90)\\
&= \sum_{\hat{x}} p(\hat{x}) Ed(X, \hat{x}) && (9.91)\\
&\geq \sum_{\hat{x}} p(\hat{x}) Ed(X, \hat{x}^*) && (9.92)\\
&= \sum_{\hat{x}} p(\hat{x}) D_{max} && (9.93)\\
&= D_{max}, && (9.94)
\end{aligned}$$

where $\hat{x}^*$ and D_{max} are defined in Definition 9.7. This leads to a contradiction because we have assumed that $0 \leq D' < D_{max}$. Therefore, we conclude that $R_I(D) > 0$ for $0 \leq D < D_{max}$.

Since $R_I(0) > 0$ and $R_I(D_{max}) = 0$, and $R_I(D)$ is non-increasing and convex from the above theorem, $R_I(D)$ must be strictly decreasing for $0 \leq D \leq D_{max}$. We now prove by contradiction that the inequality constraint in Definition 9.16 for $R_I(D)$ can be replaced by an equality constraint. Assume that $R_I(D)$ is achieved by some $\hat{X}^*$ such that

$$Ed(X, \hat{X}^*) = D'' < D. \tag{9.95}$$

Then

$$R_I(D'') = \min_{\hat{X}:Ed(X,\hat{X})\le D''} I(X;\hat{X}) \le I(X;\hat{X}^*) = R_I(D). \tag{9.96}$$

This is a contradiction because $R_I(D)$ is strictly decreasing for $0 \le D \le D_{max}$. Hence,

$$Ed(X,\hat{X}^*) = D. \tag{9.97}$$

This implies that the inequality constraint in Definition 9.16 for $R_I(D)$ can be replaced by an equality constraint. □

Remark In all problems of interest, $R(0) = R_I(0) > 0$. Otherwise, $R(D) = 0$ for all $D \ge 0$ because $R(D)$ is nonnegative and non-increasing.

EXAMPLE 9.20 (BINARY SOURCE) *Let X be a binary random variable with*

$$\Pr\{X = 0\} = 1 - \gamma \text{ and } \Pr\{X = 1\} = \gamma. \tag{9.98}$$

Let $\hat{\mathcal{X}} = \{0,1\}$ be the reproduction alphabet for X, and let d be the Hamming distortion measure. We first consider the case that $0 \le \gamma \le \frac{1}{2}$. Then if we make a guess on the value of X, we should guess 0 in order to minimize the expected distortion. Therefore, $\hat{x}^* = 0$ and

$$\begin{aligned} D_{max} &= Ed(X,0) & (9.99)\\ &= \Pr\{X = 1\} & (9.100)\\ &= \gamma. & (9.101)\end{aligned}$$

We will show that for $0 \le \gamma \le \frac{1}{2}$,

$$R_I(D) = \begin{cases} h_b(\gamma) - h_b(D) & \text{if } 0 \le D < \gamma \\ 0 & \text{if } D \ge \gamma. \end{cases} \tag{9.102}$$

Let $\hat{X}$ be an estimate of X taking values in $\hat{\mathcal{X}}$, and let Y be the Hamming distortion measure between X and $\hat{X}$, i.e.,

$$Y = d(X,\hat{X}). \tag{9.103}$$

Observe that conditioning on $\hat{X}$, X and Y determine each other. Therefore,

$$H(X|\hat{X}) = H(Y|\hat{X}). \tag{9.104}$$

Then for $D < \gamma = D_{max}$ and any $\hat{X}$ such that

$$Ed(X,\hat{X}) \le D, \tag{9.105}$$

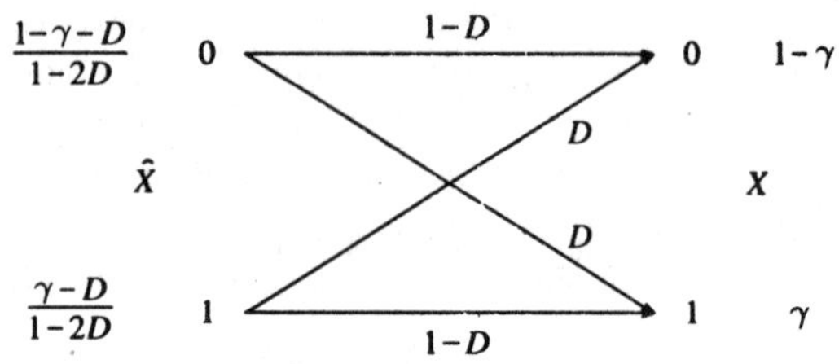

Figure 9.3. Achieving $R_I(D)$ for a binary source via a reverse binary symmetric channel.

we have

$$\begin{aligned} I(X;\hat{X}) &= H(X) - H(X|\hat{X}) & (9.106)\\ &= h_b(\gamma) - H(Y|\hat{X}) & (9.107)\\ &\geq h_b(\gamma) - H(Y) & (9.108)\\ &= h_b(\gamma) - h_b(\Pr\{X \neq \hat{X}\}) & (9.109)\\ &\geq h_b(\gamma) - h_b(D), & (9.110)\end{aligned}$$

where the last inequality is justified because

$$\Pr\{X \neq \hat{X}\} = Ed(X,\hat{X}) \leq D \tag{9.111}$$

and $h_b(a)$ is increasing for $0 \leq a \leq \frac{1}{2}$. Minimizing over all $\hat{X}$ satisfying (9.105) in (9.110), we obtain the lower bound

$$R_I(D) \geq h_b(\gamma) - h_b(D). \tag{9.112}$$

To show that this lower bound is achievable, we need to construct an $\hat{X}$ such that the inequalities in both (9.108) and (9.110) are tight. The tightness of the inequality in (9.110) simply says that

$$\Pr\{X \neq \hat{X}\} = D, \tag{9.113}$$

while the tightness of the inequality in (9.108) says that Y should be independent of $\hat{X}$.

It would be more difficult to make Y independent of $\hat{X}$ if we specify $\hat{X}$ by $p(\hat{x}|x)$. Instead, we specify the joint distribution of X and $\hat{X}$ by means of a reverse binary symmetric channel (BSC) with crossover probability D as the shown in Figure 9.3. Here, we regard $\hat{X}$ as the input and X as the output of the BSC. Then Y is independent of the input $\hat{X}$ because the error event is independent of the input for a BSC, and (9.113) is satisfied by setting the crossover probability to D. However, we need to ensure that the marginal distribution of X so specified is equal to $p(x)$. Toward this end, we let

$$\Pr\{\hat{X} = 1\} = \alpha, \tag{9.114}$$

and consider

$$\begin{aligned}\Pr\{X=1\} &= \Pr\{\hat{X}=0\}\Pr\{X=1|\hat{X}=0\} \\ &\quad +\Pr\{\hat{X}=1\}\Pr\{X=1|\hat{X}=1\},\end{aligned} \tag{9.115}$$

or

$$\gamma = (1-\alpha)D + \alpha(1-D), \tag{9.116}$$

which gives

$$\alpha = \frac{\gamma - D}{1 - 2D}. \tag{9.117}$$

Since

$$D < D_{max} = \gamma \leq \frac{1}{2}, \tag{9.118}$$

we have $\alpha \geq 0$. On the other hand,

$$\gamma, D \leq \frac{1}{2} \tag{9.119}$$

gives

$$\gamma + D \leq 1. \tag{9.120}$$

This implies

$$\gamma - D \leq 1 - 2D, \tag{9.121}$$

or $\alpha \leq 1$. Therefore,

$$0 \leq \alpha = \Pr\{\hat{X}=1\} \leq 1 \tag{9.122}$$

and

$$0 \leq 1 - \alpha = \Pr\{\hat{X}=0\} \leq 1. \tag{9.123}$$

Hence, we have shown that the lower bound on $R_I(D)$ in (9.110) can be achieved, and $R_I(D)$ is as given in (9.102).

For $\frac{1}{2} \leq \gamma \leq 1$, by exchanging the roles of the symbols 0 and 1 in the above argument, we obtain $R_I(D)$ as in (9.102) except that γ is replaced by $1-\gamma$. Combining the two cases, we have

$$R_I(D) = \begin{cases} h_b(\gamma) - h_b(D) & \text{if } 0 \leq D < \min(\gamma, 1-\gamma) \\ 0 & \text{if } D \geq \min(\gamma, 1-\gamma). \end{cases} \tag{9.124}$$

for $0 \leq \gamma \leq 1$. The function $R_I(D)$ for $\gamma = \frac{1}{2}$ is illustrated in Figure 9.4.

Remark In the above example, we see that $R_I(0) = h_b(\gamma) = H(X)$. Then by the rate-distortion theorem, $H(X)$ is the minimum rate of a rate-distortion code which achieves an arbitrarily small average Hamming distortion. It is tempting to regarding this special case of the rate-distortion theorem as a version of the source coding theorem and conclude that the rate-distortion theorem

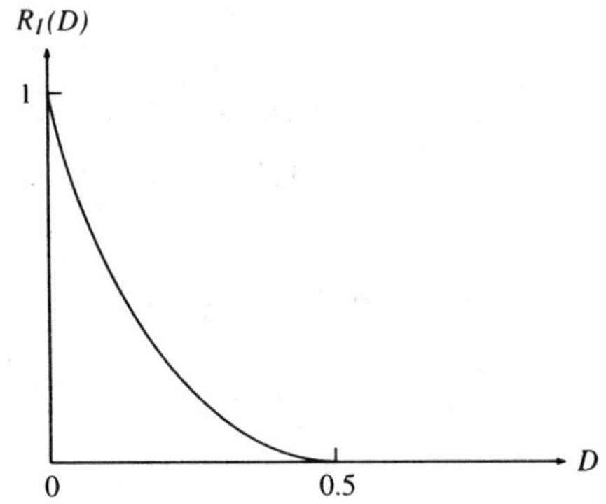

Figure 9.4. The function $R_I(D)$ for the uniform binary source with the Hamming distortion measure.

is a generalization of the source coding theorem. However, this is incorrect because the rate-distortion theorem only guarantees that the *average* Hamming distortion between $\mathbf{X}$ and $\hat{\mathbf{X}}$ is small with probability arbitrarily close to 1, but the source coding theorem guarantees that $\mathbf{X} = \hat{\mathbf{X}}$ with probability arbitrarily close to 1, which is much stronger.

It is in general not possible to obtain the rate-distortion function in closed form, and we have to resort to numerical computation. In Chapter 10, we will discuss the Blahut-Arimoto algorithm for computing the rate-distortion function.

9.4 THE CONVERSE

In this section, we prove that the rate-distortion function $R(D)$ is lower bounded by the information rate-distortion function $R_I(D)$, i.e., $R(D) \geq R_I(D)$. Specifically, we will prove that for any achievable rate-distortion pair (R, D), $R \geq R_I(D)$. Then by fixing D and minimizing R over all achievable pairs (R, D), we conclude that $R(D) \geq R_I(D)$.

Let (R, D) be any achievable rate-distortion pair. Then for any $\epsilon > 0$, there exists for sufficiently large n an (n, M) code such that

$$\frac{1}{n} \log M \leq R + \epsilon \tag{9.125}$$

and

$$\Pr\{d(\mathbf{X}, \hat{\mathbf{X}}) > D + \epsilon\} \leq \epsilon, \tag{9.126}$$

where $\hat{\mathbf{X}} = g(f(\mathbf{X}))$. Then

$$n(R + \epsilon) \stackrel{a)}{\geq} \log M \tag{9.127}$$

$$
\begin{aligned}
&\geq H(f(\mathbf{X})) && (9.128)\\
&\geq H(g(f(\mathbf{X}))) && (9.129)\\
&= H(\hat{\mathbf{X}}) && (9.130)\\
&= H(\hat{\mathbf{X}}) - H(\hat{\mathbf{X}}|\mathbf{X}) && (9.131)\\
&= I(\hat{\mathbf{X}};\mathbf{X}) && (9.132)\\
&= H(\mathbf{X}) - H(\mathbf{X}|\hat{\mathbf{X}}) && (9.133)\\
&= \sum_{k=1}^{n} H(X_k) - \sum_{k=1}^{n} H(X_k|\hat{\mathbf{X}}, X_1, X_2, \cdots, X_{k-1}) && (9.134)\\
&\overset{b)}{\geq} \sum_{k=1}^{n} H(X_k) - \sum_{k=1}^{n} H(X_k|\hat{X}_k) && (9.135)\\
&= \sum_{k=1}^{n} [H(X_k) - H(X_k|\hat{X}_k)] && (9.136)\\
&= \sum_{k=1}^{n} I(X_k;\hat{X}_k) && (9.137)\\
&\overset{c)}{\geq} \sum_{k=1}^{n} R_I(Ed(X_k,\hat{X}_k)) && (9.138)\\
&= n\left[\frac{1}{n}\sum_{k=1}^{n} R_I(Ed(X_k,\hat{X}_k))\right] && (9.139)\\
&\overset{d)}{\geq} nR_I\left(\frac{1}{n}\sum_{k=1}^{n} Ed(X_k,\hat{X}_k)\right) && (9.140)\\
&= nR_I(Ed(\mathbf{X},\hat{\mathbf{X}})). && (9.141)
\end{aligned}
$$

In the above,

a) follows from (9.125);

b) follows because conditioning does not increase entropy;

c) follows from the definition of $R_I(D)$ in Definition 9.16;

d) follows from the convexity of $R_I(D)$ proved in Theorem 9.18 and Jensen's inequality.

Now let

$$d_{max} = \max_{x,\hat{x}} d(x,\hat{x}) \tag{9.142}$$

be the maximum value which can be taken by the distortion measure d. The reader should not confuse d_{max} with D_{max} in Definition 9.7. Then from

(9.126), we have

$$\begin{aligned} & Ed(\mathbf{X},\hat{\mathbf{X}}) \\ &= E[d(\mathbf{X},\hat{\mathbf{X}})|d(\mathbf{X},\hat{\mathbf{X}}) > D+\epsilon]\Pr\{d(\mathbf{X},\hat{\mathbf{X}}) > D+\epsilon\} \\ &\quad + E[d(\mathbf{X},\hat{\mathbf{X}})|d(\mathbf{X},\hat{\mathbf{X}}) \le D+\epsilon]\Pr\{d(\mathbf{X},\hat{\mathbf{X}}) \le D+\epsilon\} && (9.143) \\ &\le d_{max}\cdot\epsilon + (D+\epsilon)\cdot 1 && (9.144) \\ &= D + (d_{max}+1)\epsilon. && (9.145) \end{aligned}$$

This shows that if the probability that the average distortion between $\mathbf{X}$ and $\hat{\mathbf{X}}$ exceeds $D+\epsilon$ is small, then the expected average distortion between $\mathbf{X}$ and $\hat{\mathbf{X}}$ can exceed D only by a small amount[3]. Following (9.141), we have

$$\begin{aligned} R+\epsilon &\ge R_I(Ed(\mathbf{X},\hat{\mathbf{X}})) && (9.146) \\ &\ge R_I(D+(d_{max}+1)\epsilon), && (9.147) \end{aligned}$$

where the last inequality follows from (9.145) because $R_I(D)$ is non-increasing in D. We note that the convexity of $R_I(D)$ implies that it is a continuous function of D. Then taking the limit as $\epsilon \to 0$, we obtain

$$\begin{aligned} R &\ge \lim_{\epsilon\to 0} R_I(D+(d_{max}+1)\epsilon) && (9.148) \\ &= R_I\left(D+(d_{max}+1)\lim_{\epsilon\to 0}\epsilon\right) && (9.149) \\ &= R_I(D), && (9.150) \end{aligned}$$

where we have invoked the continuity of $R_I(D)$ in obtaining (9.149). Upon minimizing R over all achievable pairs (R, D) for a fixed D in (9.150), we have proved that

$$R(D) \ge R_I(D). \tag{9.151}$$

This completes the proof for the converse of the rate-distortion theorem.

9.5 ACHIEVABILITY OF $R_I(D)$

In this section, we prove that the rate-distortion function $R(D)$ is upper bounded by the information rate-distortion function $R_I(D)$, i.e., $R(D) \le R_I(D)$. Then by combining with the result that $R(D) \ge R_I(D)$ from the last section, we conclude that $R(D) = R_I(D)$, and the rate-distortion theorem is proved.

For any $0 \le D \le D_{max}$, we will prove that for every random variable $\hat{X}$ taking values in $\hat{\mathcal{X}}$ such that

$$Ed(X,\hat{X}) \le D, \tag{9.152}$$

[3]The converse is not true.

the rate-distortion pair $(I(X;\hat{X}), D)$ is achievable. This will be proved by showing for sufficiently large n the existence of a rate-distortion code such that

1. the rate of the code is not more than $I(X;\hat{X}) + \epsilon$;
2. $d(\mathbf{X}, \hat{\mathbf{X}}) \leq D + \epsilon$ with probability almost 1.

Then by minimizing $I(X;\hat{X})$ over all $\hat{X}$ satisfying (9.152), we conclude that the rate-distortion pair $(R_I(D), D)$ is achievable, which implies $R_I(D) \geq R(D)$ because $R(D)$ is the minimum of all R such that (R, D) is achievable.

Fix any $0 \leq D \leq D_{max}$ and any $\epsilon > 0$, and let δ be a small positive quantity to be specified later. Toward proving the existence of a desired code, we fix a random variable $\hat{X}$ which satisfies (9.152) and let M be an integer satisfying

$$I(X;\hat{X}) + \frac{\epsilon}{2} \leq \frac{1}{n} \log M \leq I(X;\hat{X}) + \epsilon, \tag{9.153}$$

where n is sufficiently large.

We now describe a random coding scheme in the following steps:

1. Construct a codebook $\mathcal{C}$ of an (n, M) code by randomly generating M codewords in $\hat{\mathcal{X}}^n$ independently and identically according to $p(\hat{x})^n$. Denote these codewords by $\hat{\mathbf{X}}(1), \hat{\mathbf{X}}(2), \cdots, \hat{\mathbf{X}}(M)$.
2. Reveal the codebook $\mathcal{C}$ to both the encoder and the decoder.
3. The source sequence $\mathbf{X}$ is generated according to $p(x)^n$.
4. The encoder encodes the source sequence $\mathbf{X}$ into an index K in the set $\mathcal{I} = \{1, 2, \cdots, M\}$. The index K takes the value i if
   a) $(\mathbf{X}, \hat{\mathbf{X}}(i)) \in T^n_{[X\hat{X}]\delta}$,
   b) for all $i' \in \mathcal{I}$, if $(\mathbf{X}, \hat{\mathbf{X}}(i')) \in T^n_{[X\hat{X}]\delta}$, then $i' \leq i$;

   otherwise, K takes the constant value 1.
5. The index K is delivered to the decoder.
6. The decoder outputs $\hat{\mathbf{X}}(K)$ as the reproduction sequence $\hat{\mathbf{X}}$.

Remark Strong typicality is used in defining the encoding function in Step 4. This is made possible by the assumption that both the source alphabet $\mathcal{X}$ and the reproduction alphabet $\hat{\mathcal{X}}$ are finite.

Let us further explain the encoding scheme described in Step 4. After the source sequence $\mathbf{X}$ is generated, we search through all the codewords in the codebook $\mathcal{C}$ for those which are jointly typical with $\mathbf{X}$ with respect to $p(x, \hat{x})$. If there is at least one such codeword, we let i be the largest index of such codewords and let $K = i$. If such a codeword does not exist, we let $K = 1$.

The event $\{K = 1\}$ occurs in one of the following two scenarios:

1. $\hat{X}(1)$ is the only codeword in $\mathcal{C}$ which is jointly typical with $\mathbf{X}$.
2. No codeword in $\mathcal{C}$ is jointly typical with $\mathbf{X}$.

In either scenario, $\mathbf{X}$ is not jointly typical with the codewords $\hat{X}(2), \hat{X}(3), \cdots, \hat{X}(M)$. In other words, if $K = 1$, then $\mathbf{X}$ is jointly typical with none of the codewords $\hat{X}(2), \hat{X}(3), \cdots, \hat{X}(M)$.

Define

$$E_i = \left\{(\mathbf{X}, \hat{\mathbf{X}}(i)) \in T^n_{[X\hat{X}]\delta}\right\} \tag{9.154}$$

to be the event that $\mathbf{X}$ is jointly typical with the codeword $\hat{X}(i)$. We see from the above discussion that

$$\{K = 1\} \subset E_2^c \cap E_3^c \cap \cdots \cap E_M^c. \tag{9.155}$$

Since the codewords are generated i.i.d., conditioning on $\{\mathbf{X} = \mathbf{x}\}$ for any $\mathbf{x} \in \mathcal{X}^n$, the events E_i are mutually independent[4], and they all have the same probability. Then for any $\mathbf{x} \in \mathcal{X}^n$,

$$\begin{aligned}
\Pr\{K = 1|\mathbf{X} = \mathbf{x}\} &\leq \Pr\{E_2^c \cap E_3^c \cap \cdots \cap E_M^c|\mathbf{X} = \mathbf{x}\} && (9.156)\\
&= \prod_{i=2}^{M} \Pr\{E_i^c|\mathbf{X} = \mathbf{x}\} && (9.157)\\
&= (\Pr\{E_1^c|\mathbf{X} = \mathbf{x}\})^{M-1} && (9.158)\\
&= (1 - \Pr\{E_1|\mathbf{X} = \mathbf{x}\})^{M-1}. && (9.159)
\end{aligned}$$

We now obtain a lower bound on $\Pr\{E_1|\mathbf{X} = \mathbf{x}\}$ for $\mathbf{x} \in S^n_{[X]\delta}$, where

$$S^n_{[X]\delta} = \{\mathbf{x} \in T^n_{[X]\delta} : |T^n_{[\hat{X}|X]\delta}(\mathbf{x})| \geq 1\} \tag{9.160}$$

(cf. Section 5.3). Consider

$$\begin{aligned}
\Pr\{E_1|\mathbf{X} = \mathbf{x}\} &= \Pr\left\{(\mathbf{x}, \hat{\mathbf{X}}(1)) \in T^n_{[X\hat{X}]\delta}\right\} && (9.161)\\
&= \sum_{\hat{\mathbf{x}}:(\mathbf{x},\hat{\mathbf{x}}) \in T^n_{[X\hat{X}]\delta}} p(\hat{\mathbf{x}}). && (9.162)
\end{aligned}$$

[4]Without conditioning on $\{\mathbf{X} = \mathbf{x}\}$, the events E_i are not mutually independent because they depend on each other through $\mathbf{X}$.

The summation above is over all $\hat{\mathbf{x}}$ such that $(\mathbf{x}, \hat{\mathbf{x}}) \in T^n_{[X\hat{X}]\delta}$. From the consistency of strong typicality (Theorem 5.7), if $(\mathbf{x}, \hat{\mathbf{x}}) \in T^n_{[X\hat{X}]\delta}$, then $\hat{\mathbf{x}} \in T^n_{[\hat{X}]\delta}$. By the strong AEP (Theorem 5.2), all $p(\hat{\mathbf{x}})$ in the above summation satisfy

$$p(\hat{\mathbf{x}}) \geq 2^{-n(H(\hat{X})+\eta)}, \tag{9.163}$$

where $\eta \to 0$ as $\delta \to 0$. By Theorem 5.9,

$$|T^n_{[\hat{X}|X]\delta}(\mathbf{x})| \geq 2^{n(H(\hat{X}|X)-\xi)}, \tag{9.164}$$

where $\xi \to 0$ as $\delta \to 0$. Then from (9.162), we have

$$\begin{aligned} \Pr\{E_1|\mathbf{X} = \mathbf{x}\} &\geq 2^{n(H(\hat{X}|X)-\xi)} 2^{-n(H(\hat{X})+\eta)} & (9.165)\\ &= 2^{-n(H(\hat{X})-H(\hat{X}|X)+\xi+\eta)} & (9.166)\\ &= 2^{-n(I(X;\hat{X})+\zeta)}, & (9.167) \end{aligned}$$

where

$$\zeta = \xi + \eta \to 0 \tag{9.168}$$

as $\delta \to 0$. Following (9.159), we have

$$\Pr\{K = 1|\mathbf{X} = \mathbf{x}\} \leq \left[1 - 2^{-n(I(X;\hat{X})+\zeta)}\right]^{M-1} \tag{9.169}$$

The lower bound in (9.153) implies

$$M \geq 2^{n(I(X;\hat{X})+\frac{\epsilon}{2})}. \tag{9.170}$$

Then upon taking natural logarithm in (9.169), we obtain

$$\begin{aligned} &\ln \Pr\{K = 1|\mathbf{X} = \mathbf{x}\} \\ &\leq (M-1) \ln\left[1 - 2^{-n(I(X;\hat{X})+\zeta)}\right] & (9.171)\\ &\overset{a)}{\leq} \left(2^{n(I(X;\hat{X})+\frac{\epsilon}{2})} - 1\right) \ln\left[1 - 2^{-n(I(X;\hat{X})+\zeta)}\right] & (9.172)\\ &\overset{b)}{\leq} -\left(2^{n(I(X;\hat{X})+\frac{\epsilon}{2})} - 1\right) 2^{-n(I(X;\hat{X})+\zeta)} & (9.173)\\ &= -\left[2^{n(\frac{\epsilon}{2}-\zeta)} - 2^{-n(I(X;\hat{X})+\zeta)}\right]. & (9.174) \end{aligned}$$

In the above, a) follows from (9.170) by noting that the logarithm in (9.171) is negative, and b) follows from the fundamental inequality $\ln a \leq a - 1$. By letting δ be sufficiently small so that

$$\frac{\epsilon}{2} - \zeta > 0, \tag{9.175}$$

the above upper bound on $\ln \Pr\{K = 1|\mathbf{X} = \mathbf{x}\}$ tends to $-\infty$ as $n \to \infty$, i.e., $\Pr\{K = 1|\mathbf{X} = \mathbf{x}\} \to 0$ as $n \to \infty$. This implies

$$\Pr\{K = 1|\mathbf{X} = \mathbf{x}\} \leq \frac{\epsilon}{2} \tag{9.176}$$

for sufficiently large n. It then follows that

$$\begin{aligned}
&\Pr\{K = 1\} && (9.177)\\
&= \sum_{\mathbf{x} \in S^n_{[X]\delta}} \Pr\{K = 1|\mathbf{X} = \mathbf{x}\}\Pr\{\mathbf{X} = \mathbf{x}\} \\
&\quad + \sum_{\mathbf{x} \notin S^n_{[X]\delta}} \Pr\{K = 1|\mathbf{X} = \mathbf{x}\}\Pr\{\mathbf{X} = \mathbf{x}\} && (9.178)\\
&\leq \sum_{\mathbf{x} \in S^n_{[X]\delta}} \frac{\epsilon}{2} \cdot \Pr\{\mathbf{X} = \mathbf{x}\} + \sum_{\mathbf{x} \notin S^n_{[X]\delta}} 1 \cdot \Pr\{\mathbf{X} = \mathbf{x}\} && (9.179)\\
&= \frac{\epsilon}{2} \cdot \Pr\{\mathbf{X} \in S^n_{[X]\delta}\} + \Pr\{\mathbf{X} \notin S^n_{[X]\delta}\} && (9.180)\\
&\leq \frac{\epsilon}{2} \cdot 1 + (1 - \Pr\{\mathbf{X} \in S^n_{[X]\delta}\}) && (9.181)\\
&< \frac{\epsilon}{2} + \delta, && (9.182)
\end{aligned}$$

where we have invoked Proposition 5.12 in the last step. By letting δ be sufficiently small so that

$$\delta < \frac{\epsilon}{2} \tag{9.183}$$

and (9.175) is satisfied, we obtain

$$\Pr\{K = 1\} < \epsilon. \tag{9.184}$$

The main idea of the above upper bound on $\Pr\{K = 1\}$ for sufficiently large n is the following. In constructing the codebook, we randomly generate M codewords in $\hat{\mathcal{X}}^n$ according to $p(\hat{x})^n$. If M grows with n at a rate higher than $I(X; \hat{X})$, then the probability that there exists at least one codeword which is jointly typical with the source sequence $\mathbf{X}$ with respect to $p(x, \hat{x})$ is very high when n is large. Further, the average distortion between $\mathbf{X}$ and such a codeword is close to $Ed(X, \hat{X})$ because the empirical joint distribution of the symbol pairs in $\mathbf{X}$ and such a codeword is close to $p(x, \hat{x})$. Then by letting the reproduction sequence $\hat{\mathbf{X}}$ be such a codeword, the average distortion between $\mathbf{X}$ and $\hat{\mathbf{X}}$ is less than $D + \epsilon$ with probability arbitrarily close to 1 since $Ed(X, \hat{X}) \leq D$. These will be formally shown in the rest of the proof.

Now for sufficiently large n, consider

$$\Pr\{d(\mathbf{X}, \hat{\mathbf{X}}) > D + \epsilon\}$$

$$\begin{aligned} &= \Pr\{d(\mathbf{X},\hat{\mathbf{X}}) > D+\epsilon | K=1\}\Pr\{K=1\} \\ &\quad + \Pr\{d(\mathbf{X},\hat{\mathbf{X}}) > D+\epsilon | K\neq 1\}\Pr\{K\neq 1\} && (9.185) \\ &\leq 1\cdot\epsilon + \Pr\{d(\mathbf{X},\hat{\mathbf{X}}) > D+\epsilon | K\neq 1\}\cdot 1 && (9.186) \\ &= \epsilon + \Pr\{d(\mathbf{X},\hat{\mathbf{X}}) > D+\epsilon | K\neq 1\}. && (9.187) \end{aligned}$$

We will show that by choosing the value of δ carefully, it is possible to make $d(\mathbf{X},\hat{\mathbf{X}})$ always less than or equal to $D+\epsilon$ provided $K \neq 1$. Since $(\mathbf{X},\hat{\mathbf{X}}) \in T^n_{[X\hat{X}]\delta}$ conditioning on $\{K \neq 1\}$, we have

$$\begin{aligned} & d(\mathbf{X},\hat{\mathbf{X}}) \\ &= \frac{1}{n}\sum_{k=1}^{n} d(X_k,\hat{X}_k) && (9.188) \\ &= \frac{1}{n}\sum_{x,\hat{x}} d(x,\hat{x})N(x,\hat{x}|\mathbf{X},\hat{\mathbf{X}}) && (9.189) \\ &= \frac{1}{n}\sum_{x,\hat{x}} d(x,\hat{x})(np(x,\hat{x}) + N(x,\hat{x}|\mathbf{X},\hat{\mathbf{X}}) - np(x,\hat{x})) && (9.190) \\ &= \left[\sum_{x,\hat{x}} p(x,\hat{x})d(x,\hat{x})\right] + \left[\sum_{x,\hat{x}} d(x,\hat{x})\left(\frac{1}{n}N(x,\hat{x}|\mathbf{X},\hat{\mathbf{X}}) - p(x,\hat{x})\right)\right] && (9.191) \\ &= Ed(X,\hat{X}) + \sum_{x,\hat{x}} d(x,\hat{x})\left(\frac{1}{n}N(x,\hat{x}|\mathbf{X},\hat{\mathbf{X}}) - p(x,\hat{x})\right) && (9.192) \\ &\leq Ed(X,\hat{X}) + \sum_{x,\hat{x}} d(x,\hat{x})\left|\frac{1}{n}N(x,\hat{x}|\mathbf{X},\hat{\mathbf{X}}) - p(x,\hat{x})\right| && (9.193) \\ &\stackrel{a)}{\leq} Ed(X,\hat{X}) + d_{max}\sum_{x,\hat{x}}\left|\frac{1}{n}N(x,\hat{x}|\mathbf{X},\hat{\mathbf{X}}) - p(x,\hat{x})\right| && (9.194) \\ &\stackrel{b)}{\leq} Ed(X,\hat{X}) + d_{max}\delta && (9.195) \\ &\stackrel{c)}{\leq} D + d_{max}\delta, && (9.196) \end{aligned}$$

where

a) follows from the definition of d_{max} in (9.142);

b) follows because $(\mathbf{X},\hat{\mathbf{X}}) \in T^n_{[X\hat{X}]\delta}$;

c) follows from (9.152).

Therefore, by taking

$$\delta \leq \frac{\epsilon}{d_{max}}, \tag{9.197}$$

we obtain

$$d(\mathbf{X}, \hat{\mathbf{X}}) \leq D + d_{max}\left(\frac{\epsilon}{d_{max}}\right) = D + \epsilon \tag{9.198}$$

if $K \neq 1$. Therefore,

$$\Pr\{d(\mathbf{X}, \hat{\mathbf{X}}) > D + \epsilon | K \neq 1\} = 0, \tag{9.199}$$

and it follows that from (9.187) that

$$\Pr\{d(\mathbf{X}, \hat{\mathbf{X}}) > D + \epsilon\} \leq \epsilon. \tag{9.200}$$

Thus we have shown that for sufficiently large n, there exists an (n, M) random code which satisfies

$$\frac{1}{n}\log M \leq I(X; \hat{X}) + \epsilon \tag{9.201}$$

(this follows from the upper bound in (9.153)) and (9.200). This implies the existence of an (n, M) rate-distortion code which satisfies (9.201) and (9.200). Therefore, the rate-distortion pair $(I(X; \hat{X}), D)$ is achievable. Then upon minimizing over all $\hat{X}$ which satisfy (9.152), we conclude that the rate-distortion pair $(R_I(D), D)$ is achievable, which implies $R_I(D) \geq R(D)$. The proof is completed.

PROBLEMS

1. Obtain the forward channel description of $R(D)$ for the binary source with the Hamming distortion measure.

2. *Binary covering radius* The Hamming ball with center $\mathbf{c} = (c_1, c_2, \cdots, c_n) \in \{0,1\}^n$ and radius r is the set

$$S_r(\mathbf{c}) = \left\{\mathbf{x} \in \{0,1\}^n : \sum_{i=1}^{n} |x_i - c_i| \leq r\right\}.$$

Let $M_{r,n}$ be the minimum number M such that there exists Hamming balls $S_r(\mathbf{c}_j)$, $j = 1, 2, \cdots, M$ such that for all $\mathbf{x} \in \{0,1\}^n$, $\mathbf{x} \in S_r(\mathbf{c}_j)$ for some j.

a) Show that

$$M_{r,n} \geq \frac{2^n}{\sum_{k=0}^{r}\binom{n}{k}}.$$

b) What is the relation between $M_{r,n}$ and the rate-distortion function for the binary source with the Hamming distortion measure?

3. Consider a source random variable X with the Hamming distortion measure.

   a) Prove that

   $$R(D) \geq H(X) - D\log(|\mathcal{X}| - 1) - h_b(D)$$

   for $0 \leq D \leq D_{max}$.

   b) Show that the above lower bound on $R(D)$ is tight if X distributes uniformly on $\mathcal{X}$.

   See Jerohin [109] (also see [55], p.133) for the tightness of this lower bound for a general source. This bound is a special case of the Shannon lower bound for the rate-distortion function [188] (also see [52], p.369).

4. *Product source* Let X and Y be two independent source random variables with reproduction alphabets $\hat{\mathcal{X}}$ and $\hat{\mathcal{Y}}$ and distortion measures d_x and d_y, and the rate-distortion functions for X and Y are denoted by $R_x(D_x)$ and $R_y(D_y)$, respectively. Now for the product source (X, Y), define a distortion measure $d : \mathcal{X} \times \mathcal{Y} \to \hat{\mathcal{X}} \times \hat{\mathcal{Y}}$ by

   $$d((x,y),(\hat{x},\hat{y})) = d_x(x,\hat{x}) + d_y(y,\hat{y}).$$

   Prove that the rate-distortion function $R(D)$ for (X, Y) with distortion measure d is given by

   $$R(D) = \min_{D_x+D_y=D}(R_x(D_x) + R_y(D_y)).$$

   Hint: Prove that $I(X, Y; \hat{X}, \hat{Y}) \geq I(X; \hat{X}) + I(Y; \hat{Y})$ if X and Y are independent. (Shannon [188].)

5. *Compound source* Let Θ be an index set and $\mathcal{Z}_\Theta = \{X_\theta : \theta \in \Theta\}$ be a collection of source random variables. The random variables in $\mathcal{Z}_\Theta$ have a common source alphabet $\mathcal{X}$, a common reproduction alphabet $\hat{\mathcal{X}}$, and a common distortion measure d. A compound source is an i.i.d. information source whose generic random variable is X_Φ, where Φ is equal to some $\theta \in \Theta$ but we do not know which one it is. The rate-distortion function $R_\Phi(D)$ for X_Φ has the same definition as the rate-distortion function defined in this chapter except that (9.23) is replaced by

   $$\Pr\{d(\mathbf{X}_\theta, \hat{\mathbf{X}}) > D + \epsilon\} \leq \epsilon \quad \text{for all } \theta \in \Theta.$$

Show that

$$R_\Phi(D) = \sup_{\theta \in \Theta} R_\theta(D),$$

where $R_\theta(D)$ is the rate-distortion function for X_θ.

6. Show that asymptotic optimality can always be achieved by separating rate-distortion coding and channel coding when the information source is i.i.d. (with a single-letter distortion measure) and the channel is memoryless.

7. *Slepian-Wolf coding* Let ϵ, γ, and δ be small positive quantities. For $1 \le i \le 2^{n(H(Y|X)+\epsilon)}$, randomly and independently select with replacement $2^{n(I(X;Y)-\gamma)}$ sequences from $T^n_{[Y]\delta}$ according to the uniform distribution to form a bin B_i. Let $(\mathbf{x}, \mathbf{y})$ be a fixed pair of sequences in $T^n_{[XY]\delta}$. Prove the following by choosing ϵ, γ, and δ appropriately:

   a) the probability that $\mathbf{y}$ is in some B_i tends to 1 as $n \to \infty$;

   b) given that $\mathbf{y} \in B_i$, the probability that there exists another $\mathbf{y}' \in B_i$ such that $(\mathbf{x}, \mathbf{y}') \in T^n_{[XY]\delta}$ tends to 0 as $n \to \infty$.

   Let $(\mathbf{X}, \mathbf{Y}) \sim p^n(x, y)$. The results in a) and b) say that if $(\mathbf{X}, \mathbf{Y})$ is jointly typical, which happens with probability close to 1 for large n, then it is very likely that $\mathbf{Y}$ is in some bin B_i, and that $\mathbf{Y}$ is the unique vector in B_i which is jointly typical with $\mathbf{X}$. If $\mathbf{X}$ is available as side-information, then by specifying the index of the bin containing $\mathbf{Y}$, which takes about $2^{nH(Y|X)}$ bits, $\mathbf{Y}$ can be uniquely specified. Note that no knowledge about $\mathbf{X}$ is involved in specifying the index of the bin containing $\mathbf{Y}$. This is the basis of the Slepian-Wolf coding [196] which launched the whole area of multiterminal source coding (see Berger [21]).

HISTORICAL NOTES

Transmission of an information source with distortion was first conceived by Shannon in his 1948 paper [184]. He returned to the problem in 1959 and proved the rate-distortion theorem. The normalization of the rate-distortion function is due to Pinkston [163]. The rate-distortion theorem proved here is a stronger version of the original theorem. Extensions of the theorem to more general sources were proved in the book by Berger [20]. An iterative algorithm for computing the rate-distortion function developed by Blahut [28] will be discussed in Chapter 10. Rose [179] has developed an algorithm for the same purpose based on a mapping approach.

Chapter 10

THE BLAHUT-ARIMOTO ALGORITHMS

For a discrete memoryless channel $p(y|x)$, the capacity

$$C = \max_{r(x)} I(X;Y), \tag{10.1}$$

where X and Y are respectively the input and the output of the generic channel and $r(x)$ is the input distribution, characterizes the maximum asymptotically achievable rate at which information can be transmitted through the channel reliably. The expression for C in (10.1) is called a *single-letter characterization* because it depends only the transition matrix of the generic channel but not on the block length n of a code for the channel. When both the input alphabet $\mathcal{X}$ and the output alphabet $\mathcal{Y}$ are finite, the computation of C becomes a finite-dimensional maximization problem.

For an i.i.d. information source $\{X_k, k \geq 1\}$ with generic random variable X, the rate-distortion function

$$R(D) = \min_{Q(\hat{x}|x):Ed(X,\hat{X}) \leq D} I(X;\hat{X}) \tag{10.2}$$

characterizes the minimum asymptotically achievable rate of a rate-distortion code which reproduces the information source with an average distortion no more than D with respect to a single-letter distortion measure d. Again, the expression for $R(D)$ in (10.2) is a single-letter characterization because it depends only on the generic random variable X but not on the block length n of a rate-distortion code. When both the source alphabet $\mathcal{X}$ and the reproduction alphabet $\hat{\mathcal{X}}$ are finite, the computation of $R(D)$ becomes a finite-dimensional minimization problem.

Unless for very special cases, it is not possible to obtain an expression for C or $R(D)$ in closed form, and we have to resort to numerical computation.

However, computing these quantities is not straightforward because the associated optimization problem is nonlinear. In this chapter, we discuss the *Blahut-Arimoto algorithms* (henceforth the BA algorithms), which is an iterative algorithm devised for this purpose.

In order to better understand how and why the BA algorithm works, we will first describe the algorithm in a general setting in the next section. Specializations of the algorithm for the computation of C and $R(D)$ will be discussed in Section 10.2, and convergence of the algorithm will be proved in Section 10.3.

10.1 ALTERNATING OPTIMIZATION

In this section, we describe an alternating optimization algorithm. This algorithm will be specialized in the next section for computing the channel capacity and the rate-distortion function.

Consider the double supremum

$$\sup_{\mathbf{u}_1 \in A_1} \sup_{\mathbf{u}_2 \in A_2} f(\mathbf{u}_1, \mathbf{u}_2), \tag{10.3}$$

where A_i is a convex subset of $\Re^{n_i}$ for $i = 1, 2$, and f is a function defined on $A_1 \times A_2$. The function f is bounded from above, and is continuous and has continuous partial derivatives on $A_1 \times A_2$. Further assume that for all $\mathbf{u}_2 \in A_2$, there exists a unique $c_1(\mathbf{u}_2) \in A_1$ such that

$$f(c_1(\mathbf{u}_2), \mathbf{u}_2) = \max_{\mathbf{u}'_1 \in A_1} f(\mathbf{u}'_1, \mathbf{u}_2), \tag{10.4}$$

and for all $\mathbf{u}_1 \in A_1$, there exists a unique $c_2(\mathbf{u}_1) \in A_2$ such that

$$f(\mathbf{u}_1, c_2(\mathbf{u}_1)) = \max_{\mathbf{u}'_2 \in A_2} f(\mathbf{u}_1, \mathbf{u}'_2). \tag{10.5}$$

Let $\mathbf{u} = (\mathbf{u}_1, \mathbf{u}_2)$ and $A = A_1 \times A_2$. Then (10.3) can be written as

$$\sup_{\mathbf{u} \in A} f(\mathbf{u}). \tag{10.6}$$

In other words, the supremum of f is taken over a subset of $\Re^{n_1+n_2}$ which is equal to the Cartesian product of two convex subsets of $\Re^{n_1}$ and $\Re^{n_2}$, respectively.

We now describe an alternating optimization algorithm for computing f^*, the value of the double supremum in (10.3). Let $\mathbf{u}^{(k)} = (\mathbf{u}_1^{(k)}, \mathbf{u}_2^{(k)})$ for $k \geq 0$ which are defined as follows. Let $\mathbf{u}_1^{(0)}$ be an arbitrarily chosen vector in A_1, and let $\mathbf{u}_2^{(0)} = c_2(\mathbf{u}_1^{(0)})$. For $k \geq 1$, $\mathbf{u}^{(k)}$ is defined by

$$\mathbf{u}_1^{(k)} = c_1(\mathbf{u}_2^{(k-1)}) \tag{10.7}$$

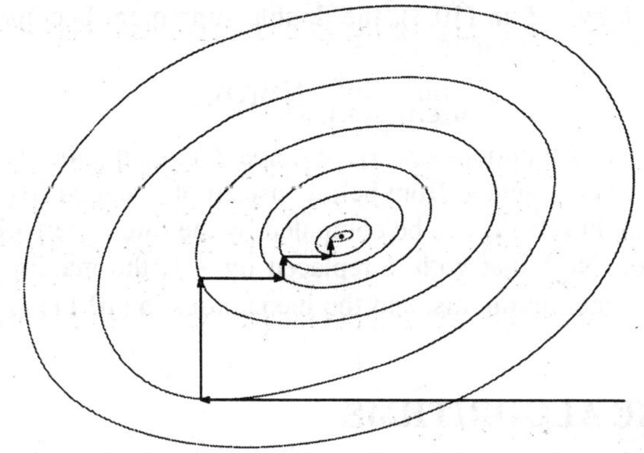

Figure 10.1. Alternating optimization.

and

$$\mathbf{u}_2^{(k)} = c_2(\mathbf{u}_1^{(k)}). \tag{10.8}$$

In other words, $\mathbf{u}_1^{(k)}$ and $\mathbf{u}_2^{(k)}$ are generated in the order $\mathbf{u}_1^{(0)}, \mathbf{u}_2^{(0)}, \mathbf{u}_1^{(1)}, \mathbf{u}_2^{(1)}, \mathbf{u}_1^{(2)}, \mathbf{u}_2^{(2)}, \cdots$, where each vector in the sequence is a function of the previous vector except that $\mathbf{u}_1^{(0)}$ is arbitrarily chosen in A_1. Let

$$f^{(k)} = f(\mathbf{u}^{(k)}). \tag{10.9}$$

Then from (10.4) and (10.5),

$$\begin{aligned} f^{(k)} &= f(\mathbf{u}_1^{(k)}, \mathbf{u}_2^{(k)}) & (10.10)\\ &\geq f(\mathbf{u}_1^{(k)}, \mathbf{u}_2^{(k-1)}) & (10.11)\\ &\geq f(\mathbf{u}_1^{(k-1)}, \mathbf{u}_2^{(k-1)}) & (10.12)\\ &= f^{(k-1)} & (10.13)\end{aligned}$$

for $k \geq 1$. Since the sequence $f^{(k)}$ is non-decreasing, it must converge because f is bounded from above. We will show in Section 10.3 that $f^{(k)} \to f^*$ if f is concave. Figure 10.1 is an illustration of the alternating maximization algorithm, where in this case both n_1 and n_2 are equal to 1, and $f^{(k)} \to f^*$.

The alternating optimization algorithm can be explained by the following analogy. Suppose a hiker wants to reach the summit of a mountain. Starting from a certain point in the mountain, the hiker moves north-south and east-west alternately. (In our problem, the north-south and east-west directions can be multi-dimensional.) In each move, the hiker moves to the highest possible point. The question is whether the hiker can eventually approach the summit starting from any point in the mountain.

Replacing f by $-f$ in (10.3), the double supremum becomes the double infimum

$$\inf_{\mathbf{u}_1 \in A_1} \inf_{\mathbf{u}_2 \in A_2} f(\mathbf{u}_1, \mathbf{u}_2). \tag{10.14}$$

All the previous assumptions on A_1, A_2, and f remain valid except that f is now assumed to be bounded from below instead of bounded from above. The double infimum in (10.14) can be computed by the same alternating optimization algorithm. Note that with f replaced by $-f$, the maximums in (10.4) and (10.5) become minimums, and the inequalities in (10.11) and (10.12) are reversed.

10.2 THE ALGORITHMS

In this section, we specialize the alternating optimization algorithm described in the last section to compute the channel capacity and the rate-distortion function. The corresponding algorithms are known as the BA algorithms.

10.2.1 CHANNEL CAPACITY

We will use $\mathbf{r}$ to denote an input distribution $r(x)$, and we write $\mathbf{r} > 0$ if $\mathbf{r}$ is strictly positive, i.e., $r(x) > 0$ for all $x \in \mathcal{X}$. If $\mathbf{r}$ is not strictly positive, we write $\mathbf{r} \geq 0$. Similar notations will be introduced as appropriate.

LEMMA 10.1 *Let $r(x)p(y|x)$ be a given joint distribution on $\mathcal{X} \times \mathcal{Y}$ such that $\mathbf{r} > 0$, and let $\mathbf{q}$ be a transition matrix from $\mathcal{Y}$ to $\mathcal{X}$. Then*

$$\max_{\mathbf{q}} \sum_x \sum_y r(x)p(y|x) \log \frac{q(x|y)}{r(x)} = \sum_x \sum_y r(x)p(y|x) \log \frac{q^*(x|y)}{r(x)}, \tag{10.15}$$

where the maximization is taken over all $\mathbf{q}$ such that

$$q(x|y) = 0 \quad \text{if and only if} \quad p(y|x) = 0, \tag{10.16}$$

and

$$q^*(x|y) = \frac{r(x)p(y|x)}{\sum_{x'} r(x')p(y|x')}, \tag{10.17}$$

i.e., the maximizing $\mathbf{q}$ is the which corresponds to the input distribution $\mathbf{r}$ and the transition matrix $p(y|x)$.

In (10.15) and the sequel, we adopt the convention that the summation is taken over all x and y such that $r(x) > 0$ and $p(y|x) > 0$. Note that the right hand side of (10.15) gives the mutual information $I(X;Y)$ when $\mathbf{r}$ is the input distribution for the generic channel $p(y|x)$.

Proof Let

$$w(y) = \sum_{x'} r(x')p(y|x') \tag{10.18}$$

in (10.17). We assume with loss of generality that for all $y \in \mathcal{Y}$, $p(y|x) > 0$ for some $x \in \mathcal{X}$. Since $\mathbf{r} > 0$, $w(y) > 0$ for all y, and hence $q^*(x|y)$ is well-defined. Rearranging (10.17), we have

$$r(x)p(y|x) = w(y)q^*(x|y). \tag{10.19}$$

Consider

$$\begin{aligned}
&\sum_x \sum_y r(x)p(y|x) \log \frac{q^*(x|y)}{r(x)} - \sum_x \sum_y r(x)p(y|x) \log \frac{q(x|y)}{r(x)} \\
&= \sum_x \sum_y r(x)p(y|x) \log \frac{q^*(x|y)}{q(x|y)} && (10.20) \\
&= \sum_y \sum_x w(y)q^*(x|y) \log \frac{q^*(x|y)}{q(x|y)} && (10.21) \\
&= \sum_y w(y) \sum_x q^*(x|y) \log \frac{q^*(x|y)}{q(x|y)} && (10.22) \\
&= \sum_y w(y) D(q^*(x|y) \| q(x|y)) && (10.23) \\
&\geq 0, && (10.24)
\end{aligned}$$

where (10.21) follows from (10.19), and the last step is an application of the divergence inequality. Then the proof is completed by noting in (10.17) that $\mathbf{q}^*$ satisfies (10.16) because $\mathbf{r} > 0$. □

THEOREM 10.2 *For a discrete memoryless channel $p(y|x)$,*

$$C = \sup_{\mathbf{r}>0} \max_{\mathbf{q}} \sum_x \sum_y r(x)p(y|x) \log \frac{q(x|y)}{r(x)}, \tag{10.25}$$

where the maximization is taken over all $\mathbf{q}$ which satisfies (10.16).

Proof Let $I(\mathbf{r}, \mathbf{p})$ denote the mutual information $I(X;Y)$ when $\mathbf{r}$ is the input distribution for the generic channel $p(y|x)$. Then we can write

$$C = \max_{\mathbf{r} \geq 0} I(\mathbf{r}, \mathbf{p}). \tag{10.26}$$

Let $\mathbf{r}^*$ achieves C. If $\mathbf{r}^* > 0$, then

$$\begin{aligned}
C &= \max_{\mathbf{r} \geq 0} I(\mathbf{r}, \mathbf{p}) && (10.27) \\
&= \max_{\mathbf{r} > 0} I(\mathbf{r}, \mathbf{p}) && (10.28) \\
&= \max_{\mathbf{r} > 0} \max_{\mathbf{q}} \sum_x \sum_y r(x)p(y|x) \log \frac{q(x|y)}{r(x)} && (10.29) \\
&= \sup_{\mathbf{r} > 0} \max_{\mathbf{q}} \sum_x \sum_y r(x)p(y|x) \log \frac{q(x|y)}{r(x)}, && (10.30)
\end{aligned}$$

where (10.29) follows from Lemma 10.1 (and the maximization is over all $\mathbf{q}$ which satisfies (10.16)).

Next, we consider the case when $\mathbf{r}^* \geq 0$. Since $I(\mathbf{r}, \mathbf{p})$ is continuous in $\mathbf{r}$, for any $\epsilon > 0$, there exists $\delta > 0$ such that if

$$\|\mathbf{r} - \mathbf{r}^*\| < \delta, \tag{10.31}$$

then

$$C - I(\mathbf{r}, \mathbf{p}) < \epsilon, \tag{10.32}$$

where $\|\mathbf{r} - \mathbf{r}^*\|$ denotes the Euclidean distance between $\mathbf{r}$ and $\mathbf{r}^*$. In particular, there exists $\tilde{\mathbf{r}} > 0$ which satisfies (10.31) and (10.32). Then

$$\begin{aligned} C &= \max_{\mathbf{r} \geq 0} I(\mathbf{r}, \mathbf{p}) && (10.33) \\ &\geq \sup_{\mathbf{r} > 0} I(\mathbf{r}, \mathbf{p}) && (10.34) \\ &\geq I(\tilde{\mathbf{r}}, \mathbf{p}) && (10.35) \\ &> C - \epsilon, && (10.36) \end{aligned}$$

where the last step follows because $\tilde{\mathbf{r}}$ satisfies (10.32). Thus we have

$$C - \epsilon < \sup_{\mathbf{r} > 0} I(\mathbf{r}, \mathbf{p}) \leq C. \tag{10.37}$$

Finally, by letting $\epsilon \to 0$, we conclude that

$$C = \sup_{\mathbf{r} > 0} I(\mathbf{r}, \mathbf{p}) = \sup_{\mathbf{r} > 0} \max_{\mathbf{q}} \sum_x \sum_y r(x) p(y|x) \log \frac{q(x|y)}{r(x)}. \tag{10.38}$$

This accomplishes the proof. □

Now for the double supremum in (10.3), let

$$f(\mathbf{r}, \mathbf{q}) = \sum_x \sum_y r(x) p(y|x) \log \frac{q(x|y)}{r(x)}, \tag{10.39}$$

with $\mathbf{r}$ and $\mathbf{q}$ playing the roles of $\mathbf{u}_1$ and $\mathbf{u}_2$, respectively. Let

$$A_1 = \left\{ (r(x), x \in \mathcal{X}) : r(x) > 0 \text{ and } \textstyle\sum_x r(x) = 1 \right\}, \tag{10.40}$$

and

$$\begin{aligned} A_2 = \{ & (q(x|y), (x,y) \in \mathcal{X} \times \mathcal{Y}) : q(x|y) > 0 \\ & \text{if } p(x|y) > 0, \ q(x|y) = 0 \text{ if } p(y|x) = 0, \\ & \text{and } \textstyle\sum_x q(x|y) = 1 \text{ for all } y \in \mathcal{Y} \}. \end{aligned} \tag{10.41}$$

Then A_1 is a subset of $\Re^{|\mathcal{X}|}$ and A_2 is a subset of $\Re^{|\mathcal{X}||\mathcal{Y}|}$, and it is readily checked that both A_1 and A_2 are convex. For all $\mathbf{r} \in A_1$ and $\mathbf{q} \in A_2$, by Lemma 10.1,

$$\begin{aligned} f(\mathbf{r}, \mathbf{q}) &= \sum_x \sum_y r(x)p(y|x) \log \frac{q(x|y)}{r(x)} &\quad (10.42)\\ &\le \sum_x \sum_y r(x)p(y|x) \log \frac{q^*(x|y)}{r(x)} &\quad (10.43)\\ &= I(X;Y) &\quad (10.44)\\ &\le H(X) &\quad (10.45)\\ &\le \log |\mathcal{X}|. &\quad (10.46) \end{aligned}$$

Thus f is bounded from above. Since for all $\mathbf{q} \in A_2$, $q(x|y) = 0$ for all x and y such that $p(x|y) = 0$, these components of $\mathbf{q}$ are degenerated. In fact, these components of $\mathbf{q}$ do not appear in the definition of $f(\mathbf{r}, \mathbf{q})$ in (10.39), which can be seen as follows. Recall the convention that the double summation in (10.39) is over all x and y such that $r(x) > 0$ and $p(y|x) > 0$. If $q(x|y) = 0$, then $p(y|x) = 0$, and hence the corresponding term is not included in the double summation. Therefore, it is readily seen that f is continuous and has continuous partial derivatives on A because all the probabilities involved in the double summation in (10.39) are strictly positive. Moreover, for any given $\mathbf{r} \in A_1$, by Lemma 10.1, there exists a unique $\mathbf{q} \in A_2$ which maximizes f. It will be shown shortly that for any given $\mathbf{q} \in A_2$, there also exists a unique $\mathbf{r} \in A_1$ which maximizes f.

The double supremum in (10.3) now becomes

$$\sup_{\mathbf{r}\in A_1} \sup_{\mathbf{q}\in A_2} \sum_x \sum_y r(x)p(y|x) \log \frac{q(x|y)}{r(x)}, \quad (10.47)$$

which by Theorem 10.2 is equal to C, where the supremum over all $\mathbf{q} \in A_2$ is in fact a maximum. We then apply the alternating optimization algorithm in the last section to compute C. First, we arbitrarily choose a *strictly positive* input distribution in A_1 and let it be $\mathbf{r}^{(0)}$. Then we define $\mathbf{q}^{(0)}$ and in general $\mathbf{q}^{(k)}$ for $k \ge 0$ by

$$q^{(k)}(x|y) = \frac{r^{(k)}(x)p(y|x)}{\sum_{x'} r^{(k)}(x')p(y|x')} \quad (10.48)$$

in view of Lemma 10.1. In order to define $\mathbf{r}^{(1)}$ and in general $\mathbf{r}^{(k)}$ for $k \ge 1$, we need to find the $\mathbf{r} \in A_1$ which maximizes f for a given $\mathbf{q} \in A_2$, where the constraints on $\mathbf{r}$ are

$$\sum_x r(x) = 1 \quad (10.49)$$

and

$$r(x) > 0 \quad \text{for all } x \in \mathcal{X}. \tag{10.50}$$

We now use the method of Lagrange multipliers to find the best $\mathbf{r}$ by ignoring temporarily the positivity constraints in (10.50). Let

$$J = \sum_x \sum_y r(x)p(y|x) \log \frac{q(x|y)}{r(x)} - \lambda \sum_x r(x). \tag{10.51}$$

For convenience sake, we assume that the logarithm is the natural logarithm. Differentiating with respect to $r(x)$ gives

$$\frac{\partial J}{\partial r(x)} = \sum_y p(y|x) \log q(x|y) - \log r(x) - 1 - \lambda. \tag{10.52}$$

Upon setting $\frac{\partial J}{\partial r(x)} = 0$, we have

$$\log r(x) = \sum_y p(y|x) \log q(x|y) - 1 - \lambda, \tag{10.53}$$

or

$$r(x) = e^{-(\lambda+1)} \prod_y q(x|y)^{p(y|x)}. \tag{10.54}$$

By considering the normalization constraint in (10.49), we can eliminate λ and obtain

$$r(x) = \frac{\prod_y q(x|y)^{p(y|x)}}{\sum_{x'} \prod_y q(x'|y)^{p(y|x')}}. \tag{10.55}$$

The above product is over all y such that $p(y|x) > 0$, and $q(x|y) > 0$ for all such y. This implies that both the numerator and the denominator on the right hand side above are positive, and therefore $r(x) > 0$. In other words, the $\mathbf{r}$ thus obtained happen to satisfy the positivity constraints in (10.50) although these constraints were ignored when we set up the Lagrange multipliers. We will show in Section 10.3.2 that f is concave. Then $\mathbf{r}$ as given in (10.55), which is unique, indeed achieves the maximum of f for a given $\mathbf{q} \in A_2$ because $\mathbf{r}$ is in the interior of A_1. In view of (10.55), we define $\mathbf{r}^{(k)}$ for $k \geq 1$ by

$$r^{(k)}(x) = \frac{\prod_y q^{(k-1)}(x|y)^{p(y|x)}}{\sum_{x'} \prod_y q^{(k-1)}(x'|y)^{p(y|x')}}. \tag{10.56}$$

The vectors $\mathbf{r}^{(k)}$ and $\mathbf{q}^{(k)}$ are defined in the order $\mathbf{r}^{(0)}, \mathbf{q}^{(0)}, \mathbf{r}^{(1)}, \mathbf{q}^{(1)}, \mathbf{r}^{(2)}, \mathbf{q}^{(2)}, \cdots$, where each vector in the sequence is a function of the previous vector except that $\mathbf{r}^{(0)}$ is arbitrarily chosen in A_1. It remains to show by induction that $\mathbf{r}^{(k)} \in A_1$ for $k \geq 1$ and $\mathbf{q}^{(k)} \in A_2$ for $k \geq 0$. If $\mathbf{r}^{(k)} \in A_1$, i.e., $\mathbf{r}^{(k)} > 0$,

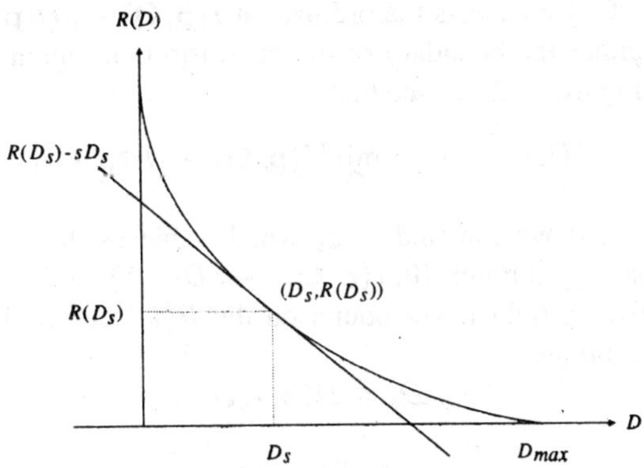

Figure 10.2. A tangent to the $R(D)$ curve with slope equal to s.

then we see from (10.48) that $q^{(k)}(x|y) = 0$ if and only if $p(x|y) = 0$, i.e., $\mathbf{q}^{(k)} \in A_2$. On the other hand, if $\mathbf{q}^{(k)} \in A_2$, then we see from (10.56) that $\mathbf{r}^{(k+1)} > 0$, i.e., $\mathbf{r}^{(k+1)} \in A_2$. Therefore, $\mathbf{r}^{(k)} \in A_1$ and $\mathbf{q}^{(k)} \in A_2$ for all $k \geq 0$. Upon determining $(\mathbf{r}^{(k)}, \mathbf{q}^{(k)})$, we can compute $f^{(k)} = f(\mathbf{r}^{(k)}, \mathbf{q}^{(k)})$ for all k. It will be shown in Section 10.3 that $f^{(k)} \to C$.

10.2.2 THE RATE-DISTORTION FUNCTION

This discussion in this section is analogous to the discussion in Section 10.2.1. Some of the details will be omitted for brevity.

For all problems of interest, $R(0) > 0$. Otherwise, $R(D) = 0$ for all $D \geq 0$ since $R(D)$ is nonnegative and non-increasing. Therefore, we assume without loss of generality that $R(0) > 0$.

We have shown in Corollary 9.19 that if $R(0) > 0$, then $R(D)$ is strictly decreasing for $0 \leq D \leq D_{max}$. Since $R(D)$ is convex, for any $s \leq 0$, there exists a point on the $R(D)$ curve for $0 \leq D \leq D_{max}$ such that the slope of a tangent[1] to the $R(D)$ curve at that point is equal to s. Denote such a point on the $R(D)$ curve by $(D_s, R(D_s))$, which is not necessarily unique. Then this tangent intersects with the ordinate at $R(D_s) - sD_s$. This is illustrated in Figure 10.2.

Let $I(\mathbf{p}, \mathbf{Q})$ denote the mutual information $I(X, \hat{X})$ and $D(\mathbf{p}, \mathbf{Q})$ denote the expected distortion $Ed(X, \hat{X})$ when $\mathbf{p}$ is the distribution for X and $\mathbf{Q}$ is the transition matrix from $\mathcal{X}$ to $\hat{\mathcal{X}}$ defining $\hat{X}$. Then for any $\mathbf{Q}$, $(I(\mathbf{p}, \mathbf{Q}), D(\mathbf{p}, \mathbf{Q}))$ is a point in the rate-distortion region, and the line with slope s passing through

[1] We say that a line is a tangent to the $R(D)$ curve if it touches the $R(D)$ curve from below.

$(I(\mathbf{p},\mathbf{Q}),D(\mathbf{p},\mathbf{Q}))$ intersects the ordinate at $I(\mathbf{p},\mathbf{Q})-sD(\mathbf{p},\mathbf{Q})$. Since the $R(D)$ curve defines the boundary of the rate-distortion region and it is above the tangent in Figure 10.2, we see that

$$R(D_s)-sD_s=\min_{\mathbf{Q}}[I(\mathbf{p},\mathbf{Q})-sD(\mathbf{p},\mathbf{Q})]. \tag{10.57}$$

For each $s\le 0$, if we can find a $\mathbf{Q}_s$ which achieves the above minimum, then the line passing through $(0,I(\mathbf{p},\mathbf{Q}_s)-sD(\mathbf{p},\mathbf{Q}_s))$, i.e., the tangent in Figure 10.2, gives a tight lower bound on the $R(D)$ curve. In particular, if $(R(D_s),D_s)$ is unique,

$$D_s=D(\mathbf{p},\mathbf{Q}_s) \tag{10.58}$$

and

$$R(D_s)=I(\mathbf{p},\mathbf{Q}_s). \tag{10.59}$$

By varying over all $s\le 0$, we can then trace out the whole $R(D)$ curve. In the rest of the section, we will devise an iterative algorithm for the minimization problem in (10.57).

LEMMA 10.3 *Let $p(x)Q(\hat{x}|x)$ be a given joint distribution on $\mathcal{X}\times\hat{\mathcal{X}}$ such that $\mathbf{Q}>0$, and let $\mathbf{t}$ be any distribution on $\hat{\mathcal{X}}$ such that $\mathbf{t}>0$. Then*

$$\min_{\mathbf{t}>0}\sum_x\sum_{\hat{x}}p(x)Q(\hat{x}|x)\log\frac{Q(\hat{x}|x)}{t(\hat{x})}=\sum_x\sum_{\hat{x}}p(x)Q(\hat{x}|x)\log\frac{Q(\hat{x}|x)}{t^*(\hat{x})}, \tag{10.60}$$

where

$$t^*(\hat{x})=\sum_x p(x)Q(\hat{x}|x), \tag{10.61}$$

i.e., the minimizing $t(\hat{x})$ is the distribution on $\hat{\mathcal{X}}$ corresponding to the input distribution $\mathbf{p}$ and the transition matrix $\mathbf{Q}$.

Proof It suffices to prove that

$$\sum_x\sum_{\hat{x}}p(x)Q(\hat{x}|x)\log\frac{Q(\hat{x}|x)}{t(\hat{x})}\ge\sum_x\sum_{\hat{x}}p(x)Q(\hat{x}|x)\log\frac{Q(\hat{x}|x)}{t^*(\hat{x})} \tag{10.62}$$

for all $\mathbf{t}>0$. The details are left as an exercise. Note in (10.61) that $\mathbf{t}^*>0$ because $\mathbf{Q}>0$. □

Since $I(\mathbf{p},\mathbf{Q})$ and $D(\mathbf{p},\mathbf{Q})$ are continuous in $\mathbf{Q}$, via an argument similar to the one we used in the proof of Theorem 10.2, we can replace the minimum over all $\mathbf{Q}$ in (10.57) by the infimum over all $\mathbf{Q}>0$. By noting that the right hand side of (10.60) is equal to $I(\mathbf{p},\mathbf{Q})$ and

$$D(\mathbf{p},\mathbf{Q})=\sum_x\sum_{\hat{x}}p(x)Q(\hat{x}|x)d(x,\hat{x}), \tag{10.63}$$

we can apply Lemma 10.3 to obtain

$$
\begin{aligned}
& R(D_s) - sD_s \\
= & \inf_{\mathbf{Q}>0} \left[\min_{\mathbf{t}>0} \sum_{x,\hat{x}} p(x)Q(\hat{x}|x) \log \frac{Q(\hat{x}|x)}{t(\hat{x})} - s \sum_{x,\hat{x}} p(x)Q(\hat{x}|x)d(x,\hat{x}) \right] \qquad (10.64) \\
= & \inf_{\mathbf{Q}>0} \min_{\mathbf{t}>0} \left[\sum_{x,\hat{x}} p(x)Q(\hat{x}|x) \log \frac{Q(\hat{x}|x)}{t(\hat{x})} - s \sum_{x,\hat{x}} p(x)Q(\hat{x}|x)d(x,\hat{x}) \right]. \qquad (10.65)
\end{aligned}
$$

Now in the double infimum in (10.14), let

$$
\begin{aligned}
f(\mathbf{Q}, \mathbf{t}) &= \sum_x \sum_{\hat{x}} p(x)Q(\hat{x}|x) \log \frac{Q(\hat{x}|x)}{t(\hat{x})} \\
&\quad - s \sum_x \sum_{\hat{x}} p(x)Q(\hat{x}|x)d(x,\hat{x}), \qquad (10.66)
\end{aligned}
$$

$$
\begin{aligned}
A_1 = \Bigg\{ & (Q(\hat{x}|x), (x,\hat{x}) \in \mathcal{X} \times \hat{\mathcal{X}}) : Q(\hat{x}|x) > 0, \\
& \sum_{\hat{x}} Q(\hat{x}|x) = 1 \text{ for all } x \in \mathcal{X} \Bigg\}, \qquad (10.67)
\end{aligned}
$$

and

$$
A_2 = \{(t(\hat{x}), \hat{x} \in \hat{\mathcal{X}}) : t(\hat{x}) > 0 \text{ and } \textstyle\sum_{\hat{x}} t(\hat{x}) = 1\}, \qquad (10.68)
$$

with $\mathbf{Q}$ and $\mathbf{t}$ playing the roles of $\mathbf{u}_1$ and $\mathbf{u}_2$, respectively. Then A_1 is a subset of $\Re^{|\mathcal{X}||\hat{\mathcal{X}}|}$ and A_2 is a subset of $\Re^{|\hat{\mathcal{X}}|}$, and it is readily checked that both A_1 and A_2 are convex. Since $s \le 0$,

$$
\begin{aligned}
& f(\mathbf{Q}, \mathbf{t}) \\
= & \sum_x \sum_{\hat{x}} p(x)Q(\hat{x}|x) \log \frac{Q(\hat{x}|x)}{t(\hat{x})} - s \sum_x \sum_{\hat{x}} p(x)Q(\hat{x}|x)d(x,\hat{x}) \qquad (10.69) \\
\ge & \sum_x \sum_{\hat{x}} p(x)Q(\hat{x}|x) \log \frac{Q(\hat{x}|x)}{t^*(\hat{x})} + 0 \qquad (10.70) \\
= & I(X;\hat{X}) \qquad (10.71) \\
\ge & 0. \qquad (10.72)
\end{aligned}
$$

Therefore, f is bounded from below.

The double infimum in (10.14) now becomes

$$
\inf_{\mathbf{Q} \in A_1} \inf_{\mathbf{t} \in A_2} \left[\sum_x \sum_{\hat{x}} p(x)Q(\hat{x}|x) \log \frac{Q(\hat{x}|x)}{t(\hat{x})} - s \sum_x \sum_{\hat{x}} p(x)Q(\hat{x}|x)d(x,\hat{x}) \right], \qquad (10.73)
$$

where the infimum over all $\mathbf{t} \in A_2$ is in fact a minimum. We then apply the alternating optimization algorithm described in Section 10.2 to compute f^*, the value of (10.73). First, we arbitrarily choose a *strictly positive* transition matrix in A_1 and let it be $\mathbf{Q}^{(0)}$. Then we define $\mathbf{t}^{(0)}$ and in general $\mathbf{t}^{(k)}$ for $k \geq 1$ by

$$t^{(k)}(\hat{x}) = \sum_x p(x) Q^{(k)}(\hat{x}|x) \tag{10.74}$$

in view of Lemma 10.3. In order to define $\mathbf{Q}^{(1)}$ and in general $\mathbf{Q}^{(k)}$ for $k \geq 1$, we need to find the $\mathbf{Q} \in A_1$ which minimizes f for a given $\mathbf{t} \in A_2$, where the constraints on $\mathbf{Q}$ are

$$Q(\hat{x}|x) > 0 \quad \text{for all } (x, \hat{x}) \in \mathcal{X} \times \hat{\mathcal{X}}, \tag{10.75}$$

and

$$\sum_{\hat{x}} Q(\hat{x}|x) = 1 \quad \text{for all } x \in \mathcal{X}. \tag{10.76}$$

As we did for the computation of the channel capacity, we first ignore the positivity constraints in (10.75) when setting up the Lagrange multipliers. Then we obtain

$$Q(\hat{x}|x) = \frac{t(\hat{x}) e^{sd(x,\hat{x})}}{\sum_{\hat{x}'} t(\hat{x}') e^{sd(x,\hat{x}')}} > 0. \tag{10.77}$$

The details are left as an exercise. We then define $\mathbf{Q}^{(k)}$ for $k \geq 1$ by

$$Q^{(k)}(\hat{x}|x) = \frac{t^{(k-1)}(\hat{x}) e^{sd(x,\hat{x})}}{\sum_{\hat{x}'} t^{(k-1)}(\hat{x}') e^{sd(x,\hat{x}')}}. \tag{10.78}$$

It will be shown in the next section that $f^{(k)} = f(\mathbf{Q}^{(k)}, \mathbf{t}^{(k)}) \to f^*$ as $k \to \infty$. If there exists a unique point $(R(D_s), D_s)$ on the $R(D)$ curve such that the slope of a tangent at that point is equal to s, then

$$(I(\mathbf{p}, \mathbf{Q}^{(k)}), D(\mathbf{p}, \mathbf{Q}^{(k)})) \to (R(D_s), D_s). \tag{10.79}$$

Otherwise, $(I(\mathbf{p}, \mathbf{Q}^{(k)}), D(\mathbf{p}, \mathbf{Q}^{(k)}))$ is arbitrarily close to the segment of the $R(D)$ curve at which the slope is equal to s when k is sufficiently large. These facts are easily shown to be true.

10.3 CONVERGENCE

In this section, we first prove that if f is concave, then $f^{(k)} \to f^*$. We then apply this sufficient condition to prove the convergence of the BA algorithm for computing the channel capacity. The convergence of the BA algorithm for computing the rate-distortion function can be proved likewise. The details are omitted.

10.3.1 A SUFFICIENT CONDITION

In the alternating optimization algorithm in Section 10.1, we see from (10.7) and (10.8) that

$$\mathbf{u}^{(k+1)} = (\mathbf{u}_1^{(k+1)}, \mathbf{u}_2^{(k+1)}) = (c_1(\mathbf{u}_2^{(k)}), c_2(c_1(\mathbf{u}_2^{(k)}))) \tag{10.80}$$

for $k \geq 0$. Define

$$\Delta f(\mathbf{u}) = f(c_1(\mathbf{u}_2), c_2(c_1(\mathbf{u}_2))) - f(\mathbf{u}_1, \mathbf{u}_2). \tag{10.81}$$

Then

$$\begin{aligned} f^{(k+1)} - f^{(k)} &= f(\mathbf{u}^{(k+1)}) - f(\mathbf{u}^{(k)}) & (10.82)\\ &= f(c_1(\mathbf{u}_2^{(k)}), c_2(c_1(\mathbf{u}_2^{(k)}))) - f(\mathbf{u}_1^{(k)}, \mathbf{u}_2^{(k)}) & (10.83)\\ &= \Delta f(\mathbf{u}^{(k)}). & (10.84) \end{aligned}$$

We will prove that f being concave is sufficient for $f^{(k)} \to f^*$. To this end, we first prove that if f is concave, then the algorithm cannot be trapped at $\mathbf{u}$ if $f(\mathbf{u}) < f^*$.

LEMMA 10.4 *Let f be concave. If $f^{(k)} < f^*$, then $f^{(k+1)} > f^{(k)}$.*

Proof We will prove that $\Delta f(\mathbf{u}) > 0$ for any $\mathbf{u} \in A$ such that $f(\mathbf{u}) < f^*$. Then if $f^{(k)} = f(\mathbf{u}^{(k)}) < f^*$, we see from (10.84) that

$$f^{(k+1)} - f^{(k)} = \Delta f(\mathbf{u}^{(k)}) > 0, \tag{10.85}$$

and the lemma is proved.

Consider any $\mathbf{u} \in A$ such that $f(\mathbf{u}) < f^*$. We will prove by contradiction that $\Delta f(\mathbf{u}) > 0$. Assume $\Delta f(\mathbf{u}) = 0$. Then it follows from (10.81) that

$$f(c_1(\mathbf{u}_2), c_2(c_1(\mathbf{u}_2))) = f(\mathbf{u}_1, \mathbf{u}_2). \tag{10.86}$$

Now we see from (10.5) that

$$f(c_1(\mathbf{u}_2), c_2(c_1(\mathbf{u}_2))) \geq f(c_1(\mathbf{u}_2), \mathbf{u}_2). \tag{10.87}$$

If $c_1(\mathbf{u}_2) \neq \mathbf{u}_1$, then

$$f(c_1(\mathbf{u}_2), \mathbf{u}_2) > f(\mathbf{u}_1, \mathbf{u}_2) \tag{10.88}$$

because $c_1(\mathbf{u}_2)$ is unique. Combining (10.87) and (10.88), we have

$$f(c_1(\mathbf{u}_2), c_2(c_1(\mathbf{u}_2))) > f(\mathbf{u}_1, \mathbf{u}_2), \tag{10.89}$$

which is a contradiction to (10.86). Therefore,

$$\mathbf{u}_1 = c_1(\mathbf{u}_2). \tag{10.90}$$

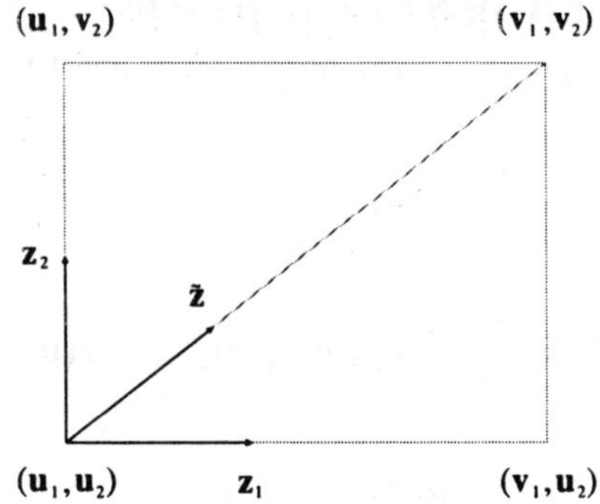

Figure 10.3. The vectors $\mathbf{u}$, $\mathbf{v}$, $\tilde{\mathbf{z}}, \mathbf{z}_1$, and $\mathbf{z}_2$.

Using this, we see from (10.86) that

$$f(\mathbf{u}_1, c_2(\mathbf{u}_1)) = f(\mathbf{u}_1, \mathbf{u}_2), \tag{10.91}$$

which implies

$$\mathbf{u}_2 = c_2(\mathbf{u}_1). \tag{10.92}$$

because $c_2(c_1(\mathbf{u}_2))$ is unique.

Since $f(\mathbf{u}) < f^*$, there exists $\mathbf{v} \in A$ such that

$$f(\mathbf{u}) < f(\mathbf{v}). \tag{10.93}$$

Consider

$$\mathbf{v} - \mathbf{u} = (\mathbf{v}_1 - \mathbf{u}_1, 0) + (0, \mathbf{v}_2 - \mathbf{u}_2). \tag{10.94}$$

Let $\tilde{\mathbf{z}}$ be the unit vector in the direction of $\mathbf{v} - \mathbf{u}$, $\mathbf{z}_1$ be the unit vector in the direction of $(\mathbf{v}_1 - \mathbf{u}_1, 0)$, and $\mathbf{z}_2$ be the unit vector in the direction of $(\mathbf{v}_2 - \mathbf{u}_2, 0)$. Then

$$\|\mathbf{v} - \mathbf{u}\|\tilde{\mathbf{z}} = \|\mathbf{v}_1 - \mathbf{u}_1\|\mathbf{z}_1 + \|\mathbf{v}_2 - \mathbf{u}_2\|\mathbf{z}_2, \tag{10.95}$$

or

$$\tilde{\mathbf{z}} = \alpha_1 \mathbf{z}_1 + \alpha_2 \mathbf{z}_2, \tag{10.96}$$

where

$$\alpha_i = \frac{\|\mathbf{v}_i - \mathbf{u}_i\|}{\|\mathbf{v} - \mathbf{u}\|}, \tag{10.97}$$

$i = 1, 2$. Figure 10.3 is an illustration of the vectors $\mathbf{u}$, $\mathbf{v}$, $\tilde{\mathbf{z}}, \mathbf{z}_1$, and $\mathbf{z}_2$.

We see from (10.90) that f attains its maximum value at $\mathbf{u} = (\mathbf{u}_1, \mathbf{u}_2)$ when $\mathbf{u}_2$ is fixed. In particular, f attains its maximum value at $\mathbf{u}$ alone the line passing through $(\mathbf{u}_1, \mathbf{u}_2)$ and $(\mathbf{v}_1, \mathbf{u}_2)$. Let ∇f denotes the gradient of

f. Since f is continuous and has continuous partial derivatives, the directional derivative of f at $\mathbf{u}$ in the direction of $\mathbf{z}_1$ exists and is given by $\nabla f \cdot \mathbf{z}_1$. It follows from the concavity of f that f is concave along the line passing through $(\mathbf{u}_1, \mathbf{u}_2)$ and $(\mathbf{v}_1, \mathbf{u}_2)$. Since f attains its maximum value at $\mathbf{u}$, the derivative of f along the line passing through $(\mathbf{u}_1, \mathbf{u}_2)$ and $(\mathbf{v}_1, \mathbf{u}_2)$ vanishes. Then we see that

$$\nabla f \cdot \mathbf{z}_1 = 0. \tag{10.98}$$

Similarly, we see from (10.92) that

$$\nabla f \cdot \mathbf{z}_2 = 0. \tag{10.99}$$

Then from (10.96), the directional derivative of f at $\mathbf{u}$ in the direction of $\tilde{\mathbf{z}}$ is given by

$$\nabla f \cdot \tilde{\mathbf{z}} = \alpha_1(\nabla f \cdot \mathbf{z}_1) + \alpha_2(\nabla f \cdot \mathbf{z}_2) = 0. \tag{10.100}$$

Since f is concave along the line passing through $\mathbf{u}$ and $\mathbf{v}$, this implies

$$f(\mathbf{u}) \geq f(\mathbf{v}), \tag{10.101}$$

which is a contradiction to (10.93). Hence, we conclude that $\Delta f(\mathbf{u}) > 0$. □

Although we have proved that the algorithm cannot be trapped at $\mathbf{u}$ if $f(\mathbf{u}) < f^*$, $f^{(k)}$ does not necessarily converge to f^* because the increment in $f^{(k)}$ in each step may be arbitrarily small. In order to prove the desired convergence, we will show in next theorem that this cannot be the case.

THEOREM 10.5 *If f is concave, then $f^{(k)} \to f^*$.*

Proof We have already shown in Section 10.1 that $f^{(k)}$ necessarily converges, say to f'. Hence, for any $\epsilon > 0$ and all sufficiently large k,

$$f' - \epsilon \leq f^{(k)} \leq f'. \tag{10.102}$$

Let

$$\gamma = \min_{\mathbf{u} \in A'} \Delta f(\mathbf{u}), \tag{10.103}$$

where

$$A' = \{\mathbf{u} \in A : f' - \epsilon \leq f(\mathbf{u}) \leq f'\}. \tag{10.104}$$

Since f has continuous partial derivatives, $\Delta f(\mathbf{u})$ is a continuous function of $\mathbf{u}$. Then the minimum in (10.103) exists because A' is compact[2].

[2] A' is compact because it is the inverse image of a closed interval under a continuous function and A is bounded.

We now show that $f' < f^*$ will lead to a contradiction if f is concave. If $f' < f^*$, then from Lemma 10.4, we see that $\Delta f(\mathbf{u}) > 0$ for all $\mathbf{u} \in A'$ and hence $\gamma > 0$. Since $f^{(k)} = f(\mathbf{u}^{(k)})$ satisfies (10.102), $\mathbf{u}^{(k)} \in A'$, and

$$f^{(k+1)} - f^{(k)} = \Delta f(\mathbf{u}^{(k)}) \geq \gamma \tag{10.105}$$

for all sufficiently large k. Therefore, no matter how smaller γ is, $f^{(k)}$ will eventually be greater than f', which is a contradiction to $f^{(k)} \to f'$. Hence, we conclude that $f^{(k)} \to f^*$. □

10.3.2 CONVERGENCE TO THE CHANNEL CAPACITY

In order to show that the BA algorithm for computing the channel capacity converges as intended, i.e., $f^{(k)} \to C$, we only need to show that the function f defined in (10.39) is concave. Toward this end, for

$$f(\mathbf{r}, \mathbf{q}) = \sum_x \sum_y r(x) p(y|x) \log \frac{q(x|y)}{r(x)} \tag{10.106}$$

defined in (10.39), we consider two ordered pairs $(\mathbf{r}_1, \mathbf{q}_1)$ and $(\mathbf{r}_2, \mathbf{q}_2)$ in A, where A_1 and A_2 are defined in (10.40) and (10.41), respectively. For any $0 \leq \lambda \leq 1$ and $\bar{\lambda} = 1 - \lambda$, an application of the log-sum inequality (Theorem 2.31) gives

$$\begin{aligned} &(\lambda r_1(x) + \bar{\lambda} r_2(x)) \log \frac{\lambda r_1(x) + \bar{\lambda} r_2(x)}{\lambda q_1(x|y) + \bar{\lambda} q_2(x|y)} \\ &\leq \lambda r_1(x) \log \frac{r_1(x)}{q_1(x|y)} + \bar{\lambda} r_2(x) \log \frac{r_2(x)}{q_2(x|y)}. \end{aligned} \tag{10.107}$$

Taking reciprocal in the logarithms yields

$$\begin{aligned} &(\lambda r_1(x) + \bar{\lambda} r_2(x)) \log \frac{\lambda q_1(x|y) + \bar{\lambda} q_2(x|y)}{\lambda r_1(x) + \bar{\lambda} r_2(x)} \\ &\geq \lambda r_1(x) \log \frac{q_1(x|y)}{r_1(x)} + \bar{\lambda} r_2(x) \log \frac{q_2(x|y)}{r_2(x)}, \end{aligned} \tag{10.108}$$

and upon multiplying by $p(y|x)$ and summing over all x and y, we obtain

$$f(\lambda \mathbf{r}_1 + \bar{\lambda} \mathbf{r}_2, \lambda \mathbf{q}_1 + \bar{\lambda} \mathbf{q}_2) \geq \lambda f(\mathbf{r}_1, \mathbf{q}_1) + \bar{\lambda} f(\mathbf{r}_2, \mathbf{q}_2). \tag{10.109}$$

Therefore, f is concave. Hence, we have shown that $f^{(k)} \to C$.

PROBLEMS

1. Implement the BA algorithm for computing channel capacity.

2. Implement the BA algorithm for computing the rate-distortion function.

3. Explain why in the BA Algorithm for computing channel capacity, we should not choose an initial input distribution which contains zero probability masses.

4. Prove Lemma 10.3.

5. Consider $f(\mathbf{Q}, \mathbf{t})$ in the BA algorithm for computing the rate-distortion function.

   a) Show that for fixed s and $\mathbf{t}$, $f(\mathbf{Q}, \mathbf{t})$ is minimized by

   $$Q(\hat{x}|x) = \frac{t(\hat{x})e^{sd(x,\hat{x})}}{\sum_{\hat{x}'} t(\hat{x}')e^{sd(x,\hat{x}')}}.$$

   b) Show that $f(\mathbf{Q}, \mathbf{t})$ is convex.

HISTORICAL NOTES

An iterative algorithm for computing the channel capacity was developed by Arimoto [14], where the convergence of the algorithm was proved. Blahut [28] independently developed two similar algorithms, the first for computing the channel capacity and the second for computing the rate-distortion function. The convergence of Blahut's second algorithm was proved by Csiszár [54]. These two algorithms are now commonly referred to as the Blahut-Arimoto algorithms. The simplified proof of convergence in this chapter is based on Yeung and Berger [232].

The Blahut-Arimoto algorithms are special cases of a general iterative algorithm due to Csiszár and Tusnády [58] which also include the EM algorithm [62] for fitting models from incomplete data and the algorithm for finding the log-optimal portfolio for a stock market due to Cover [48].

Chapter 11

SINGLE-SOURCE NETWORK CODING

In a *point-to-point communication system*, the transmitting point and the receiving point are connected by a communication channel. An information source is generated at the transmission point, and the purpose of the communication system is to deliver the information generated at the transmission point to the receiving point via the channel.

In a point-to-point network communication system, there are a number of points, called *nodes*. Between certain pair of nodes, there exist point-to-point communication channels on which information can be transmitted. On these channels, information can be sent only in the specified direction. At a node, one or more information sources may be generated, and each of them is multicast[1] to a set of destination nodes on the network. Examples of point-to-point network communication systems include the telephone network and certain computer networks such as the Internet backbone. For simplicity, we will refer to a point-to-point network communication system as a *point-to-point communication network*. It is clear that a point-to-point communication system is a special case of a point-to-point communication network.

Point-to-point communication systems have been thoroughly studied in classical information theory. However, point-to-point communication networks have been studied in depth only during the last ten years, and there are a lot of problems yet to be solved. In this chapter and Chapter 15, we focus on point-to-point communication networks satisfying the following:

1. the communication channels are free of error;
2. the information is received at the destination nodes with zero error or almost perfectly.

[1]Multicast means to send information to a specified set of destinations.

In this chapter, we consider networks with one information source. Networks with multiple information sources will be discussed in Chapter 15 after we have developed the necessary tools.

11.1 A POINT-TO-POINT NETWORK

A point-to-point network is represented by a directed graph $G = (V, E)$, where V is the set of nodes in the network and E is the set of edges in G which represent the communication channels. An edge from node i to node j, which represents the communication channel from node i to node j, is denoted by (i, j). We assume that G is finite, i.e., $|V| < \infty$. For a network with one information source, which is the subject of discussion in this chapter, the node at which information is generated is referred to as the *source node*, denoted by s, and the destination nodes are referred to as the *sink nodes*, denoted by $t_1, t_2, \cdots, t_L$.

For a communication channel from node i to node j, let R_{ij} be the *rate constraint*, i.e., the maximum number of bits which can be sent per unit time on the channel. R_{ij} is also referred to as the *capacity*[2] of edge (i, j). Let

$$\mathbf{R} = [R_{ij} : (i, j) \in E] \tag{11.1}$$

be the rate constraints for graph G. To simplify our discussion, we assume that R_{ij} are (nonnegative) integers for all $(i, j) \in E$.

In the following, we introduce some notions in graph theory which will facilitate the description of a point-to-point network. Temporarily regard an edge in G as a water pipe and the graph G as a network of water pipes. Fix a sink node t_l and assume that all the sink nodes except for t_l is blocked. Suppose water is generated at a constant rate at node s. We assume that the rate of water flow in each pipe does not exceed its capacity. We also assume that there is no leakage in the network, so that water is conserved at every node other than s and t_l in the sense that the total rate of water flowing into the node is equal to the total rate of water flowing out of the node. The water generated at node s is eventually drained at node t_l.

A *flow*

$$\mathbf{F} = [F_{ij} : (i, j) \in E] \tag{11.2}$$

in G from node s to node t_l with respect to rate constraints $\mathbf{R}$ is a valid assignment of a nonnegative integer F_{ij} to every edge $(i, j) \in E$ such that F_{ij} is equal to the rate of water flow in edge (i, j) under all the above assumptions. F_{ij} is referred to as the value of $\mathbf{F}$ on edge (i, j). Specifically, $\mathbf{F}$ is a flow in G from node s to node t_l if for all $(i, j) \in E$,

$$0 \leq F_{ij} \leq R_{ij}, \tag{11.3}$$

---

[2]Here the term "capacity" is used in the sense of graph theory.

and for all $i \in V$ except for s and t_l,

$$F_+(i) = F_-(i), \tag{11.4}$$

where

$$F_+(i) = \sum_{j:(j,i)\in E} F_{ji} \tag{11.5}$$

and

$$F_-(i) = \sum_{j:(i,j)\in E} F_{ij}. \tag{11.6}$$

In the above, $F_+(i)$ is the total flow into node i and $F_-(i)$ is the total flow out of node i, and (11.4) are called the *conservation* conditions.

Since the conservation conditions require that the resultant flow out of any node other than s and t_l is zero, it is intuitively clear and not difficult to show that the resultant flow out of node s is equal to the resultant flow into node t_l. This common value is called the *value* of $\mathbf{F}$. $\mathbf{F}$ is a *max-flow* from node s to node t_l in G with respect to rate constraints $\mathbf{R}$ if $\mathbf{F}$ is a flow from s to t_l whose value is greater than or equal to the value of any other flow from s to t_l.

A *cut* between node s and node t_l is a subset U of V such that $s \in U$ and $t_l \notin U$. Let

$$E_U = \{(i,j) \in E : i \in U \text{ and } j \notin U\} \tag{11.7}$$

be the set of edges across the cut U. The *capacity* of the cut U with respect to rate constraints $\mathbf{R}$ is defined as the sum of the capacities of all the edges across the cut, i.e.,

$$\sum_{(i,j)\in E_U} R_{ij}. \tag{11.8}$$

U is a *min-cut* between node s and node t_l if U is a cut between s and t_l whose capacity is less than or equal to the capacity of any other cut between s and t_l.

A min-cut between node s and node t_l can be thought of as a *bottleneck* between s and t_l. Therefore, it is intuitively clear that the value of a max-flow from node s to node t_l cannot exceed the capacity of a min-cut between node s and node t_l. The following theorem, known as the *max-flow min-cut theorem*, states that the capacity of a min-cut is always achievable. This theorem will play a key role in subsequent discussions in this chapter.

THEOREM 11.1 (MAX-FLOW MIN-CUT THEOREM [72]) *Let G be a graph with source node s and sink nodes $t_1, t_2, \cdots, t_L$, and $\mathbf{R}$ be the rate constraints. Then for $l = 1, 2, \cdots, L$, the value of a max-flow from node s to node t_l is equal to the capacity of a min-cut between node s and node t_l.*

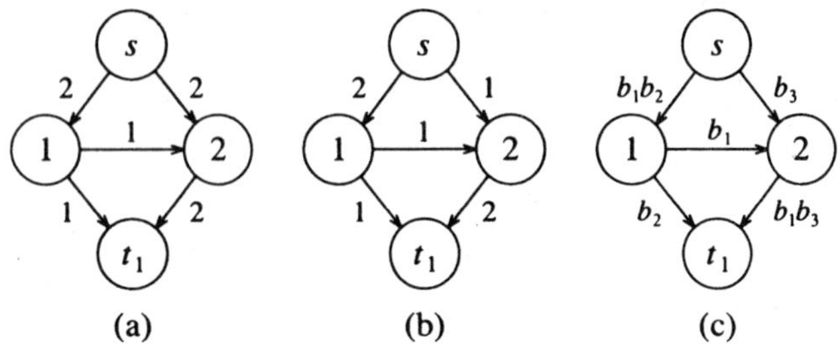

Figure 11.1. A one-sink network.

11.2 WHAT IS NETWORK CODING?

Let ω be the rate at which information is multicast from the source node s to the sink nodes $t_1, t_2, \cdots, t_L$ in a network G with rate constraints $\mathbf{R}$. We are naturally interested in the maximum possible value of ω. With a slight abuse of notation, we denote the value of a max-flow from the source node s to a sink node t_l by $\text{maxflow}(s, t_l)$. It is intuitive that

$$\omega \leq \text{maxflow}(s, t_l) \tag{11.9}$$

for all $l = 1, 2, \cdots, L$, i.e.,

$$\omega \leq \min_l \text{maxflow}(s, t_l). \tag{11.10}$$

This is called the *max-flow bound*, which will be proved in Section 11.4. Further, it will be proved in Section 11.5 that the max-flow bound can always be achieved. In this section, we first show that the max-flow bound can be achieved in a few simple examples. In these examples, the unit of information is the bit.

First, we consider the network in Figure 11.1 which has one sink node. Figure 11.1(a) shows the capacity of each edge. By identifying the min-cut to be $\{s, 1, 2\}$ and applying max-flow min-cut theorem, we see that

$$\text{maxflow}(s, t_1) = 3. \tag{11.11}$$

Therefore the flow in Figure 11.1(b) is a max-flow. In Figure 11.1(c), we show how we can send three bits b_1, b_2, and b_3 from node s to node t_1 based on the max-flow in Figure 11.1(b). Evidently, the max-flow bound is achieved.

In fact, we can easily see that the max-flow bound can always be achieved when there is only one sink node in the network. In this case, we only need to treat the information bits constituting the message as *physical entities* and route them through the network according to any fixed routing scheme. Eventually, all the bits will arrive at the sink node. Since the routing scheme is fixed, the

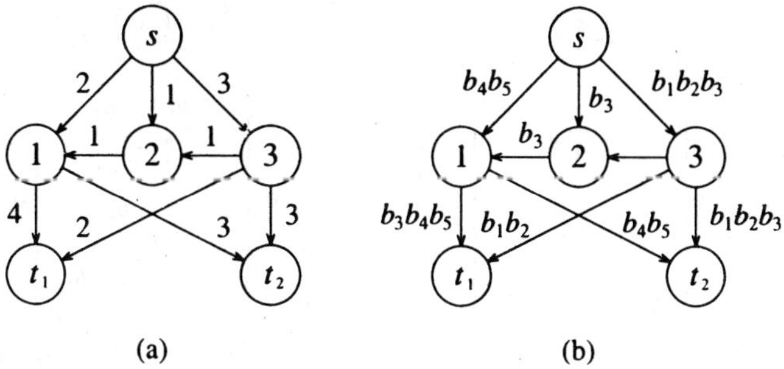

Figure 11.2. A two-sink network without coding.

sink node knows which bit is coming in from which edge, and the message can be recovered accordingly.

Next, we consider the network in Figure 11.2 which has two sink nodes. Figure 11.2(a) shows the capacity of each edge. It is easy to see that

$$\text{maxflow}(s, t_1) = 5 \tag{11.12}$$

and

$$\text{maxflow}(s, t_2) = 6. \tag{11.13}$$

So the max-flow bound asserts that we cannot send more than 5 bits to both t_1 and t_2. Figure 11.2(b) shows a scheme which sends 5 bits b_1, b_2, b_3, b_4, and b_5 to both t_1 and t_2. Therefore, the max-flow bound is achieved. In this scheme, b_1 and b_2 are replicated at node 3, b_3 is replicated at node s, while b_4 and b_5 are replicated node 1. Note that each bit is replicated exactly once in the network because two copies of each bit are needed to send to the two sink nodes.

We now consider the network in Figure 11.3 which again has two sink nodes. Figure 11.3(a) shows the capacity of each edge. It is easy to see that

$$\text{maxflow}(s, t_l) = 2, \tag{11.14}$$

$l = 1, 2$. So the max-flow bound asserts that we cannot send more than 2 bits to both t_1 and t_2. In Figure 11.3(b), we try to devise a routing scheme which sends 2 bits b_1 and b_2 to both t_1 and t_2. By symmetry, we send one bit on each output channel at node s. In this case, b_1 is sent on channel $(s, 1)$ and b_2 is sent on channel $(s, 2)$. At node i, $i = 1, 2$, b_i is replicated and the copies are sent on the two output channels. At node 3, since both b_1 and b_2 are received but there is only one output channel, we have to choose one of the two bits to send on the output channel $(3, 4)$. Suppose we choose b_1 as in Figure 11.3(b). Then the bit is replicated at node 4 and the copies are sent to nodes t_1 and t_2. At node t_2, both b_1 and b_2 are received. However, at node t_1, two copies of b_1 are

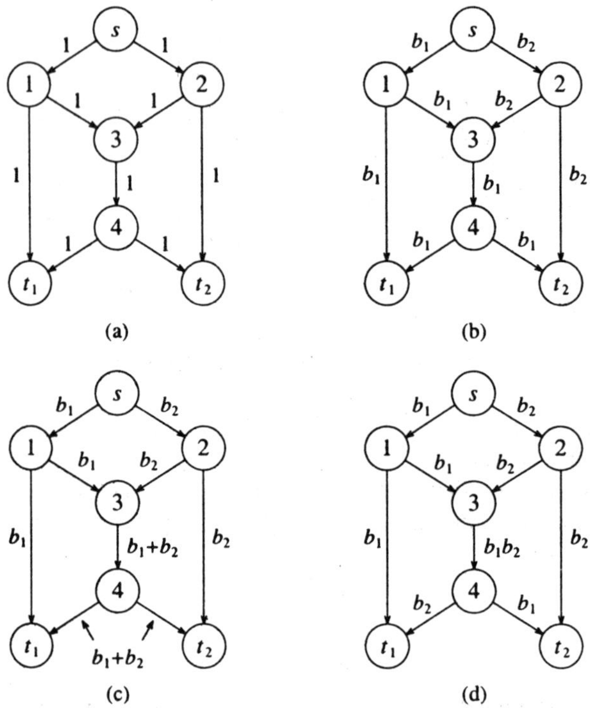

Figure 11.3. A two-sink network with coding.

received but b_2 cannot be recovered. Thus this routing scheme does not work. Similarly, if b_2 instead of b_1 is sent on channel $(3, 4)$, b_1 cannot be recovered at node t_2. Therefore, we conclude that for this network, the max-flow bound cannot be achieved by *routing* and *replication* of bits.

However, if coding is allowed at the nodes, it is actually possible to achieve the max-flow bound. Figure 11.3(c) shows a scheme which sends 2 bits b_1 and b_2 to both nodes t_1 and t_2, where '+' denotes modulo 2 addition. At node t_1, b_1 is received, and b_2 can be recovered by adding b_1 and $b_1 + b_2$, because

$$b_2 = b_1 + (b_1 + b_2). \tag{11.15}$$

Similarly, b_2 is received at node t_2, and b_1 can be recovered by adding b_2 and $b_1 + b_2$. Therefore, the max-flow bound is achieved. In this scheme, b_1 and b_2 are encoded into the codeword $b_1 + b_2$ which is then sent on channel $(3, 4)$. If coding at a node is not allowed, in order to send both b_1 and b_2 to nodes t_1 and t_2, at least one more bit has to be sent. Figure 11.3(d) shows such a scheme. In this scheme, however, the capacity of channel $(3, 4)$ is exceeded by 1 bit.

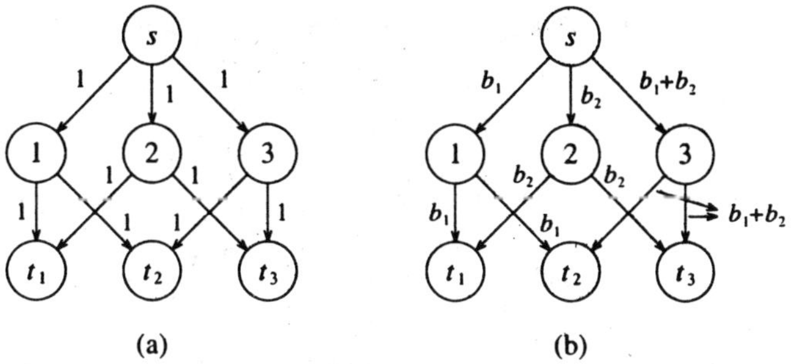

Figure 11.4. A diversity coding scheme.

Finally, we consider the network in Figure 11.4 which has three sink nodes. Figure 11.4(a) shows the capacity of each edge. It is easy to see that

$$\text{maxflow}(s, t_l) = 2 \tag{11.16}$$

for all l. In Figure 11.4(b), we show how to multicast 2 bits b_1 and b_2 to all the sink nodes. Therefore, the max-flow bound is achieved. Again, it is necessary to code at the nodes in order to multicast the maximum number of bits to all the sink nodes.

The network in Figure 11.4 is of special interest in practice because it is a special case of the *diversity coding* scheme used in commercial *disk arrays*, which are a kind of fault-tolerant data storage system. For simplicity, assume the disk array has three disks which are represented by nodes 1, 2, and 3 in the network, and the information to be stored are the bits b_1 and b_2. The information is encoded into three pieces, namely b_1, b_2, and $b_1 + b_2$, which are stored on the disks represented by nodes 1, 2, and 3, respectively. In the system, there are three decoders, represented by sink nodes t_1, t_2, and t_3, such that each of them has access to a distinct set of two disks. The idea is that when any one disk is out of order, the information can still be recovered from the remaining two disks. For example, if the disk represented by node 1 is out of order, then the information can be recovered by the decoder represented by the sink node t_3 which has access to the disks represented by node 2 and node 3. When all the three disks are functioning, the information can be recovered by any decoder.

The above examples reveal the surprising fact that although information may be generated at the source node in the form of raw bits[3] which are incompressible (see Section 4.3), coding within the network plays an essential role in

[3]Raw bits refer to i.i.d. bits, each distributing uniformly on $\{0, 1\}$.

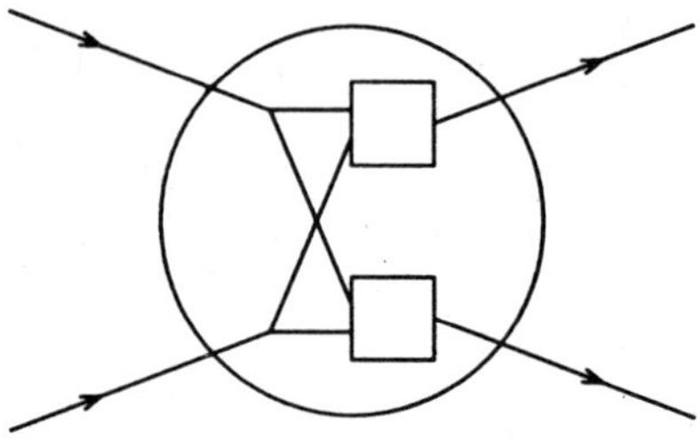

Figure 11.5. A node with two input channels and two output channels.

multicasting the information to two or more sink nodes. In particular, unless there is only one sink node in the network, it is generally not valid to regard information bits as physical entities.

We refer to coding at the nodes in a network as *network coding*. In the paradigm of network coding, a node consists of a set of encoders which are dedicated to the individual output channels. Each of these encoders encodes the information received from all the input channels (together with the information generated by the information source if the node is the source node) into codewords and sends them on the corresponding output channel. At a sink node, there is also a decoder which decodes the multicast information. Figure 11.5 shows a node with two input channels and two output channels.

In existing computer networks, each node functions as a *switch* in the sense that a data packet received from an input channel is replicated if necessary and sent on one or more output channels. When a data packet is multicast in the network, the packet is replicated and routed at the intermediate nodes according to a certain scheme so that eventually a copy of the packet is received by every sink node. Such a scheme can be regarded as a degenerate special case of network coding.

11.3 A NETWORK CODE

In this section, we describe a coding scheme for the network defined in the previous sections. Such a coding scheme is referred to as a *network code*.

Since the max-flow bound concerns only the values of max-flows from the source node s to the sink nodes, we assume without loss of generality that there is no loop in G, i.e., $(i, i) \notin E$ for all $i \in V$, because such edges do not increase the value of a max-flow from node s to a sink node. For the same reason, we assume that there is no input edge at node s, i.e., $(i, s) \notin E$ for all $i \in V \backslash \{s\}$.

We consider a block code of length n. Let X denote the information source and assume that x, the outcome of X, is obtained by selecting an index from a set $\mathcal{X}$ according to the uniform distribution. The elements in $\mathcal{X}$ are called messages. For $(i, j) \in E$, node i can send information to node j which depends only on the information previously received by node i.

An $(n, (\eta_{ij} : (i, j) \in E), \tau)$ α*-code* on a graph G is defined by the components listed below. The construction of an α-code from these components will be described after their definitions are given.

1) A positive integer K.

2)
$$u : \{1, 2, \cdots, K\} \to V \tag{11.17}$$
and
$$v : \{1, 2, \cdots, K\} \to V \tag{11.18}$$
such that $(u(k), v(k)) \in E$.

3) $A_k = \{1, 2, \cdots, |A_k|\}$, $1 \le k \le K$, such that
$$\prod_{k \in T_{ij}} |A_k| = \eta_{ij}, \tag{11.19}$$
where
$$T_{ij} = \{1 \le k \le K : (u(k), v(k)) = (i, j)\}. \tag{11.20}$$

4) If $u(k) = s$, then
$$f_k : \mathcal{X} \to A_k, \tag{11.21}$$
where
$$\mathcal{X} = \{1, 2, \cdots, \lceil 2^{n\tau} \rceil\}. \tag{11.22}$$
If $u(k) \neq s$, if
$$Q_k = \{1 \le k' < k : v(k') = u(k)\} \tag{11.23}$$
is nonempty, then
$$f_k : \prod_{k' \in Q_k} A_{k'} \to A_k; \tag{11.24}$$
otherwise, let f_k be an arbitrary constant taken from A_k.

5)
$$g_l : \prod_{k' \in W_l} A_{k'} \to \mathcal{X}, \tag{11.25}$$
$l = 1, 2, \cdots, L$, where
$$W_l = \{1 \le k \le K : v(k) = t_l\} \tag{11.26}$$
such that for all $l = 1, 2, \cdots, L$,
$$\tilde{g}_l(x) = x \tag{11.27}$$

for all $x \in \mathcal{X}$, where $\tilde{g}_l$ is the function from $\mathcal{X}$ to $\mathcal{X}$ induced inductively by $f_k, 1 \le k \le K$, and g_l such that $\tilde{g}_l(x)$ denotes the value of g_l as a function of x.

The quantity τ is the rate of the information source X, which is also the rate at which information is multicast from the source node to all the sink nodes. The $(n, (\eta_{ij} : (i,j) \in E), \tau)$ α-code is constructed from these components as follows. At the beginning of a coding session, the value of X is available to node s. During the coding session, there are K transactions which take place in chronological order, where each transaction refers to a node sending information to another node. In the kth transaction, node $u(k)$ encodes according to encoding function f_k and sends an index in A_k to node $v(k)$. The domain of f_k is the information received by node $u(k)$ so far, and we distinguish two cases. If $u(k) = s$, the domain of f_k is $\mathcal{X}$. If $u(k) \neq s$, Q_k gives the indices of all the previous transactions for which information was sent to node $u(k)$, so the domain of f_k is $\prod_{k' \in Q_k} A_{k'}$. The set T_{ij} gives the indices of all the transactions for which information is sent from node i to node j, so η_{ij} is the number of possible index tuples that can be sent from node i to node j during the coding session. Finally, W_l gives the indices of all the transactions for which information is sent to node t_l, and g_l is the decoding function at node t_l which recovers x with zero error.

11.4 THE MAX-FLOW BOUND

In this section, we formally state and prove the max-flow bound discussed in Section 11.2. The achievability of this bound will be proved in the next section.

DEFINITION 11.2 *For a graph G with rate constraints $\mathbf{R}$, an information rate $\omega \ge 0$ is asymptotically achievable if for any $\epsilon > 0$, there exists for sufficiently large n an $(n, (\eta_{ij} : (i,j) \in E), \tau)$ α-code on G such that*

$$n^{-1} \log_2 \eta_{ij} \le R_{ij} + \epsilon \tag{11.28}$$

for all $(i,j) \in E$, where $n^{-1} \log_2 \eta_{ij}$ is the average bit rate of the code on channel (i,j), and

$$\tau \ge \omega - \epsilon. \tag{11.29}$$

For brevity, an asymptotically achievable information rate will be referred to as an achievable information rate.

Remark It follows from the above definition that if $\omega \ge 0$ is achievable, then ω' is also achievable for all $0 \le \omega' \le \omega$. Also, if $\omega^{(k)}, k \ge 1$ is achievable, then $\omega = \lim_{k \to \infty} \omega^{(k)}$ is also achievable. Therefore, the set of all achievable information rates is closed and fully characterized by the maximum value in the set.

THEOREM 11.3 (MAX-FLOW BOUND) *For a graph G with rate constraints $\mathbf{R}$, if ω is achievable, then*

$$\omega \leq \min_l \text{maxflow}(s, t_l). \tag{11.30}$$

Proof It suffices to prove that for a graph G with rate constraints $\mathbf{R}$, if for any $\epsilon > 0$ there exists for sufficiently large n an $(n, (\eta_{ij} : (i,j) \in E), \tau)$ α-code on G such that

$$n^{-1} \log_2 \eta_{ij} \leq R_{ij} + \epsilon \tag{11.31}$$

for all $(i,j) \in E$, and

$$\tau \geq \omega - \epsilon, \tag{11.32}$$

then ω satisfies (11.30).

Consider such a code for a fixed ϵ and a sufficiently large n, and consider any $l = 1, 2, \cdots, L$ and any cut U between node s and node t_l. Let

$$w_j(x) = (\tilde{f}_k(x) : k \in \cup_{i \in V} T_{ij}), \tag{11.33}$$

where $x \in \mathcal{X}$ and $\tilde{f}_k$ is the function from $\mathcal{X}$ to A_k induced inductively by $f_{k'}, 1 \leq k' \leq k$, such that $\tilde{f}_k(x)$ denotes the value of f_k as a function of x. $w_j(x)$ is all the information known by node j during the whole coding session when the message is x. Since $\tilde{f}_k(x)$ is a function of the information previously received by node $u(k)$, we see inductively that $w_{t_l}(x)$ is a function of $\tilde{f}_k(x), k \in \cup_{(i,j) \in E_U} T_{ij}$, where

$$E_U = \{(i,j) \in E : i \in U \text{ and } j \notin U\} \tag{11.34}$$

is the set of edges across the cut U as previously defined. Since x can be determined at node t_l, we have

$$\begin{aligned}
H(X) &\leq H(X, w_{t_l}(X)) && (11.35)\\
&= H(w_{t_l}(X)) && (11.36)\\
&\leq H\left(\tilde{f}_k(X), k \in \bigcup_{(i,j) \in E_U} T_{ij}\right) && (11.37)\\
&\leq \sum_{(i,j) \in E_U} \sum_{k \in T_{ij}} H(\tilde{f}_k(X)) && (11.38)\\
&\leq \sum_{(i,j) \in E_U} \sum_{k \in T_{ij}} \log_2 |A_k| && (11.39)\\
&= \sum_{(i,j) \in E_U} \log_2 \left(\prod_{k \in T_{ij}} |A_k| \right) && (11.40)\\
&= \sum_{(i,j) \in E_U} \log_2 \eta_{ij}. && (11.41)
\end{aligned}$$

Thus

$$\begin{aligned}
\omega - \epsilon &\leq \tau && (11.42)\\
&\leq n^{-1}\log_2\lceil 2^{n\tau}\rceil && (11.43)\\
&= n^{-1}\log_2|\mathcal{X}| && (11.44)\\
&\leq n^{-1}H(X) && (11.45)\\
&\leq \sum_{(i,j)\in E_U} n^{-1}\log_2\eta_{ij} && (11.46)\\
&\leq \sum_{(i,j)\in E_U} (R_{ij}+\epsilon) && (11.47)\\
&\leq \sum_{(i,j)\in E_U} R_{ij} + |E|\epsilon, && (11.48)
\end{aligned}$$

where (11.46) follows from (11.41). Minimizing the right hand side over all U, we have

$$\omega - \epsilon \leq \min_U \sum_{(i,j)\in E_U} R_{ij} + |E|\epsilon. \tag{11.49}$$

The first term on the right hand side is the capacity of a min-cut between node s and node t_l. By the max-flow min-cut theorem, it is equal to the value of a max-flow from node s to node t_l, i.e., $\text{maxflow}(s, t_l)$. Letting $\epsilon \to 0$, we obtain

$$\omega \leq \text{maxflow}(s, t_l). \tag{11.50}$$

Since this upper bound on ω holds for all $l = 1, 2, \cdots, L$,

$$\omega \leq \min_l \text{maxflow}(s, t_l). \tag{11.51}$$

The theorem is proved. □

Remark 1 The max-flow bound cannot be proved by a straightforward application of the data processing theorem because for a cut U between node s and node t_l, the random variables X, $(w_i(X) : i \in B_U)$, $(w_j(X) : j \in B'_U)$, and $w_{t_l}(X)$ in general do not form a Markov chain in this order, where

$$B_U = \{i \in V : (i, j) \in E_U \text{ for some } j \in V\} \tag{11.52}$$

and

$$B'_U = \{j \in V : (i, j) \in E_U \text{ for some } i \in V\} \tag{11.53}$$

are the two sets of nodes on the boundary of the cut U. The time parameter and the causality of an α-code must be taken into account in proving the max-flow bound.

Remark 2 Even if we allow an arbitrarily small probability of decoding error in the usual Shannon sense, by modifying our proof by means of a standard application of Fano's inequality, it can be shown that it is still necessary for ω to satisfy (11.51). The details are omitted here.

11.5 ACHIEVABILITY OF THE MAX-FLOW BOUND

The max-flow bound has been proved in the last section. The achievability of this bound is stated in the following theorem. This theorem will be proved after the necessary preliminaries are presented.

THEOREM 11.4 *For a graph G with rate constraints $\mathbf{R}$, if*

$$\omega \leq \min_l \text{maxflow}(s, t_l), \tag{11.54}$$

then ω is achievable.

A *directed path* in a graph G is a finite non-null sequence of nodes

$$v_1, v_2, \cdots, v_{m-1}, v_m \tag{11.55}$$

such that $(v_i, v_{i+1}) \in E$ for all $i = 1, 2, \cdots, m-1$. The edges (v_i, v_{i+1}), $i = 1, 2, \cdots, m-1$, are referred to as the edges on the directed path. Such a sequence is called a directed path from v_1 to v_m. If $v_1 = v_m$, then it is called a *directed cycle*. If there exists a directed cycle in G, then G is *cyclic*, otherwise G is *acyclic*.

In Section 11.5.1, we will first prove Theorem 11.4 for the special case when the network is acyclic. Acyclic networks are easier to handle because the nodes in the network can be ordered in a way which allows coding at the nodes to be done in a sequential and consistent manner. The following proposition describes such an order, and the proof shows how it can be obtained. In Section 11.5.2, Theorem 11.4 will be proved in full generality.

PROPOSITION 11.5 *If $G = (V, E)$ is a finite acyclic graph, then it is possible to order the nodes of G in a sequence such that if there is an edge from node i to node j, then node i is before node j in the sequence.*

Proof We partition V into subsets $V_1, V_2, \cdots$, such that node i is in V_k if and only if the length of a longest directed path ending at node i is equal to k. We claim that if node i is in $V_{k'}$ and node j is in V_k such that $k \leq k'$, then there exists no directed path from node i to node j. This is proved by contradiction as follows. Assume that there exists a directed path from node i to node j. Since there exists a directed path ending at node i of length k', there exists a directed path ending at node j containing node i whose length is at least $k' + 1$.

Since node j is in V_k, the length of a longest directed path ending at node j is equal to k. Then

$$k' + 1 \leq k. \tag{11.56}$$

However, this is a contradiction because

$$k' + 1 > k' \geq k. \tag{11.57}$$

Therefore, we conclude that there exists a directed path from a node in $V_{k'}$ to a node in V_k, then $k' < k$.

Hence, by listing the nodes of G in a sequence such that the nodes in V_k appear before the nodes in $V_{k'}$ if $k < k'$, where the order of the nodes within each V_k is arbitrary, we obtain an order of the nodes of G with the desired property. □

EXAMPLE 11.6 *Consider ordering the nodes in the acyclic graph in Figure 11.3 by the sequence*

$$s, 2, 1, 3, 4, t_2, t_1. \tag{11.58}$$

It is easy to check that in this sequence, if there is a directed path from node i to node j, then node i appears before node j.

11.5.1 ACYCLIC NETWORKS

In this section, we prove Theorem 11.4 for the special case when the graph G is acyclic. Let the vertices in G be labeled by $0, 1, \cdots, |V| - 1$ in the following way. The source s has the label 0. The other vertices are labeled in a way such that for $1 \leq j \leq |V| - 1$, $(i, j) \in E$ implies $i < j$. Such a labeling is possible by Proposition 11.5. We regard $s, t_1, \cdots, t_L$ as aliases of the corresponding vertices.

We will consider an $(n, (\eta_{ij} : (i,j) \in E), \tau)$ *β-code* on the graph G defined by

1) for all $(s, j) \in E$, an encoding function

$$f_{sj} : \mathcal{X} \to \{1, 2, \cdots, \eta_{sj}\}, \tag{11.59}$$

where

$$\mathcal{X} = \{1, 2, \cdots, \lceil 2^{n\tau} \rceil\}; \tag{11.60}$$

2) for all $(i, j) \in E$ such that $i \neq s$, an encoding function

$$f_{ij} : \prod_{i':(i',i)\in E} \{1, 2, \cdots, \eta_{i'i}\} \to \{1, 2, \cdots, \eta_{ij}\} \tag{11.61}$$

(if $\{i' : (i', i) \in E\}$ is empty, we adopt the convention that f_{ij} is an arbitrary constant taken from $\{1, 2, \cdots, \eta_{ij}\}$);

3) for all $l = 1, 2, \cdots, L$, a decoding function

$$g_l : \prod_{i:(i,t_l)\in E} \{1, 2, \cdots, \eta_{it_l}\} \to \mathcal{X} \tag{11.62}$$

such that

$$\tilde{g}_l(x) = x \tag{11.63}$$

for all $x \in \mathcal{X}$. (Recall that $\tilde{g}_l(x)$ denotes the value of g_l as a function of x.)

In the above, f_{ij} is the encoding function for edge (i, j), and g_l is the decoding function for sink node t_l. In a coding session, f_{ij} is applied before $f_{i'j'}$ if $i < i'$, and f_{ij} is applied before $f_{ij'}$ if $j < j'$. This defines the order in which the encoding functions are applied. Since $i' < i$ if $(i', i) \in E$, a node does not encode until all the necessary information is received on the input channels. A β-code is a special case of an α-code defined in Section 11.3.

Assume that ω satisfies (11.54) with respect to rate constraints $\mathbf{R}$. It suffices to show that for any $\epsilon > 0$, there exists for sufficiently large n an $(n, (\eta_{ij} : (i, j) \in E), \omega - \epsilon)$ β-code on G such that

$$n^{-1} \log_2 \eta_{ij} \leq R_{ij} + \epsilon \tag{11.64}$$

for all $(i, j) \in E$. Instead, we will show the existence of an $(n, (\eta_{ij} : (i, j) \in E), \omega)$ β-code satisfying the same set of conditions. This will be done by constructing a random code. In constructing this code, we temporarily replace $\mathcal{X}$ by

$$\mathcal{X}' = \{1, 2, \cdots, \lceil C2^{n\omega} \rceil\}, \tag{11.65}$$

where C is any constant greater than 1. Thus the domain of f_{sj} is expanded from $\mathcal{X}$ to $\mathcal{X}'$ for all $(s, j) \in E$.

We now construct the encoding functions as follows. For all $j \in V$ such that $(s, j) \in E$, for all $x \in \mathcal{X}'$, let $f_{sj}(x)$ be a value selected independently from the set $\{1, 2, \cdots, \eta_{sj}\}$ according to the uniform distribution. For all $(i, j) \in E, i \neq s$, and for all

$$z \in \prod_{i':(i',i)\in E} \{1, 2, \cdots, \eta_{i'i}\}, \tag{11.66}$$

let $f_{ij}(z)$ be a value selected independently from the set $\{1, 2, \cdots, \eta_{ij}\}$ according to the uniform distribution.

Let

$$z_s(x) = x, \tag{11.67}$$

and for $j \in V, j \neq s$, let

$$z_j(x) = (\tilde{f}_{ij}(x), (i, j) \in E), \tag{11.68}$$

where $x \in \mathcal{X}'$ and $\tilde{f}_{ij}(x)$ denotes the value of f_{ij} as a function of x. $z_j(x)$ is all the information received by node j during the coding session when the message is x. For distinct $x, x' \in \mathcal{X}$, x and x' are indistinguishable at sink t_l if and only if $z_{t_l}(x) = z_{t_l}(x')$. For all $x \in \mathcal{X}$, define

$$F(x) = \begin{cases} 1 & \text{if for some } l = 1, 2, \cdots, L, \text{ there exists } x' \in \mathcal{X}, \\ & x' \neq x, \text{ such that } z_{t_l}(x) = z_{t_l}(x'), \\ 0 & \text{otherwise.} \end{cases} \tag{11.69}$$

$F(x)$ is equal to 1 if and only if x cannot be uniquely determined at at least one of the sink nodes. Now fix $x \in \mathcal{X}$ and $1 \leq l \leq L$. Consider any $x' \in \mathcal{X}$ not equal to x and define the sets

$$U_0 = \{i \in V : z_i(x) \neq z_i(x')\} \tag{11.70}$$

and

$$U_1 = \{i \in V : z_i(x) = z_i(x')\}. \tag{11.71}$$

U_0 is the set of nodes at which the two messages x and x' are distinguishable, and U_1 is the set of nodes at which x and x' are indistinguishable. Obviously, $s \in U_0$.

Now suppose $z_{t_l}(x) = z_{t_l}(x')$. Then $U_0 = U$ for some $U \subset V$, where $s \in U$ and $t_l \notin U$, i.e., U is a cut between node s and node t_l. For any $(i,j) \in E$,

$$\Pr\{\tilde{f}_{ij}(x) = \tilde{f}_{ij}(x') | z_i(x) \neq z_i(x')\} = \eta_{ij}^{-1}. \tag{11.72}$$

Therefore,

$$\begin{aligned} & \Pr\{U_0 = U\} \\ &= \Pr\{U_0 = U, U_0 \supset U\} && (11.73) \\ &= \Pr\{U_0 = U | U_0 \supset U\} \Pr\{U_0 \supset U\} && (11.74) \\ &\leq \Pr\{U_0 = U | U_0 \supset U\} && (11.75) \\ &= \prod_{(i,j) \in E_U} \Pr\{\tilde{f}_{ij}(x) = \tilde{f}_{ij}(x') | z_i(x) \neq z_i(x')\} && (11.76) \\ &= \prod_{(i,j) \in E_U} \eta_{ij}^{-1}, && (11.77) \end{aligned}$$

where

$$E_U = \{(i,j) \in E : i \in U, j \notin U\} \tag{11.78}$$

is the set of all the edges across the cut U as previously defined.

Let ϵ be any fixed positive real number. For all $(i,j) \in E$, take η_{ij} such that

$$R_{ij} + \zeta \leq n^{-1} \log \eta_{ij} \leq R_{ij} + \epsilon \tag{11.79}$$

for some $0 < \zeta < \epsilon$. Then

$$\begin{aligned}
\Pr\{U_0 = U\} &\leq \prod_{(i,j)\in E_U} \eta_{ij}^{-1} && (11.80)\\
&\leq \prod_{(i,j)\in E_U} 2^{-n(R_{ij}+\zeta)} && (11.81)\\
&= 2^{-n\left(|E_U|\zeta+\sum_{(i,j)\in E_U} R_{ij}\right)} && (11.82)\\
&\overset{a)}{\leq} 2^{-n\left(\zeta+\sum_{(i,j)\in E_U} R_{ij}\right)} && (11.83)\\
&\overset{b)}{\leq} 2^{-n(\zeta+\text{maxflow}(s,t_l))} && (11.84)\\
&\overset{c)}{\leq} 2^{-n(\omega+\zeta)}, && (11.85)
\end{aligned}$$

where

a) follows because $|E_U| \geq 1$;

b) follows because

$$\sum_{(i,j)\in E_U} R_{ij} \geq \min_{U'} \sum_{(i,j)\in E_U} R_{ij} = \text{maxflow}(s, t_l), \qquad (11.86)$$

where U' is a cut between node s and node t_l, by the max-flow min-cut theorem;

c) follows from (11.54).

Note that this upper bound does not depend on U. Since U is some subset of V and V has $2^{|V|}$ subsets,

$$\begin{aligned}
&\Pr\{z_{t_l}(x) = z_{t_l}(x')\}\\
&\quad= \Pr\{U_0 = U \text{ for some cut } U \text{ between node } s \text{ and node } t_l\} && (11.87)\\
&\quad\leq 2^{|V|}2^{-n(\omega+\zeta)} && (11.88)
\end{aligned}$$

by the union bound. Further,

$$\begin{aligned}
&\Pr\{z_{t_l}(x) = z_{t_l}(x') \text{ for some } x' \in \mathcal{X}', x' \neq x\}\\
&\quad\leq (|\mathcal{X}'| - 1)2^{|V|}2^{-n(\omega+\zeta)} && (11.89)\\
&\quad< C2^{n\omega}2^{|V|}2^{-n(\omega+\zeta)} && (11.90)\\
&\quad= C2^{|V|}2^{-n\zeta}, && (11.91)
\end{aligned}$$

where (11.89) follows from the union bound and (11.90) follows from (11.65). Therefore,

$E[F(x)]$

$$
\begin{aligned}
&= \Pr\{F(x) = 1\} && (11.92)\\
&= \Pr\left\{\bigcup_{l=1}^{L}\{z_{t_l}(x) = z_{t_l}(x') \text{ for some } x' \in \mathcal{X}', x' \neq x\}\right\} && (11.93)\\
&< LC2^{|V|}2^{-n\zeta} && (11.94)\\
&= \delta(n, \zeta) && (11.95)
\end{aligned}
$$

by the union bound, where

$$\delta(n, \zeta) = LC2^{|V|}2^{-n\zeta}. \tag{11.96}$$

Now the total number of messages which can be uniquely determined at all the sink nodes is equal to

$$\sum_{x \in \mathcal{X}'} (1 - F(x)). \tag{11.97}$$

By taking expectation for the random code we have constructed, we have

$$
\begin{aligned}
E \sum_{x \in \mathcal{X}'} (1 - F(x)) &= \sum_{x \in \mathcal{X}'} (1 - E[F(x)]) && (11.98)\\
&> \sum_{x \in \mathcal{X}'} (1 - \delta(n, \zeta)) && (11.99)\\
&\geq (1 - \delta(n, \zeta))C2^{n\omega}, && (11.100)
\end{aligned}
$$

where (11.99) follows from (11.95), and the last step follows from (11.65). Hence, there exists a deterministic code for which the number of messages which can be uniquely determined at all the sink nodes is at least

$$(1 - \delta(n, \zeta))C2^{n\omega}, \tag{11.101}$$

which is greater than $2^{n\omega}$ for n sufficiently large since $\delta(n, \zeta) \to 0$ as $n \to \infty$. Let $\mathcal{X}$ to be any set of $\lceil 2^{n\omega} \rceil$ such messages in $\mathcal{X}'$. For $l = 1, 2, \cdots, L$ and

$$z \in \prod_{i' : (i', t_l) \in E} \{1, 2, \cdots, \eta_{i't_l}\}, \tag{11.102}$$

upon defining

$$g_l(z) = x \tag{11.103}$$

where $x \in \mathcal{X}$ such that

$$z = z_{t_l}(x), \tag{11.104}$$

we have obtained a desired $(n, (\eta_{ij} : (i, j) \in E), \omega)$ β-code. Hence, Theorem 11.4 is proved for the special case when G is acyclic.

11.5.2 CYCLIC NETWORKS

For cyclic networks, there is no natural ordering of the nodes which allows coding in a sequential manner as for acyclic networks. In this section, we will prove Theorem 11.4 in full generality, which involves the construction of a more elaborate code on a *time-parametrized* acyclic graph.

Consider the graph G with rate constraints $\mathbf{R}$ we have defined in the previous sections but without the assumption that G is acyclic. We first construct a time-parametrized graph $G^* = (V^*, E^*)$ from the graph G. The set V^* consists of $\Lambda + 1$ layers of nodes, each of which is a copy of V. Specifically,

$$V^* = \bigcup_{\lambda=0}^{\Lambda} V^{(\lambda)}, \tag{11.105}$$

where

$$V^{(\lambda)} = \{i^{(\lambda)} : i \in V\}. \tag{11.106}$$

As we will see later, λ is interpreted as the time parameter. The set E^* consists of the following three types of edges:

1. $(s^{(0)}, s^{(\lambda)}), 1 \leq \lambda \leq \Lambda$
2. $(t_l^{(\lambda)}, t_l^{(\Lambda)}), 0 \leq \lambda \leq \Lambda - 1$
3. $(i^{(\lambda)}, j^{(\lambda+1)}), (i, j) \in E, 0 \leq \lambda \leq \Lambda - 1.$

For G^*, let $s^* = s^{(0)}$ be the source node, and let $t_l^* = t_l^{(\Lambda)}$ be a sink node which corresponds to the sink node t_l in G, $l = 1, 2, \cdots, L$. Clearly, G^* is acyclic because each edge in G^* ends at a vertex in a layer with a larger index.

Let $R^*_{uv}, (u, v) \in E^*$ be the capacity of an edge $(u, v) \in E^*$, where

$$R^*_{uv} = \begin{cases} R_{ij} & \text{if } (u,v) = (i^{(\lambda)}, j^{(\lambda+1)}) \text{ for some } (i,j) \in E \\ & \text{and } 0 \leq \lambda \leq \Lambda - 1, \\ \infty & \text{otherwise,} \end{cases} \tag{11.107}$$

and let

$$\mathbf{R}^* = [R^*_{uv}, (u, v) \in E^*], \tag{11.108}$$

which is referred to as the rate constraints for graph G^*.

EXAMPLE 11.7 *In Figure 11.6, we show the graph G^* for $\Lambda = 5$ for the graph G in Figure 11.1(a).*

LEMMA 11.8 *For $l = 1, 2, \cdots, L$, there exists a max-flow $\mathbf{F}$ in graph G from node s to node t_l which is expressible as the sum of a number of flows (from*

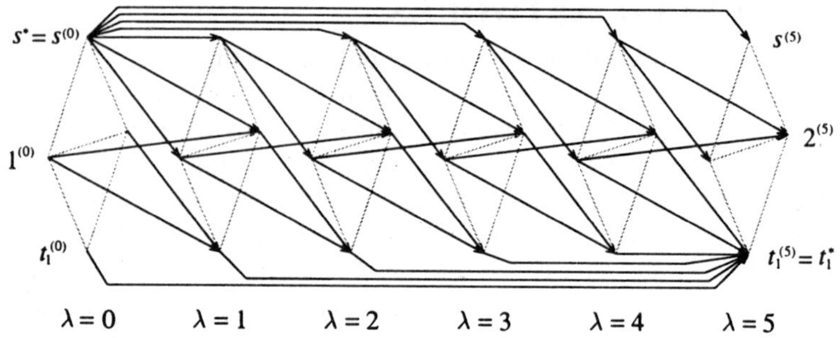

Figure 11.6. The graph G^* with $\Lambda = 5$ for the graph G in Figure 11.1(a).

node s to node t_l), each of them consisting of a simple path (i.e., a directed path without cycle) from node s to node t_l only.

Proof Let $\mathbf{F}$ be a flow from node s to node t_l in the graph G. If a directed path in G is such that the value of $\mathbf{F}$ on every edge on the path is strictly positive, then we say that the path is a positive directed path. If a positive directed path forms a cycle, it is called a positive directed cycle.

Suppose $\mathbf{F}$ contains a positive directed cycle. Let y be the minimum value of $\mathbf{F}$ on an edge in the cycle. By subtracting y from the value of $\mathbf{F}$ on every edge in the cycle, we obtain another flow $\mathbf{F}'$ from node s to node t_l such that the resultant flow out of each node is the same as $\mathbf{F}$. Consequently, the values of $\mathbf{F}$ and $\mathbf{F}'$ are the same. Note that the positive directed cycle in $\mathbf{F}$ is eliminated in $\mathbf{F}'$. By repeating this procedure if necessary, one can obtain a flow from node s to node t_l containing no positive directed cycle such that the value of this flow is the same as the value of $\mathbf{F}$.

Let $\mathbf{F}$ be a max-flow from node s to node t_l in G. From the foregoing, we assume without loss of generality that $\mathbf{F}$ does not contain a positive directed cycle. Let P_1 be any positive directed path from s to t_l in $\mathbf{F}$ (evidently P_1 is simple), and let c_1 be the minimum value of $\mathbf{F}$ on an edge along P_1. Let $\tilde{\mathbf{F}}^1$ be the flow from node s to node t_l along P_1 with value c_1. Subtracting $\tilde{\mathbf{F}}^1$ from $\mathbf{F}$, $\mathbf{F}$ is reduced to $\mathbf{F} - \tilde{\mathbf{F}}^1$, a flow from node s to node t_l which does not contain a positive directed cycle. Then we can apply the same procedure repeatedly until $\mathbf{F}$ is reduced to the *zero flow*[4], and $\mathbf{F}$ is seen to be the sum of all the flows which have been subtracted from $\mathbf{F}$. The lemma is proved. □

EXAMPLE 11.9 *The max-flow $\mathbf{F}$ from node s to node t_1 in Figure 11.1(b), whose value is 3, can be expressed as the sum of the three flows in Figure 11.7,*

[4]The process must terminate because the components of $\mathbf{F}$ are integers.

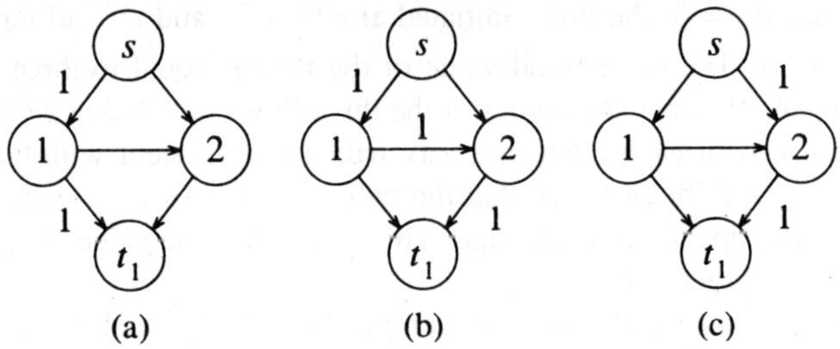

Figure 11.7. The max-flow in Figure 11.1(b) decomposed into three flows.

each of them consisting of a simple path from node s to node t_l only. The value of each of these flows is equal to 1.

LEMMA 11.10 *For $l = 1, 2, \cdots, L$, if the value of a max-flow from s to t_l in G is greater than or equal to ω, then the value of a max-flow from s^* to t_l^* in G^* is greater than or equal to $(\Lambda - d_l + 1)\omega$, where d_l is the maximum length of a simple path from s to t_l.*

We first use the last two examples to illustrate the idea of the proof of this lemma which will be given next. For the graph G in Figure 11.1(a), d_1, the maximum length of a simple path from node s to node t_1, is equal to 3. In Example 11.9, we have expressed the max-flow $\mathbf{F}$ in Figure 11.1(b), whose value is equal to 3, as the sum of the three flows in Figure 11.7. Based on these three flows, we now construct a flow from s^* to t_1^* in the graph G^* in Figure 11.8. In this figure, copies of these three flows are initiated at nodes $s^{(0)}$, $s^{(1)}$, and $s^{(2)}$, and they traverse in G^* as shown. The reader should think of the flows initiated at nodes $s^{(1)}$ and $s^{(2)}$ as being generated at node $s^* = s^{(0)}$ and delivered to nodes $s^{(1)}$ and $s^{(2)}$ via edges $(s^*, s^{(1)})$ and $(s^*, s^{(2)})$, respectively, whose capacities are infinite. These are not shown in the figure for the sake of

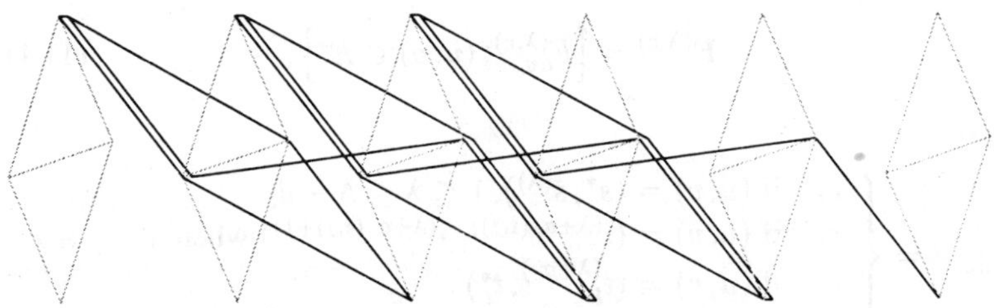

Figure 11.8. A flow from s^* to t_1^* in the graph G^* in Figure 11.6.

clarity. Since $d_1 = 3$, the flows initiated at $s^{(0)}$, $s^{(1)}$, and $s^{(2)}$ all terminate at $t_1^{(\lambda)}$ for $\lambda \leq 5$. Therefore, total value of the flow is equal to three times the value of $\mathbf{F}$, i.e., 9. Since the copies of the three flows initiated at $s^{(0)}, s^{(1)}$, and $s^{(2)}$ interleave with each other in a way which is consistent with the original flow F, it is not difficult to see that the value of the flow so constructed does not exceed the capacity in each edge. This shows that the value of a max-flow from s^* to t^* is at least 9.

We now give a formal proof for Lemma 11.10. The reader may skip this proof without affecting further reading of this section.

Proof of Lemma 11.10 For a fixed $1 \leq l \leq L$, let $\mathbf{F}$ be a max-flow from node s to node t_l in G with value ω such that $\mathbf{F}$ does not contain a positive directed cycle. Using Lemma 11.8, we can write

$$\mathbf{F} = \tilde{\mathbf{F}}^1 + \tilde{\mathbf{F}}^2 + \cdots + \tilde{\mathbf{F}}^\varphi, \tag{11.109}$$

where $\tilde{\mathbf{F}}^r$, $r = 1, 2, \cdots, \varphi$, contains a positive simple path P_r from node s to node t_l only. Specifically,

$$\tilde{\mathbf{F}}^r_{ij} = \begin{cases} c_r & \text{if } (i,j) \in P_r \\ 0 & \text{otherwise,} \end{cases} \tag{11.110}$$

where

$$c_1 + c_2 + \cdots + c_\varphi = \omega. \tag{11.111}$$

Let q_r be the length of P_r. For an edge $(i,j) \in P_r$, let $a_r(i,j)$ be the distance of node i from node s along P_r. Clearly,

$$a_r(i,j) \leq q_r - 1 \tag{11.112}$$

and

$$q_r \leq d_l. \tag{11.113}$$

Now for $0 \leq \lambda \leq \Lambda - d_l$, define

$$\mathbf{F}^{(\lambda,r)} = \left[F_{uv}^{(\lambda,r)}, (u,v) \in E^*\right], \tag{11.114}$$

where

$$F_{uv}^{(\lambda,r)} = \begin{cases} c_r & \text{if } (u,v) = (s^*, s^{(\lambda)}), 1 \leq \lambda \leq \Lambda - d_l \\ c_r & \text{if } (u,v) = (i^{(\lambda + a_r(i,j))}, j^{(\lambda + a_r(i,j)+1)}) \text{ where } (i,j) \in P_r \\ c_r & \text{if } (u,v) = (t_l^{(\lambda+q_r)}, t_l^*) \\ 0 & \text{otherwise.} \end{cases} \tag{11.115}$$

Since

$$\lambda + a_r(i,j) + 1 \leq \lambda + q_r \leq \lambda + d_l \leq \Lambda, \tag{11.116}$$

the second case and the third case in (11.115) and hence $\mathbf{F}^{(\lambda,r)}$ is well-defined for $0 \leq \lambda \leq \Lambda - d_l$. $\mathbf{F}^{(\lambda,r)}$ is a flow from node s^* to node t_l^* in G^* derived from the flow $\tilde{\mathbf{F}}^r$ in G as follows. A flow of c_r is generated at node s^* and enters the λth layer of nodes from $s^{(\lambda)}$. Then the flow traverses consecutive layers of nodes by emulating the path P_r in G until it eventually terminate at node t_l^* via node $t_l^{(\lambda+q_r)}$. Based on $\mathbf{F}^{(\lambda,r)}$, we construct

$$\mathbf{F}^{(\lambda)} = \sum_{r=1}^{\varphi} \mathbf{F}^{(\lambda,r)}, \tag{11.117}$$

and

$$\mathbf{F}^* = \sum_{\lambda=0}^{\Lambda-d_l} \mathbf{F}^{(\lambda)}. \tag{11.118}$$

We will prove that $\mathbf{F}^* \leq \mathbf{R}^*$ componentwise. Then $\mathbf{F}^*$ is a flow from node s^* to node t_l^* in G^*, and from (11.111), its value is given by

$$\sum_{\lambda=0}^{\Lambda-d_l} \sum_{r=1}^{\varphi} c_r = \sum_{\lambda=0}^{\Lambda-d_l} \omega = (\Lambda - d_l + 1)\omega. \tag{11.119}$$

This would imply that the value of a max-flow from node s^* to node t_l^* in G^* is at least $(\Lambda - d_l + 1)\omega$, and the lemma is proved.

Toward proving that $\mathbf{F}^* \leq \mathbf{R}^*$, we only need to consider $(u, v) \in E^*$ such that

$$(u, v) = (i^{(\lambda)}, j^{(\lambda+1)}) \tag{11.120}$$

for some $(i, j) \in E$ and $0 \leq \lambda \leq \Lambda - 1$, because R_{uv}^* is infinite otherwise (cf. (11.107)). For notational convenience, we adopt the convention that

$$\mathbf{F}^{(\lambda,r)} = 0 \tag{11.121}$$

for $\lambda < 0$. Now for $0 \leq \lambda \leq \Lambda - 1$ and $(i, j) \in E$,

$$\begin{aligned} F_{i^{(\lambda)}j^{(\lambda+1)}}^* &= \sum_{\mu=0}^{\Lambda-d_l} F_{i^{(\lambda)}j^{(\lambda+1)}}^{(\mu)} && (11.122) \\ &= \sum_{\mu=0}^{\Lambda-d_l} \sum_{r=1}^{\varphi} F_{i^{(\lambda)}j^{(\lambda+1)}}^{(\mu,r)} && (11.123) \\ &= \sum_{r=1}^{\varphi} \sum_{\mu=0}^{\Lambda-d_l} F_{i^{(\lambda)}j^{(\lambda+1)}}^{(\mu,r)} && (11.124) \\ &= \sum_{r=1}^{\varphi} F_{i^{(\lambda)}j^{(\lambda+1)}}^{(\lambda-a_r(i,j),r)} && (11.125) \end{aligned}$$

$$\leq \sum_{r=1}^{\varphi} \tilde{F}_{ij}^{r} \tag{11.126}$$

$$= F_{ij} \tag{11.127}$$

$$\leq R_{ij} \tag{11.128}$$

$$= R^{*}_{i(\lambda)j(\lambda+1)}. \tag{11.129}$$

In the above derivation, (11.125) follows from the second case in (11.115) because $F^{(\mu,r)}_{i(\lambda)j(\lambda+1)}$ is possibly nonzero only when

$$\lambda = \mu + a_r(i,j), \tag{11.130}$$

or

$$\mu = \lambda - a_r(i,j), \tag{11.131}$$

and (11.126) can be justified as follows. First, the inequality is justified for $\lambda < a_r(i,j)$ since by (11.121),

$$\mathbf{F}^{(\mu,r)} = 0 \tag{11.132}$$

for $\mu < 0$. For $\lambda \geq a_r(i,j)$, we distinguish two cases. From (11.115) and (11.110), if $(i,j) \in P_r$, we have

$$F^{(\lambda - a_r(i,j),r)}_{i(\lambda)j(\lambda+1)} = \tilde{F}_{ij}^{r} = c_r. \tag{11.133}$$

If $(i,j) \notin P_r$, we have

$$F^{(\lambda - a_r(i,j),r)}_{i(\lambda)j(\lambda+1)} = \tilde{F}_{ij}^{r} = 0. \tag{11.134}$$

Thus the inequality is justified for all cases. Hence we conclude that $\mathbf{F}^* \leq \mathbf{R}^*$, and the lemma is proved. □

From Lemma 11.10 and the result in Section 11.5.1, we see that $(\Lambda - d + 1)\omega$ is achievable for graph G^* with rate constraints $\mathbf{R}^*$, where

$$d = \max_{l} d_l. \tag{11.135}$$

Thus, for every $\epsilon > 0$, there exists for sufficiently large ν a $(\nu, (\eta^*_{uv} : (u,v) \in E^*, (\Lambda - d + 1)\omega)$ β-code on G^* such that

$$\nu^{-1} \log_2 \eta^*_{uv} \leq R^*_{uv} + \epsilon \tag{11.136}$$

for all $(u,v) \in E^*$. For this β-code on G^*, we denote the encoding function for an edge $(u,v) \in E^*$ by f^*_{uv}, and denote the decoding function at the sink node t_l^* by g_l^*, $l = 1, 2, \cdots, L$. Without loss of generality, we assume that for $1 \leq \lambda \leq \Lambda$,

$$f^*_{s^* s^{(\lambda)}}(x) = x \tag{11.137}$$

for all x in

$$\mathcal{X} = \{1, \cdots, \lceil 2^{\nu(\Lambda-d+1)\omega} \rceil\}, \tag{11.138}$$

and for $0 \leq \lambda \leq \Lambda - 1$,

$$f^*_{t_l^{(\lambda)} t_l^*}(y) = y \tag{11.139}$$

for all y in

$$\prod_{k:(k,t_l)\in E} \{1, \cdots, \eta^*_{k^{(\lambda-1)} t_l^{(\lambda)}}\}, \tag{11.140}$$

$1 \leq l \leq L$. Note that if the β-code does not satisfy these assumptions, it can readily be converted into one because the capacities of the edges $(s^*, s^{(\lambda)})$, $1 \leq \lambda \leq \Lambda$, and the edges $(t_l^{(\lambda)}, t_l^*)$, $0 \leq \lambda \leq \Lambda - 1$, are infinite.

Let Λ be a positive integer, and let $n = \Lambda\nu$. Using the β-code on G^*, we now construct an $(n, (\eta_{ij} : (i,j) \in E), (\Lambda - d + 1)\omega/\Lambda)$ *γ-code* on G, where

$$\eta_{ij} = \prod_{\lambda=0}^{\Lambda-1} \eta^*_{i^{(\lambda)} j^{(\lambda+1)}} \tag{11.141}$$

for $(i,j) \in E$. The code is defined by the following components:

1) for $(i,j) \in E$ such that $i \neq s$, an arbitrary constant $f_{ij}^{(1)}$ taken from the set

$$\{1, 2, \cdots, \eta^*_{i^{(0)} j^{(1)}}\}; \tag{11.142}$$

2) for $1 \leq \lambda \leq \Lambda$, an encoding function

$$f_{sj}^{(\lambda)} : \mathcal{X} \to \{1, 2, \cdots, \eta^*_{s^{(\lambda-1)} j^{(\lambda)}}\} \tag{11.143}$$

for all $j \in V$ such that $(s,j) \in E$, where

$$\mathcal{X} = \{1, 2, \cdots, \lceil 2^{\nu(\Lambda-d+1)\omega} \rceil\}, \tag{11.144}$$

and for $2 \leq \lambda \leq \Lambda$, an encoding function

$$f_{ij}^{(\lambda)} : \prod_{k:(k,i)\in E} \{1, 2, \cdots, \eta^*_{k^{(\lambda-2)} i^{(\lambda-1)}}\} \to \{1, 2, \cdots, \eta^*_{i^{(\lambda-1)} j^{(\lambda)}}\} \tag{11.145}$$

for all $(i,j) \in E$ such that $i \neq s$ (if $\{k : (k,i) \in E\}$ is empty, we adopt the convention that $f_{ij}^{(\lambda)}$ is an arbitrary constant taken from the set $\{1, 2, \cdots, \eta^*_{i^{(\lambda-1)} j^{(\lambda)}}\}$);

3) for $l = 1, 2, \cdots, L$, a decoding function

$$g_l : \prod_{\lambda=0}^{\Lambda-1} \prod_{i:(i,t_l)\in E} \{1, 2, \cdots, \eta^*_{i^{(\lambda)}t_l^{(\lambda+1)}}\} \to \mathcal{X} \tag{11.146}$$

such that $\tilde{g}_l(x) = x$ for all $x \in \mathcal{X}$ (recall that $\tilde{g}_l(x)$ denotes the value of g_l as a function of x);

where

i) for $(i, j) \in E$ such that $i \neq s$,

$$f_{ij}^{(1)} = f^*_{i^{(0)}j^{(1)}} \tag{11.147}$$

($f^*_{i^{(0)}j^{(1)}}$ is an arbitrary constant in $\{1, 2, \cdots, \eta^*_{i^{(0)}j^{(1)}}\}$ since $\{u \in V^* : (u, i^{(0)}) \in E^*\}$ is empty);

ii) for $1 \leq \lambda \leq \Lambda$, for all $x \in \mathcal{X}$,

$$f_{sj}^{(\lambda)}(x) = f^*_{s^{(\lambda-1)}j^{(\lambda)}}(x), \tag{11.148}$$

and for $2 \leq \lambda \leq \Lambda$ and all $(i, j) \in E$ such that $i \neq s$,

$$f_{ij}^{(\lambda)}(y) = f^*_{i^{(\lambda-1)}j^{(\lambda)}}(y) \tag{11.149}$$

for all y in

$$\prod_{k:(k,i)\in E} \{1, 2, \cdots, \eta^*_{k^{(\lambda-2)}i^{(\lambda-1)}}\}; \tag{11.150}$$

iii) for $l = 1, 2, \cdots, L$,

$$g_l(z) = g_l^*(z) \tag{11.151}$$

for all z in

$$\prod_{\lambda=0}^{\Lambda-1} \prod_{i:(i,t_l)\in E} \{1, 2, \cdots, \eta^*_{i^{(\lambda)}t_l^{(\lambda+1)}}\}. \tag{11.152}$$

The coding process of the γ-code consists of $\Lambda + 1$ phases:

1. In Phase 1, for all $(i, j) \in E$ such that $i \neq s$, node i sends $f_{ij}^{(1)}$ to node j, and for all $j \in V$ such that $(s, j) \in E$, node s sends $f_{sj}^{(1)}(x)$ to node j.

2. In Phase λ, $2 \leq \lambda \leq \Lambda$, for all $(i, j) \in E$, node i sends $\tilde{f}_{ij}^{(\lambda)}(x)$ to node j, where $\tilde{f}_{ij}^{(\lambda)}(x)$ denotes the value of $f_{ij}^{(\lambda)}$ as a function of x, and it depends

only on $\tilde{f}_{ki}^{(\lambda-1)}(x)$ for all $k \in V$ such that $(k,i) \in E$, i.e., the information received by node i during Phase $\lambda - 1$.

3. In Phase $\Lambda + 1$, for $l = 1, 2, \cdots, L$, the sink node t_l uses g_l to decode x.

From the definitions, we see that an $(n, (\eta_{ij} : (i,j) \in E), (\Lambda - d + 1)\omega/\Lambda)$ γ-code on G is a special case of an $(n, (\eta_{ij} : (i,j) \in E), (\Lambda - d + 1)\omega/\Lambda)$ α-code on G. For the γ-code we have constructed,

$$
\begin{aligned}
n^{-1}\log_2 \eta_{ij} &= (\Lambda\nu)^{-1}\log_2\left(\prod_{\lambda=0}^{\Lambda-1}\eta^*_{i(\lambda)j(\lambda+1)}\right) && (11.153)\\
&= \Lambda^{-1}\sum_{\lambda=0}^{\Lambda-1}\nu^{-1}\log_2 \eta^*_{i(\lambda)j(\lambda+1)} && (11.154)\\
&\le \Lambda^{-1}\sum_{\lambda=0}^{\Lambda-1}(R^*_{i(\lambda)j(\lambda+1)} + \epsilon) && (11.155)\\
&= \Lambda^{-1}\sum_{\lambda=0}^{\Lambda-1}(R_{ij} + \epsilon) && (11.156)\\
&= R_{ij} + \epsilon && (11.157)
\end{aligned}
$$

for all $(i,j) \in E$, where (11.155) follows from (11.136), and (11.156) follows from (11.107). Finally, for any $\epsilon > 0$, by taking a sufficiently large Λ, we have

$$\frac{(\Lambda - d + 1)\omega}{\Lambda} > \omega - \epsilon. \tag{11.158}$$

Hence, we conclude that ω is achievable for the graph G with rate constraints $\mathbf{R}$.

PROBLEMS

In the following problems, the rate constraint for an edge is in bits per unit time.

1. Consider the following network.

   We want to multicast information to the sink nodes at the maximum rate without using network coding. Let $B = \{b_1, b_2, \cdots, b_\kappa\}$ be the set of bits to be multicast. Let B_i be the set of bits sent in edge (s,i), where $|B_i| = 2$, $i = 1, 2, 3$. At node i, the received bits are duplicated and sent in the two out-going edges. Thus two bits are sent in each edge in the network.

   a) Show that $B = B_i \cup B_j$ for any $1 \le i < j \le 3$.
   b) Show that $B_3 \cup (B_1 \cap B_2) = B$.

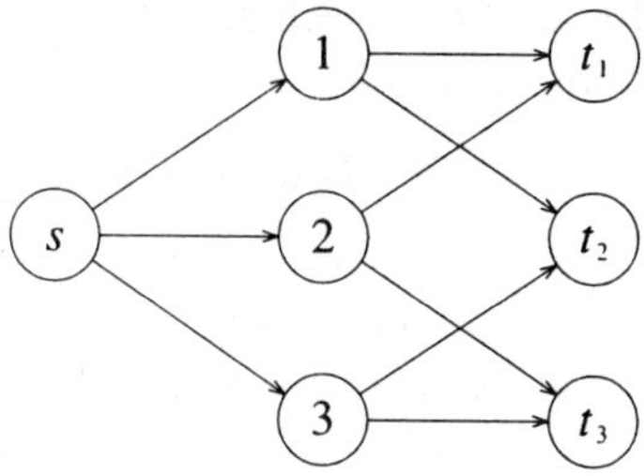

c) Show that $|B_3 \cup (B_1 \cap B_2)| \leq |B_3| + |B_1| + |B_2| - |B_1 \cup B_2|$.

d) Determine the maximum value of κ and devise a network code which achieves this maximum value.

e) What is the percentage of improvement if network coding is used?

(Ahlswede *et al.* [6].)

2. Consider the following network.

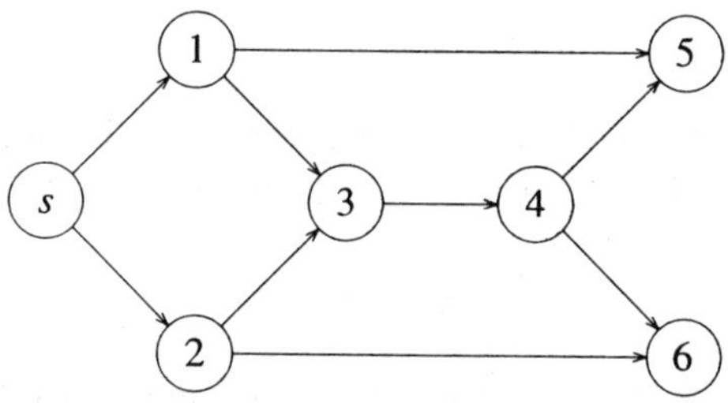

Devise a network coding scheme which multicasts two bits b_1 and b_2 from node s to all the other nodes such that nodes 3, 5, and 6 receive b_1 and b_2 after 1 unit time and nodes 1, 2, and 4 receive b_1 and b_2 after 2 units of time. In other words, node i receives information at a rate equal to $\text{maxflow}(s, i)$ for all $i \neq s$.

See Li *et al.* [132] for such a scheme for a general network.

3. Determine the maximum rate at which information can be multicast to nodes 5 and 6 only in the network in Problem 2 if network coding is not used. Devise a network coding scheme which achieves this maximum rate.

4. *Convolutional network code*
In the following network, $\text{maxflow}(s, t_l) = 3$ for $l = 1, 2, 3$. The max-flow bound asserts that 3 bits can be multicast to all the three sink nodes per unit time. We now describe a network coding scheme which achieve this. Let 3

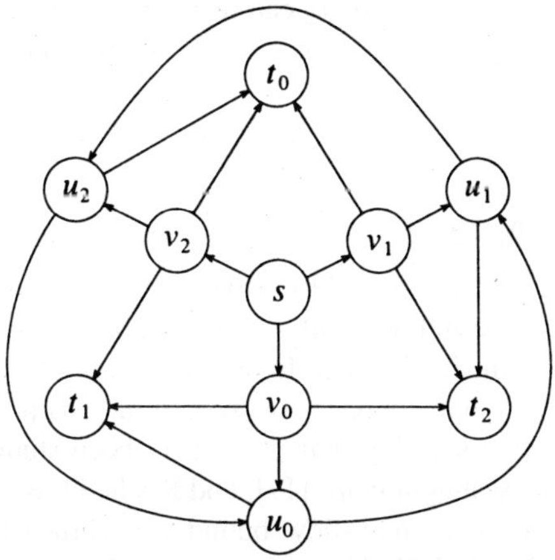

bits $b_1(k), b_2(k), b_3(k)$ be generated at node s at time $k = 1, 2, \cdots$, where we assume without loss of generality that $b_l(k)$ is an element of the finite field $GF(2)$. We adopt the convention that $b_l(k) = 0$ for $k \leq 0$. At time $k \geq 1$, information transactions T1 to T11 occur in the following order:

T1. s sends $b_l(k)$ to v_l, $l = 0, 1, 2$

T2. v_l sends $b_l(k)$ to u_l, $t_{l\oplus 1}$, and $t_{l\oplus 2}$, $l = 0, 1, 2$

T3. u_0 sends $b_0(k) + b_1(k-1) + b_2(k-1)$ to u_1

T4. u_1 sends $b_0(k) + b_1(k-1) + b_2(k-1)$ to t_2

T5. u_1 sends $b_0(k) + b_1(k) + b_2(k-1)$ to u_2

T6. u_2 sends $b_0(k) + b_1(k) + b_2(k-1)$ to t_0

T7. u_2 sends $b_0(k) + b_1(k) + b_2(k)$ to u_0

T8. u_0 sends $b_0(k) + b_1(k) + b_2(k)$ to t_1

T9. t_2 decodes $b_2(k-1)$

T10. t_0 decodes $b_0(k)$

T11 t_1 decodes $b_1(k)$

where "$\oplus$" denotes modulo 3 addition and "+" denotes modulo 2 addition.

a) Show that the information transactions T1 to T11 can be performed at time $k = 1$.

b) Show that T1 to T11 can be performed at any time $k \geq 1$ by induction on k.

c) Verify that at time k, nodes t_0 and t_1 can recover $b_0(k')$, $b_1(k')$, and $b_2(k')$ for all $k' \leq k$.

d) Verify that at time k, node t_2 can recover $b_0(k')$ and $b_1(k')$ for all $k' \leq k$, and $b_2(k')$ for all $k' \leq k-1$. Note the unit time delay for t_2 to recover $b_2(k)$.

(Ahlswede *et al.* [6].)

HISTORICAL NOTES

Network coding was introduced by Ahlswede *et al.* [6], where it was shown that information can be multicast in a point-to-point network at a higher rate by coding at the intermediate nodes. In the same paper, the achievability of the max-flow bound for single-source network coding was proved. Special cases of single-source network coding has previously been studied by Roche *et al.* [174], Rabin [167], Ayanoglu *et al.* [17], and Roche [173].

The achievability of the max-flow bound was shown in [6] by a random coding technique. Li *et al.* [132] have subsequently shown that the max-flow bound can be achieved by linear network codes. A more transparent proof of this result and its extensions can be found in Koetter and Médard [120], where they developed an algebraic approach to network coding.

Chapter 12

INFORMATION INEQUALITIES

An information expression f refers to a *linear combination*[1] of Shannon's information measures involving a finite number of random variables. For example,

$$H(X,Y) + 2I(X;Z) \tag{12.1}$$

and

$$I(X;Y) - I(X;Y|Z) \tag{12.2}$$

are information expressions. An information inequality has the form

$$f \geq c, \tag{12.3}$$

where the constant c is usually equal to zero. We consider non-strict inequalities only because these are usually the form of inequalities in information theory. Likewise, an information identity has the form

$$f = c. \tag{12.4}$$

We point out that an information identity $f = c$ is equivalent to the pair of information inequalities $f \geq c$ and $f \leq c$.

An information inequality or identity is said to *always hold* if it holds for any joint distribution for the random variables involved. For example, we say that the information inequality

$$I(X;Y) \geq 0 \tag{12.5}$$

[1]More generally, an information expression can be nonlinear, but they do not appear to be useful in information theory.

always holds because it holds for any joint distribution $p(x, y)$. On the other hand, we say that an information inequality does not always hold if there exists a joint distribution for which the inequality does not hold. Consider the information inequality

$$I(X;Y) \leq 0. \tag{12.6}$$

Since

$$I(X;Y) \geq 0 \tag{12.7}$$

always holds, (12.6) is equivalent to

$$I(X;Y) = 0, \tag{12.8}$$

which holds if and only if X and Y are independent. In other words, (12.6) does not hold if X and Y are not independent. Therefore, we say that (12.6) does not always hold.

As we have seen in the previous chapters, information inequalities are the major tools for proving converse coding theorems. These inequalities govern the impossibilities in information theory. More precisely, information inequalities imply that certain things cannot happen. As such, they are referred to as the *laws of information theory*.

The basic inequalities form the most important set of information inequalities. In fact, almost all the information inequalities known to date are implied by the basic inequalities. These are called *Shannon-type inequalities*. On the other hand, if an information inequality always holds but is not implied by the basic inequalities, then it is called a *non-Shannon-type inequality*. We have not yet explained what it means by that an inequality is or is not implied by the basic inequalities, but this will become clear later in the chapter.

Let us now rederive the inequality obtained in Example 6.15 (Imperfect secrecy theorem) without using an information diagram. In this example, three random variables X, Y, and Z are involved, and the setup of the problem is equivalent to the constraint

$$H(X|Y,Z) = 0. \tag{12.9}$$

Then

$$\begin{aligned}
I(X;Y) & \\
&= H(X) + H(Y) - H(X,Y) && (12.10)\\
&= H(X) + H(Y) - [H(X,Y,Z) - H(Z|X,Y)] && (12.11)\\
&\geq H(X) + H(Y) - H(X,Y,Z) && (12.12)\\
&= H(X) + H(Y) - [H(Z) + H(Y|Z) + H(X|Y,Z)] && (12.13)\\
&= H(X) - H(Z) + I(Y;Z) - H(X|Y,Z) && (12.14)\\
&\geq H(X) - H(Z), && (12.15)
\end{aligned}$$

where we have used

$$H(Z|X,Y) \geq 0 \tag{12.16}$$

in obtaining (12.12), and

$$I(Y;Z) \geq 0 \tag{12.17}$$

and (12.9) in obtaining (12.15). This derivation is less transparent than the one we presented in Example 6.15, but the point here is that the final inequality we obtain in (12.15) can be proved by invoking the basic inequalities (12.16) and (12.17). In other words, (12.15) is implied by the basic inequalities. Therefore, it is a (constrained) Shannon-type inequality.

We are motivated to ask the following two questions:

1. How can Shannon-type inequalities be characterized? That is, given an information inequality, how can we tell whether it is implied by the basic inequalities?
2. Are there any non-Shannon-type information inequalities?

These are two very fundamental questions in information theory. We point out that the first question naturally comes before the second question because if we cannot characterize all Shannon-type inequalities, even if we are given a non-Shannon-type inequality, we cannot tell that it actually is one.

In this chapter, we develop a geometric framework for information inequalities which allows them to be studied systematically. This framework naturally leads to an answer to the first question, which makes machine-proving of all Shannon-type inequalities possible. This will be discussed in the next chapter. The second question will be answered positively in Chapter 14. In other words, there *do* exist laws in information theory beyond those laid down by Shannon.

12.1 THE REGION Γ_n^*

Let

$$\mathcal{N}_n = \{1, 2, \cdots, n\}, \tag{12.18}$$

where $n \geq 2$, and let

$$\Theta = \{X_i, i \in \mathcal{N}_n\} \tag{12.19}$$

be any collection of n random variables. Associated with Θ are

$$k = 2^n - 1 \tag{12.20}$$

joint entropies. For example, for $n = 3$, the 7 joint entropies associated with random variables X_1, X_2, and X_3 are

$$H(X_1), H(X_2), H(X_3), H(X_1, X_2),$$
$$H(X_2, X_3), H(X_1, X_3), H(X_1, X_2, X_3). \tag{12.21}$$

Let $\Re$ denote the set of real numbers. For any nonempty subset α of $\mathcal{N}_n$, let

$$X_\alpha = (X_i, i \in \alpha) \tag{12.22}$$

and

$$H_\Theta(\alpha) = H(X_\alpha). \tag{12.23}$$

For a fixed Θ, we can then view H_Θ as a set function from $2^{\mathcal{N}_n}$ to $\Re$ with $H_\Theta(\emptyset) = 0$, i.e., we adopt the convention that the entropy of an empty set of random variable is equal to zero. For this reason, we call H_Θ the *entropy function* of Θ.

Let $\mathcal{H}_n$ be the k-dimensional Euclidean space with the coordinates labeled by $h_\alpha, \alpha \in 2^{\mathcal{N}_n}\backslash\{\emptyset\}$, where h_α corresponds to the value of $H_\Theta(\alpha)$ for any collection Θ of n random variables. We will refer to $\mathcal{H}_n$ as the *entropy space* for n random variables. Then an entropy function H_Θ can be represented by a column vector in $\mathcal{H}_n$. On the other hand, a column vector $\mathbf{h} \in \mathcal{H}_n$ is called *entropic* if $\mathbf{h}$ is equal to the entropy function H_Θ of some collection Θ of n random variables. We are motivated to define the following region in $\mathcal{H}_n$:

$$\Gamma_n^* = \{\mathbf{h} \in \mathcal{H}_n : \mathbf{h} \text{ is entropic}\}. \tag{12.24}$$

For convenience, the vectors in Γ_n^* will also be referred to as entropy functions. As an example, for $n = 3$, the coordinates of $\mathcal{H}_3$ are labeled by

$$h_1, h_2, h_3, h_{12}, h_{13}, h_{23}, h_{123}, \tag{12.25}$$

where h_{123} denotes $h_{\{1,2,3\}}$, etc, and Γ_3^* is the region in $\mathcal{H}_3$ of all entropy functions for 3 random variables.

While further characterizations of Γ_n^* will be given later, we first point out a few basic properties of Γ_n^*:

1. Γ_n^* contains the origin.

2. $\overline{\Gamma}_n^*$, the closure of Γ_n^*, is convex.

3. Γ_n^* is in the nonnegative orthant of the entropy space $\mathcal{H}_n$[2].

The origin of the entropy space corresponds to the entropy function of n degenerate random variables taking constant values. Hence, Property 1 follows. Property 2 will be proved in Chapter 14. Properties 1 and 2 imply that $\overline{\Gamma}_n^*$ is a convex cone. Property 3 is true because the coordinates in the entropy space $\mathcal{H}_n$ correspond to joint entropies, which are always nonnegative.

[2]The nonnegative orthant of $\mathcal{H}^n$ is the region $\{\mathbf{h} \in \mathcal{H}_n : h_\alpha \geq 0 \text{ for all } \alpha \in 2^{\mathcal{N}_n}\backslash\{\emptyset\}\}$.

12.2 INFORMATION EXPRESSIONS IN CANONICAL FORM

Any Shannon's information measure other than a joint entropy can be expressed as a linear combination of joint entropies by application of one of the following information identities:

$$H(X|Y) = H(X,Y) - H(Y) \tag{12.26}$$
$$I(X;Y) = H(X) + H(Y) - H(X,Y) \tag{12.27}$$
$$I(X;Y|Z) = H(X,Z) + H(Y,Z) - H(X,Y,Z) - H(Z). \tag{12.28}$$

The first and the second identity are special cases of the third identity, which has already been proved in Lemma 6.8. Thus any information expression which involves n random variables can be expressed as a linear combination of the k associated joint entropies. We call this the *canonical form* of an information expression. When we write an information expression f as $f(\mathbf{h})$, it means that f is in canonical form. Since an information expression in canonical form is a linear combination of the joint entropies, it has the form

$$\mathbf{b}^\top \mathbf{h} \tag{12.29}$$

where $\mathbf{b}^\top$ denotes the transpose of a constant column vector $\mathbf{b}$ in $\Re^k$.

The identities in (12.26) to (12.28) provide a way to express every information expression in canonical form. However, it is not clear whether such a canonical form is unique. To illustrate the point, we consider obtaining the canonical form of $H(X|Y)$ in two ways. First,

$$H(X|Y) = H(X,Y) - H(Y). \tag{12.30}$$

Second,

$$H(X|Y) = H(X) - I(X;Y) \tag{12.31}$$
$$= H(X) - (H(Y) - H(Y|X)) \tag{12.32}$$
$$= H(X) - (H(Y) - H(X,Y) + H(X)) \tag{12.33}$$
$$= H(X,Y) - H(Y). \tag{12.34}$$

Thus it turns out that we can obtain the same canonical form for $H(X|Y)$ via two different expansions. This is not accidental, as it is implied by the uniqueness of the canonical form which we will prove shortly.

Recall from the proof of Theorem 6.6 that the vector $\mathbf{h}$ represents the values of the I-Measure μ^* on the unions in $\mathcal{F}_n$. Moreover, $\mathbf{h}$ is related to the values of μ^* on the atoms of $\mathcal{F}_n$, represented as $\mathbf{u}$, by

$$\mathbf{h} = \mathbf{C}_n \mathbf{u} \tag{12.35}$$

where $\mathbf{C}_n$ is a unique $k \times k$ matrix (cf. (6.27)). We now state the following lemma which is a rephrase of Theorem 6.11. This lemma is essential for proving the next theorem which implies the uniqueness of the canonical form.

LEMMA 12.1 *Let*

$$\Psi_n^* = \{\mathbf{u} \in \Re^k : \mathbf{C}_n \mathbf{u} \in \Gamma_n^*\}. \tag{12.36}$$

Then the nonnegative orthant of $\Re^k$ *is a subset of* Ψ_n^*.

THEOREM 12.2 *Let f be an information expression. Then the unconstrained information identity $f = 0$ always holds if and only if f is the zero function.*

Proof Without loss of generality, assume f is in canonical form and let

$$f(\mathbf{h}) = \mathbf{b}^\top \mathbf{h}. \tag{12.37}$$

Assume $f = 0$ always holds and f is not the zero function, i.e., $\mathbf{b} \neq 0$. We will show that this leads to a contradiction. Now $f = 0$, or more precisely the set

$$\{\mathbf{h} : \mathbf{b}^\top \mathbf{h} = 0\}, \tag{12.38}$$

is a hyperplane[3] in the entropy space which has zero Lebesgue measure[4]. If $f = 0$ always holds, i.e., it holds for all joint distributions, then Γ_n^* must be contained in the hyperplane $f = 0$, otherwise there exists an $\mathbf{h}_0 \in \Gamma_n^*$ which is not on $f = 0$, i.e., $f(\mathbf{h}_0) \neq 0$. Since $\mathbf{h}_0 \in \Gamma_n^*$, it corresponds to the entropy function of some joint distribution. This means that there exists a joint distribution such that $f(\mathbf{h}) = 0$ does not hold, which cannot be true because $f = 0$ always holds.

If Γ_n^* has positive Lebesgue measure, it cannot be contained in the hyperplane $f = 0$ which has zero Lebesgue measure. Therefore, it suffices to show that Γ_n^* has positive Lebesgue measure. To this end, we see from Lemma 12.1 that the nonnegative orthant of $\mathcal{H}_n$, which has positive Lebesgue measure, is a subset of Ψ_n^*. Thus Ψ_n^* has positive Lebesgue measure. Since Γ_n^* is an invertible transformation of Ψ_n^*, its Lebesgue measure is also positive.

Therefore, Γ_n^* is not contained in the hyperplane $f = 0$, which implies that there exists a joint distribution for which $f = 0$ does not hold. This leads to a contradiction because we have assumed that $f = 0$ always holds. Hence, we have proved that if $f = 0$ always holds, then f must be the zero function.

Conversely, if f is the zero function, then it is trivial that $f = 0$ always holds. The theorem is proved. □

[3] If $\mathbf{b} = 0$, then $\{\mathbf{h} : \mathbf{b}^\top \mathbf{h} = 0\}$ is equal to $\mathcal{H}_n$.

[4] The Lebesque measure can be thought of as "volume" in the Euclidean space if the reader is not familiar with measure theory.

COROLLARY 12.3 *The canonical form of an information expression is unique.*

Proof Let f_1 and f_2 be canonical forms of an information expression g. Since

$$g = f_1 \tag{12.39}$$

and

$$g = f_2 \tag{12.40}$$

always hold,

$$f_1 - f_2 = 0 \tag{12.41}$$

always holds. By the above theorem, $f_1 - f_2$ is the zero function, which implies that f_1 and f_2 are identical. The corollary is proved. □

Due to the uniqueness of the canonical form of an information expression, it is an easy matter to check whether for two information expressions f_1 and f_2 the unconstrained information identity

$$f_1 = f_2 \tag{12.42}$$

always holds. All we need to do is to express $f_1 - f_2$ in canonical form. If all the coefficients are zero, then (12.42) always holds, otherwise it does not.

12.3 A GEOMETRICAL FRAMEWORK

In the last section, we have seen the role of the region Γ_n^* in proving unconstrained information identities. In this section, we explain the geometrical meanings of unconstrained information inequalities, constrained information inequalities, and constrained information identities in terms of Γ_n^*. Without loss of generality, we assume that all information expressions are in canonical form.

12.3.1 UNCONSTRAINED INEQUALITIES

Consider an unconstrained information inequality $f \geq 0$, where $f(\mathbf{h}) = \mathbf{b}^\top \mathbf{h}$. Then $f \geq 0$ corresponds to the set

$$\{\mathbf{h} \in \mathcal{H}_n : \mathbf{b}^\top \mathbf{h} \geq 0\} \tag{12.43}$$

which is a half-space in the entropy space $\mathcal{H}_n$ containing the origin. Specifically, for any $\mathbf{h} \in \mathcal{H}_n$, $f(\mathbf{h}) \geq 0$ if and only if $\mathbf{h}$ belongs to this set. For simplicity, we will refer to this set as the half-space $f \geq 0$. As an example, for $n = 2$, the information inequality

$$I(X_1; X_2) = H(X_1) + H(X_2) - H(X_1, X_2) \geq 0, \tag{12.44}$$

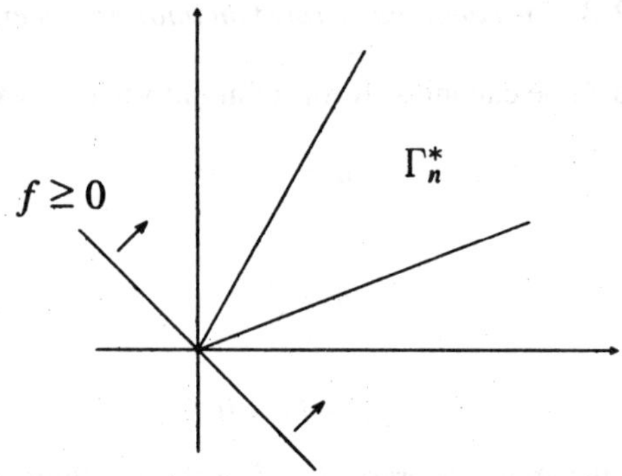

Figure 12.1. An illustration for $f \geq 0$ always holds.

written as

$$h_1 + h_2 - h_{12} \geq 0, \tag{12.45}$$

corresponds to the half-space

$$\{\mathbf{h} \in \mathcal{H}_n : h_1 + h_2 - h_{12} \geq 0\}. \tag{12.46}$$

in the entropy space $\mathcal{H}_2$.

Since an information inequality always holds if and only if it is satisfied by the entropy function of any joint distribution for the random variables involved, we have the following geometrical interpretation of an information inequality:

$f \geq 0$ always holds if and only if $\Gamma_n^* \subset \{\mathbf{h} \in \mathcal{H}_n : f(\mathbf{h}) \geq 0\}$.

This gives a complete characterization of all unconstrained inequalities in terms of Γ_n^*. If Γ_n^* is known, we in principle can determine whether any information inequality involving n random variables always holds.

The two possible cases for $f \geq 0$ are illustrated in Figure 12.1 and Figure 12.2. In Figure 12.1, Γ_n^* is completely included in the half-space $f \geq 0$, so $f \geq 0$ always holds. In Figure 12.2, there exists a vector $\mathbf{h}_0 \in \Gamma_n^*$ such that $f(\mathbf{h}_0) < 0$. Thus the inequality $f \geq 0$ does not always hold.

12.3.2 CONSTRAINED INEQUALITIES

In information theory, we very often deal with information inequalities (identities) with certain constraints on the joint distribution for the random variables involved. These are called constrained information inequalities (identities), and the constraints on the joint distribution can usually be expressed as linear constraints on the entropies. The following are such examples:

1. X_1, X_2, and X_3 are mutually independent if and only if $H(X_1, X_2, X_3) = H(X_1) + H(X_2) + H(X_3)$.

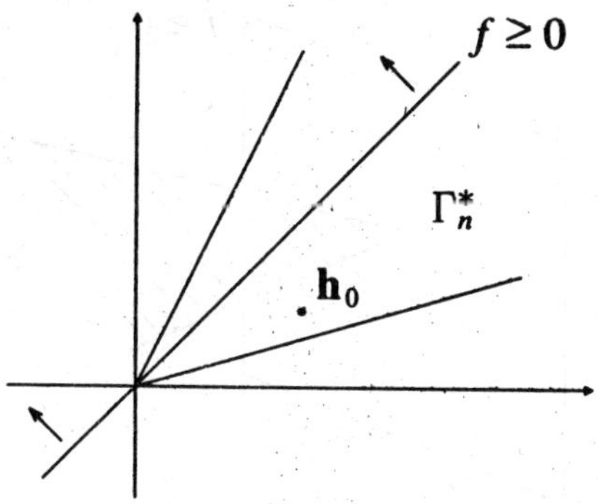

Figure 12.2. An illustration for $f \geq 0$ not always holds.

2. X_1, X_2, and X_3 are pairwise independent if and only if $I(X_1; X_2) = I(X_2; X_3) = I(X_1; X_3) = 0$.
3. X_1 is a function of X_2 if and only if $H(X_1|X_2) = 0$.
4. $X_1 \to X_2 \to X_3 \to X_4$ forms a Markov chain if and only if $I(X_1; X_3|X_2) = 0$ and $I(X_1, X_2; X_4|X_3) = 0$.

Suppose there are q linear constraints on the entropies given by

$$\mathbf{Q}\mathbf{h} = 0, \tag{12.47}$$

where $\mathbf{Q}$ is a $q \times k$ matrix. Here we do not assume that the q constraints are linearly independent, so $\mathbf{Q}$ is not necessarily full rank. Let

$$\Phi = \{\mathbf{h} \in \mathcal{H}_n : \mathbf{Q}\mathbf{h} = 0\}. \tag{12.48}$$

In other words, the q constraints confine $\mathbf{h}$ to a linear subspace Φ in the entropy space. Parallel to our discussion on unconstrained inequalities, we have the following geometrical interpretation of a constrained inequality:

> Under the constraint Φ, $f \geq 0$ always holds if and only if $(\Gamma_n^* \cap \Phi) \subset \{\mathbf{h} : f(\mathbf{h}) \geq 0\}$.

This gives a complete characterization of all constrained inequalities in terms of Γ_n^*. Note that $\Phi = \mathcal{H}_n$ when there is no constraint on the entropies. In this sense, an unconstrained inequality is a special case of a constrained inequality.

The two cases of $f \geq 0$ under the constraint Φ are illustrated in Figure 12.3 and Figure 12.4. Figure 12.3 shows the case when $f \geq 0$ always holds under the constraint Φ. Note that $f \geq 0$ may or may not always hold when there is no constraint. Figure 12.4 shows the case when $f \geq 0$ does not always hold

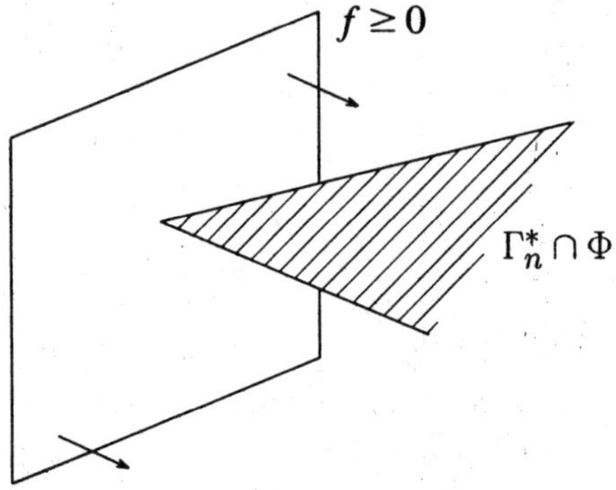

Figure 12.3. An illustration for $f \geq 0$ always holds under the constraint Φ.

under the constraint Φ. In this case, $f \geq 0$ does not always hold when there is no constraint, because

$$(\Gamma_n^* \cap \Phi) \not\subset \{\mathbf{h} : f(\mathbf{h}) \geq 0\} \tag{12.49}$$

implies

$$\Gamma_n^* \not\subset \{\mathbf{h} : f(\mathbf{h}) \geq 0\}. \tag{12.50}$$

12.3.3 CONSTRAINED IDENTITIES

As we have pointed out at the beginning of the chapter, an identity

$$f = 0 \tag{12.51}$$

always holds if and only if both the inequalities $f \geq 0$ and $f \leq 0$ always hold. Then following our discussion on constrained inequalities, we have

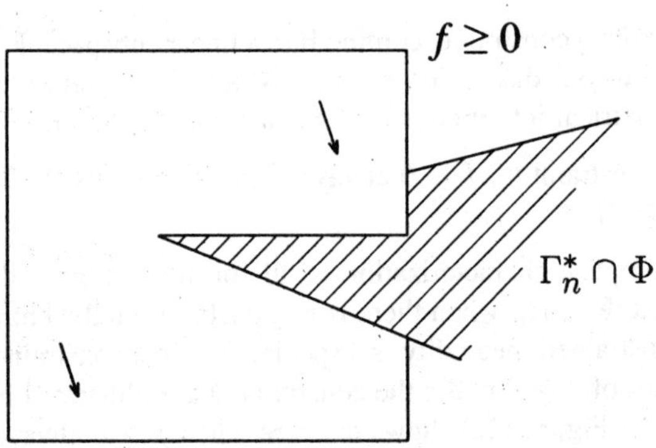

Figure 12.4. An illustration for $f \geq 0$ not always holds under the constraint Φ.

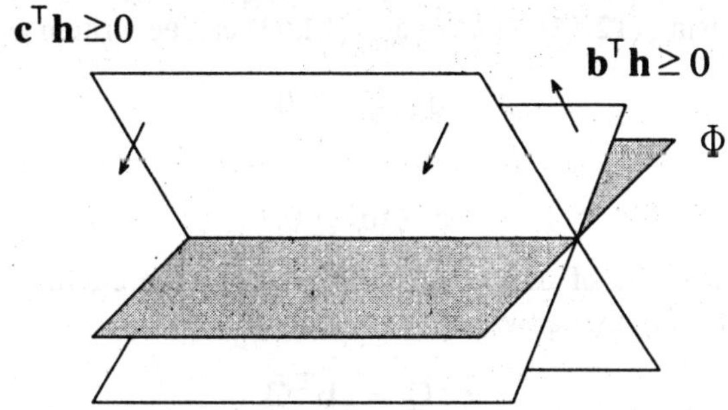

Figure 12.5. Equivalence of $\mathbf{b}^\top\mathbf{h} \geq 0$ and $\mathbf{c}^\top\mathbf{h} \geq 0$ under the constraint Φ.

Under the constraint Φ, $f = 0$ always holds if and only if $(\Gamma_n^* \cap \Phi) \subset \{\mathbf{h} : f(\mathbf{h}) \geq 0\} \cap \{\mathbf{h} : f(\mathbf{h}) \leq 0\}$,

or

Under the constraint Φ, $f = 0$ always holds if and only if $(\Gamma_n^* \cap \Phi) \subset \{\mathbf{h} : f(\mathbf{h}) = 0\}$.

This condition says that the intersection of Γ_n^* and Φ is contained in the hyperplane $f = 0$.

12.4 EQUIVALENCE OF CONSTRAINED INEQUALITIES

When there is no constraint on the entropies, two information inequalities

$$\mathbf{b}^\top\mathbf{h} \geq 0 \tag{12.52}$$

and

$$\mathbf{c}^\top\mathbf{h} \geq 0 \tag{12.53}$$

are equivalent if and only if $\mathbf{c} = a\mathbf{b}$, where a is a positive constant. However, this is not the case under a non-trivial constraint $\Phi \neq \mathcal{H}_n$. This situation is illustrated in Figure 12.5. In this figure, although the inequalities in (12.52) and (12.53) correspond to different half-spaces in the entropy space, they actually impose the same constraint on $\mathbf{h}$ when $\mathbf{h}$ is confined to Φ.

In this section, we present a characterization of (12.52) and (12.53) being equivalent under a set of linear constraint Φ. The reader may skip this section at first reading.

Let r be the rank of $\mathbf{Q}$ in (12.47). Since $\mathbf{h}$ is in the null space of $\mathbf{Q}$, we can write

$$\mathbf{h} = \tilde{\mathbf{Q}}\mathbf{h}', \tag{12.54}$$

where $\tilde{\mathbf{Q}}$ is a $k \times (k-r)$ matrix such that the rows of $\tilde{\mathbf{Q}}^\top$ form a basis of the orthogonal complement of the row space of $\mathbf{Q}$, and $\mathbf{h}'$ is a column $(k-r)$-vector. Then using (12.54), (12.52) and (12.53) can be written as

$$\mathbf{b}^\top \tilde{\mathbf{Q}} \mathbf{h}' \geq 0 \tag{12.55}$$

and

$$\mathbf{c}^\top \tilde{\mathbf{Q}} \mathbf{h}' \geq 0, \tag{12.56}$$

respectively in terms of the set of basis given by the columns of $\tilde{\mathbf{Q}}$. Then (12.55) and (12.56) are equivalent if and only if

$$\mathbf{c}^\top \tilde{\mathbf{Q}} = a\mathbf{b}^\top \tilde{\mathbf{Q}}, \tag{12.57}$$

where a is a positive constant, or

$$(\mathbf{c} - a\mathbf{b})^\top \tilde{\mathbf{Q}} = 0. \tag{12.58}$$

In other words, $(\mathbf{c} - a\mathbf{b})^\top$ is in the orthogonal complement of the row space of $\tilde{\mathbf{Q}}^\top$, i.e., $(\mathbf{c} - a\mathbf{b})^\top$ is in the row space of $\mathbf{Q}$. Let $\mathbf{Q}'$ be an $r \times k$ matrix whose row space is the same as that of $\mathbf{Q}$. ($\mathbf{Q}$ can be taken as $\mathbf{Q}'$ if $\mathbf{Q}$ is full rank.) Since the rank of $\mathbf{Q}$ is r and $\mathbf{Q}'$ has r rows, the rows of $\mathbf{Q}'$ form a basis for the row space of $\mathbf{Q}$, and $\mathbf{Q}'$ is full rank. Then from (12.58), (12.55) and (12.56) are equivalent under the constraint Φ if and only if

$$\mathbf{c} = a\mathbf{b} + (\mathbf{Q}')^\top \mathbf{e} \tag{12.59}$$

for some positive constant a and some column r-vector $\mathbf{e}$.

Suppose for given $\mathbf{b}$ and $\mathbf{c}$, we want to see whether (12.55) and (12.56) are equivalent under the constraint Φ. We first consider the case when either $\mathbf{b}^\top$ or $\mathbf{c}^\top$ is in the row space of $\mathbf{Q}$. This is actually not an interesting case because if $\mathbf{b}^\top$, for example, is in the row space of $\mathbf{Q}$, then

$$\mathbf{b}^\top \tilde{\mathbf{Q}} = 0 \tag{12.60}$$

in (12.55), which means that (12.55) imposes no additional constraint under the constraint Φ.

THEOREM 12.4 *If either $\mathbf{b}^\top$ or $\mathbf{c}^\top$ is in the row space of $\mathbf{Q}$, then $\mathbf{b}^\top \mathbf{h} \geq 0$ and $\mathbf{c}^\top \mathbf{h} \geq 0$ are equivalent under the constraint Φ if and only if both $\mathbf{b}^\top$ and $\mathbf{c}^\top$ are in the row space of $\mathbf{Q}$.*

The proof of this theorem is left as an exercise. We now turn to the more interesting case when neither $\mathbf{b}^\top$ nor $\mathbf{c}^\top$ is in the row space of $\mathbf{Q}$. The following theorem gives an explicit condition for (12.55) and (12.56) to be equivalent under the constraint Φ.

THEOREM 12.5 *If neither* $\mathbf{b}^\top$ *nor* $\mathbf{c}^\top$ *is in the row space of* $\mathbf{Q}$, *then* $\mathbf{b}^\top\mathbf{h} \geq 0$ *and* $\mathbf{c}^\top\mathbf{h} \geq 0$ *are equivalent under the constraint* Φ *if and only if*

$$\begin{bmatrix} (\mathbf{Q}')^\top & \mathbf{b} \end{bmatrix} \begin{bmatrix} \mathbf{e} \\ a \end{bmatrix} = \mathbf{c}. \tag{12.61}$$

has a unique solution with $a > 0$, *where* $\mathbf{Q}'$ *is any matrix whose row space is the same as that of* $\mathbf{Q}$.

Proof For $\mathbf{b}^\top$ and $\mathbf{c}^\top$ not in the row space of $\mathbf{Q}$, we want to see when we can find unknowns a and $\mathbf{e}$ satisfying (12.59) with $a > 0$. To this end, we write (12.59) in matrix form as (12.61). Since $\mathbf{b}$ is not in the column space of $(\mathbf{Q}')^\top$ and $(\mathbf{Q}')^\top$ is full rank, $\begin{bmatrix} (\mathbf{Q}')^\top & \mathbf{b} \end{bmatrix}$ is also full rank. Then (12.61) has either a unique solution or no solution. Therefore, the necessary and sufficient condition for (12.55) and (12.56) to be equivalent is that (12.61) has a unique solution and $a > 0$. The theorem is proved. □

EXAMPLE 12.6 *Consider three random variables* X_1, X_2, *and* X_3 *with the Markov constraint*

$$I(X_1; X_3|X_2) = 0, \tag{12.62}$$

which is equivalent to

$$H(X_1, X_2) + H(X_2, X_3) - H(X_1, X_2, X_3) - H(X_2) = 0. \tag{12.63}$$

In terms of the coordinates in the entropy space $\mathcal{H}_3$, *this constraint is written as*

$$\mathbf{Q}\mathbf{h} = 0, \tag{12.64}$$

where

$$\mathbf{Q} = [\ 0\ \ -1\ \ 0\ \ 1\ \ 1\ \ 0\ \ -1\ ] \tag{12.65}$$

and

$$\mathbf{h} = [\ h_1\ \ h_2\ \ h_3\ \ h_{12}\ \ h_{23}\ \ h_{13}\ \ h_{123}\ ]^\top. \tag{12.66}$$

We now show that under the constraint in (12.64), the inequalities

$$H(X_1|X_3) - H(X_1|X_2) \geq 0 \tag{12.67}$$

and

$$I(X_1; X_2|X_3) \geq 0 \tag{12.68}$$

are in fact equivalent. Toward this end, we write (12.67) and (12.68) as $\mathbf{b}^\top\mathbf{h} \geq 0$ *and* $\mathbf{c}^\top\mathbf{h} \geq 0$, *respectively, where*

$$\mathbf{b} = [\ 0\ \ 1\ \ -1\ \ -1\ \ 0\ \ 1\ \ 0\ ]^\top \tag{12.69}$$

and

$$\mathbf{c} = [\ 0\ \ 0\ \ -1\ \ 0\ \ 1\ \ 1\ \ -1\ ]^\top. \tag{12.70}$$

Since $\mathbf{Q}$ *is full rank, we may take* $\mathbf{Q}' = \mathbf{Q}$. *Upon solving*

$$\left[\ \mathbf{Q}^\top\ \mathbf{b}\ \right]\left[\begin{array}{c}\mathbf{e}\\ a\end{array}\right] = \mathbf{c}, \tag{12.71}$$

we obtain the unique solution $a = 1 > 0$ *and* $\mathbf{e} = 1$ *(*$\mathbf{e}$ *is a* 1×1 *matrix). Therefore, (12.67) and (12.68) are equivalent under the constraint in (12.64).*

Under the constraint Φ, if neither $\mathbf{b}^\top$ nor $\mathbf{c}^\top$ is in the row space of $\mathbf{Q}$, it can be shown that the identities

$$\mathbf{b}^\top\mathbf{h} = 0 \tag{12.72}$$

and

$$\mathbf{c}^\top\mathbf{h} = 0 \tag{12.73}$$

are equivalent if and only if (12.61) has a unique solution. We leave the proof as an exercise.

12.5 THE IMPLICATION PROBLEM OF CONDITIONAL INDEPENDENCE

We use $X_\alpha \perp X_\beta | X_\gamma$ to denote the conditional independency (CI)

X_α and X_β are conditionally independent given X_γ.

We have proved in Theorem 2.34 that $X_\alpha \perp X_\beta | X_\gamma$ is equivalent to

$$I(X_\alpha; X_\beta | X_\gamma) = 0. \tag{12.74}$$

When $\gamma = \emptyset$, $X_\alpha \perp X_\beta | X_\gamma$ becomes an unconditional independency which we regard as a special case of a conditional independency. When $\alpha = \beta$, (12.74) becomes

$$H(X_\alpha | X_\gamma) = 0, \tag{12.75}$$

which we see from Proposition 2.36 that X_α is a function of X_γ. For this reason, we also regard functional dependency as a special case of conditional independency.

In probability problems, we are often given a set of CI's and we need to determine whether another given CI is logically implied. This is called the *implication problem*, which is perhaps the most basic problem in probability theory. We have seen in Section 7.2 that the implication problem has a solution if only full conditional mutual independencies are involved. However, the general problem is extremely difficult, and it has recently been solved only up to four random variables by Matúš [146].

We end this section by explaining the relation between the implication problem and the region Γ_n^*. A CI involving random variables $X_1, X_2, \cdots, X_n$ has the form

$$X_\alpha \perp X_\beta | X_\gamma, \tag{12.76}$$

where $\alpha, \beta, \gamma \subset \mathcal{N}_n$. Since $I(X_\alpha; X_\beta | X_\gamma) = 0$ is equivalent to

$$H(X_{\alpha\cup\gamma}) + H(X_{\beta\cup\gamma}) - H(X_{\alpha\cup\beta\cup\gamma}) - H(X_\gamma) = 0, \tag{12.77}$$

$X_\alpha \perp X_\beta | X_\gamma$ corresponds to the hyperplane

$$\{\mathbf{h} : h_{\alpha\cup\gamma} + h_{\beta\cup\gamma} - h_{\alpha\cup\beta\cup\gamma} - h_\gamma = 0\}. \tag{12.78}$$

For a CI K, we denote the hyperplane in $\mathcal{H}_n$ corresponding to K by $\mathcal{E}(K)$.

Let $\Pi = \{K_l\}$ be a collection of CI's, and we want to determine whether Π implies a given CI K. This would be the case if and only if the following is true:

$$\text{For all } \mathbf{h} \in \Gamma_n^*, \text{ if } \mathbf{h} \in \bigcap_l \mathcal{E}(K_l), \text{ then } \mathbf{h} \in \mathcal{E}(K).$$

Equivalently,

$$\Pi \text{ implies } K \text{ if and only if } \left(\bigcap_l \mathcal{E}(K_l)\right) \cap \Gamma_n^* \subset \mathcal{E}(K).$$

Therefore, the implication problem can be solved if Γ_n^* can be characterized. Hence, the region Γ_n^* is not only of fundamental importance in information theory, but is also of fundamental importance in probability theory.

PROBLEMS

1. *Symmetrical information expressions* An information expression is said to be symmetrical if it is identical under every permutation of the random variables involved. However, sometimes a symmetrical information expression cannot be readily recognized symbolically. For example, $I(X_1; X_2) - I(X_1; X_2|X_3)$ is symmetrical in X_1, X_2, and X_3 but it is not symmetrical symbolically. Devise a general method for recognizing symmetrical information expressions.

2. The canonical form of an information expression is unique when there is no constraint on the random variables involved. Show by an example that this does not hold when certain constraints are imposed on the random variables involved.

3. *Alternative canonical form* Denote $\cap_{i\in G}\tilde{X}_i$ by $\check{X}_G$ and let

$$\mathcal{C} = \left\{\check{X}_G : G \text{ is a nonempty subset of } \mathcal{N}_n\right\}.$$

a) Prove that a signed measure μ on $\mathcal{F}_n$ is completely specified by $\{\mu(C), C \in \mathcal{C}\}$, which can be any set of real numbers.

b) Prove that an information expression involving $X_1, X_2, \cdots, X_n$ can be expressed uniquely as a linear combination of $\mu^*(\tilde{X}_G)$, where G are nonempty subsets of $\mathcal{N}_n$.

4. *Uniqueness of the canonical form for nonlinear information expressions* Consider a function $f : \Re^k \to \Re$, where $k = 2^n - 1$ such that $\{\mathbf{h} \in \Re^k : f(\mathbf{h}) = 0\}$ has zero Lebesgue measure.

   a) Prove that f cannot be identically zero on Γ_n^*.

   b) Use the result in a) to show the uniqueness of the canonical form for the class of information expressions of the form $g(\mathbf{h})$ where g is a polynomial.

   (Yeung [231].)

5. Prove that under the constraint $\mathbf{Qh} = 0$, if neither $\mathbf{b}^\top$ nor $\mathbf{c}^\top$ is in the row space of $\mathbf{Q}$, the identities $\mathbf{b}^\top\mathbf{h} = 0$ and $\mathbf{c}^\top\mathbf{h} = 0$ are equivalent if and only if (12.61) has a unique solution.

HISTORICAL NOTES

The uniqueness of the canonical form for linear information expressions was first proved by Han [92]. The same result was independently obtained in the book by Csiszár and Körner [55]. The geometrical framework for information inequalities is due to Yeung [231]. The characterization of equivalent constrained inequalities in Section 12.4 is previously unpublished.

Chapter 13

SHANNON-TYPE INEQUALITIES

The basic inequalities form the most important set of information inequalities. In fact, almost all the information inequalities known to date are implied by the basic inequalities. These are called *Shannon-type inequalities*. In this chapter, we show that verification of Shannon-type inequalities can be formulated as a linear programming problem, thus enabling machine-proving of all such inequalities.

13.1 THE ELEMENTAL INEQUALITIES

Consider the conditional mutual information

$$I(X, Y; X, Z, U|Z, T), \tag{13.1}$$

in which the random variables X and Z appear more than once. It is readily seen that $I(X, Y; X, Z, U|Z, T)$ can be written as

$$H(X|Z, T) + I(Y; U|X, Z, T), \tag{13.2}$$

where in both $H(X|Z, T)$ and $I(Y; U|X, Z, T)$, each random variable appears only once.

A Shannon's information measure is said to be *reducible* if there exists a random variable which appears more than once in the information measure, otherwise the information measure is said to be *irreducible*. Without loss of generality, we will consider irreducible Shannon's information measures only, because a reducible Shannon's information measure can always be written as the sum of irreducible Shannon's information measures.

The nonnegativity of all Shannon's information measures form a set of inequalities called the basic inequalities. The set of basic inequalities, however, is not minimal in the sense that some basic inequalities are implied by the others. For example,

$$H(X|Y) \geq 0 \tag{13.3}$$

and

$$I(X;Y) \geq 0, \tag{13.4}$$

which are both basic inequalities involving random variables X and Y, imply

$$H(X) = H(X|Y) + I(X;Y) \geq 0, \tag{13.5}$$

which again is a basic inequality involving X and Y.

Let $\mathcal{N}_n = \{1, 2, \cdots, n\}$, where $n \geq 2$. Unless otherwise specified, all information expressions in this chapter involve some or all of the random variables $X_1, X_2, \cdots, X_n$. The value of n will be specified when necessary. Through application of the identities

$$\begin{aligned} H(X) &= H(X|Y) + I(X;Y) & (13.6)\\ H(X,Y) &= H(X) + H(Y|X) & (13.7)\\ I(X;Y,Z) &= I(X;Y) + I(X;Z|Y) & (13.8)\\ H(X|Z) &= H(X|Y,Z) + I(X;Y|Z) & (13.9)\\ H(X,Y|Z) &= H(X|Z) + H(Y|X,Z) & (13.10)\\ I(X;Y,Z|T) &= I(X;Y|T) + I(X;Z|Y,T), & (13.11) \end{aligned}$$

any Shannon's information measure can be expressed as the sum of Shannon's information measures of the following two *elemental forms*:

i) $H(X_i|X_{\mathcal{N}_n - \{i\}}), i \in \mathcal{N}_n$

ii) $I(X_i; X_j|X_K)$, where $i \neq j$ and $K \subset \mathcal{N}_n - \{i, j\}$.

This will be illustrated in the next example. It is not difficult to check that the total number of the two elemental forms of Shannon's information measures for n random variables is equal to

$$m = n + \binom{n}{2} 2^{n-2}. \tag{13.12}$$

The proof of (13.12) is left as an exercise.

EXAMPLE 13.1 *We can expand $H(X_1, X_2)$ into a sum of elemental forms of Shannon's information measures for $n = 3$ by applying the identities in (13.6) to (13.11) as follows:*

$$\begin{aligned} &H(X_1, X_2)\\ &= H(X_1) + H(X_2|X_1) & (13.13) \end{aligned}$$

$$\begin{aligned}
&= H(X_1|X_2,X_3) + I(X_1;X_2,X_3) + H(X_2|X_1,X_3) \\
&\quad + I(X_2;X_3|X_1) && (13.14) \\
&= H(X_1|X_2,X_3) + I(X_1;X_2) + I(X_1;X_3|X_2) \\
&\quad + H(X_2|X_1,X_3) + I(X_2;X_3|X_1). && (13.15)
\end{aligned}$$

The nonnegativity of the two elemental forms of Shannon's information measures form a proper subset of the set of basic inequalities. We call the m inequalities in this smaller set the *elemental inequalities*. They are equivalent to the basic inequalities because each basic inequality which is not an elemental inequality can be obtained as the sum of a set of elemental inequalities in view of (13.6) to (13.11). This will be illustrated in the next example. The proof for the minimality of the set of elemental inequalities is deferred to Section 13.6.

EXAMPLE 13.2 *In the last example, we expressed $H(X_1, X_2)$ as*

$$\begin{aligned}
&H(X_1|X_2,X_3) + I(X_1;X_2) + I(X_1;X_3|X_2) \\
&+ H(X_2|X_1,X_3) + I(X_2;X_3|X_1). && (13.16)
\end{aligned}$$

All the five Shannon's information measures in the above expression are in elemental form for $n = 3$. Then the basic inequality

$$H(X_1, X_2) \geq 0 \tag{13.17}$$

can be obtained as the sum of the following elemental inequalities:

$$\begin{aligned}
H(X_1|X_2,X_3) &\geq 0 && (13.18) \\
I(X_1;X_2) &\geq 0 && (13.19) \\
I(X_1;X_3|X_2) &\geq 0 && (13.20) \\
H(X_2|X_1,X_3) &\geq 0 && (13.21) \\
I(X_2;X_3|X_1) &\geq 0. && (13.22)
\end{aligned}$$

13.2 A LINEAR PROGRAMMING APPROACH

Recall from Section 12.2 that any information expression can be expressed uniquely in canonical form, i.e., a linear combination of the $k = 2^n - 1$ joint entropies involving some or all of the random variables $X_1, X_2, \cdots, X_n$. If the elemental inequalities are expressed in canonical form, they become linear inequalities in the entropy space $\mathcal{H}_n$. Denote this set of inequalities by $\mathbf{Gh} \geq 0$, where $\mathbf{G}$ is an $m \times k$ matrix, and define

$$\Gamma_n = \{\mathbf{h} : \mathbf{Gh} \geq 0\}. \tag{13.23}$$

We first show that Γ_n is a pyramid in the nonnegative orthant of the entropy space $\mathcal{H}_n$. Evidently, Γ_n contains the origin. Let $\mathbf{e}_j, 1 \leq j \leq k$, be the column k-vector whose jth component is equal to 1 and all the other components are equal to 0. Then the inequality

$$\mathbf{e}_j^\top \mathbf{h} \geq 0 \tag{13.24}$$

corresponds to the nonnegativity of a joint entropy, which is a basic inequality. Since the set of elemental inequalities is equivalent to the set of basic inequalities, if $\mathbf{h} \in \Gamma_n$, i.e., $\mathbf{h}$ satisfies all the elemental inequalities, then $\mathbf{h}$ also satisfies the basic inequality in (13.24). In other words,

$$\Gamma_n \subset \{\mathbf{h} : \mathbf{e}_j^\top \mathbf{h} \geq 0\} \tag{13.25}$$

for all $1 \leq j \leq k$. This implies that Γ_n is in the nonnegative orthant of the entropy space. Since Γ_n contains the origin and the constraints $\mathbf{G}\mathbf{h} \geq 0$ are linear, we conclude that Γ_n is a pyramid in the nonnegative orthant of $\mathcal{H}_n$.

Since the elemental inequalities are satisfied by the entropy function of any n random variables $X_1, X_2, \cdots, X_n$, for any $\mathbf{h}$ in Γ_n^*, $\mathbf{h}$ is also in Γ_n, i.e.,

$$\Gamma_n^* \subset \Gamma_n. \tag{13.26}$$

Therefore, for any unconstrained inequality $f \geq 0$, if

$$\Gamma_n \subset \{\mathbf{h} : f(\mathbf{h}) \geq 0\}, \tag{13.27}$$

then

$$\Gamma_n^* \subset \{\mathbf{h} : f(\mathbf{h}) \geq 0\}, \tag{13.28}$$

i.e., $f \geq 0$ always holds. In other words, (13.27) is a sufficient condition for $f \geq 0$ to always hold. Moreover, an inequality $f \geq 0$ such that (13.27) is satisfied is implied by the basic inequalities, because if $\mathbf{h}$ satisfies the basic inequalities, i.e., $\mathbf{h} \in \Gamma_n$, then $\mathbf{h}$ satisfies $f(\mathbf{h}) \geq 0$.

For constrained inequalities, following our discussion in Section 12.3, we impose the constraint

$$\mathbf{Q}\mathbf{h} = 0 \tag{13.29}$$

and let

$$\Phi = \{\mathbf{h} : \mathbf{Q}\mathbf{h} = 0\}. \tag{13.30}$$

For an inequality $f \geq 0$, if

$$(\Gamma_n \cap \Phi) \subset \{\mathbf{h} : f(\mathbf{h}) \geq 0\}, \tag{13.31}$$

then by (13.26),

$$(\Gamma_n^* \cap \Phi) \subset \{\mathbf{h} : f(\mathbf{h}) \geq 0\}, \tag{13.32}$$

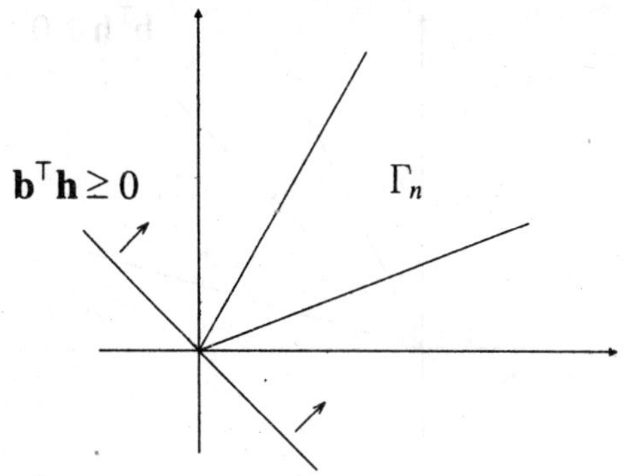

Figure 13.1. Γ_n is contained in $\{\mathbf{h} : \mathbf{b}^\top\mathbf{h} \geq 0\}$.

i.e., $f \geq 0$ always holds under the constraint Φ. In other words, (13.31) is a sufficient condition for $f \geq 0$ to always hold under the constraint Φ. Moreover, an inequality $f \geq 0$ under the constraint Φ such that (13.31) is satisfied is implied by the basic inequalities and the constraint Φ, because if $\mathbf{h} \in \Phi$ and $\mathbf{h}$ satisfies the basic inequalities, i.e., $\mathbf{h} \in \Gamma_n \cap \Phi$, then $\mathbf{h}$ satisfies $f(\mathbf{h}) \geq 0$.

13.2.1 UNCONSTRAINED INEQUALITIES

To check whether an unconstrained inequality $\mathbf{b}^\top\mathbf{h} \geq 0$ is a Shannon-type inequality, we need to check whether Γ_n is a subset of $\{\mathbf{h} : \mathbf{b}^\top\mathbf{h} \geq 0\}$. The following theorem induces a computational procedure for this purpose.

THEOREM 13.3 $\mathbf{b}^\top\mathbf{h} \geq 0$ *is a Shannon-type inequality if and only if the minimum of the problem*

$$\textit{Minimize } \mathbf{b}^\top\mathbf{h}, \textit{ subject to } \mathbf{G}\mathbf{h} \geq 0 \tag{13.33}$$

is zero. In this case, the minimum occurs at the origin.

Remark The idea of this theorem is illustrated in Figure 13.1 and Figure 13.2. In Figure 13.1, Γ_n is contained in $\{\mathbf{h} : \mathbf{b}^\top\mathbf{h} \geq 0\}$. The minimum of $\mathbf{b}^\top\mathbf{h}$ subject to Γ_n occurs at the origin with the minimum equal to 0. In Figure 13.2, Γ_n is not contained in $\{\mathbf{h} : \mathbf{b}^\top\mathbf{h} \geq 0\}$. The minimum of $\mathbf{b}^\top\mathbf{h}$ subject to Γ_n is $-\infty$. A formal proof of the theorem is given next.

Proof of Theorem 13.3 We have to prove that Γ_n is a subset of $\{\mathbf{h} : \mathbf{b}^\top\mathbf{h} \geq 0\}$ if and only if the minimum of the problem in (13.33) is zero. First of all, since $0 \in \Gamma_n$ and $\mathbf{b}^\top 0 = 0$ for any $\mathbf{b}$, the minimum of the problem in (13.33) is at most 0. Assume Γ_n is a subset of $\{\mathbf{h} : \mathbf{b}^\top\mathbf{h} \geq 0\}$ and the minimum of the

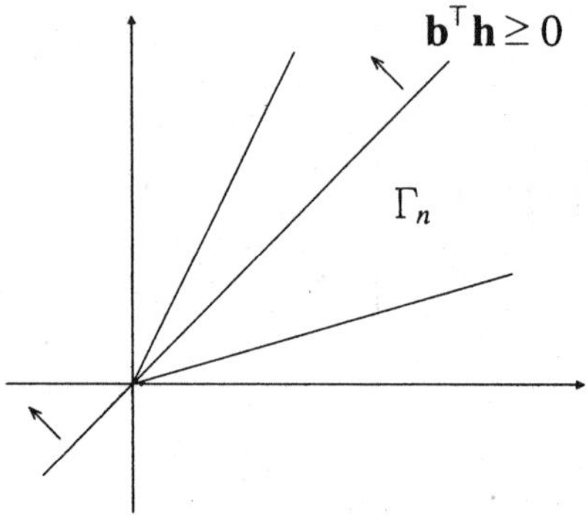

Figure 13.2. Γ_n is not contained in $\{\mathbf{h} : \mathbf{b}^\top\mathbf{h} \geq 0\}$.

problem in (13.33) is negative. Then there exists an $\mathbf{h} \in \Gamma_n$ such that

$$\mathbf{b}^\top\mathbf{h} < 0, \tag{13.34}$$

which implies

$$\Gamma_n \not\subset \{\mathbf{h} : \mathbf{b}^\top\mathbf{h} \geq 0\}, \tag{13.35}$$

which is a contradiction. Therefore, if Γ_n is a subset of $\{\mathbf{h} : \mathbf{b}^\top\mathbf{h} \geq 0\}$, then the minimum of the problem in (13.33) is zero.

To prove the converse, assume Γ_n is not a subset of $\{\mathbf{h} : \mathbf{b}^\top\mathbf{h} \geq 0\}$, i.e. (13.35) is true. Then there exists an $\mathbf{h} \in \Gamma_n$ such that

$$\mathbf{b}^\top\mathbf{h} < 0. \tag{13.36}$$

This implies that the minimum of the problem in (13.33) is negative, i.e., it is not equal to zero.

Finally, if the minimum of the problem in (13.33) is zero, since the Γ_n contains the origin and $\mathbf{b}^\top 0 = 0$, the minimum occurs at the origin. □

By virtue of this theorem, to check whether $\mathbf{b}^\top\mathbf{h} \geq 0$ is an unconstrained Shannon-type inequality, all we need to do is to apply the *optimality test* of the *simplex method* [59] to check whether the point $\mathbf{h} = 0$ is optimal for the minimization problem in (13.33). If $\mathbf{h} = 0$ is optimal, then $\mathbf{b}^\top\mathbf{h} \geq 0$ is an unconstrained Shannon-type inequality, otherwise it is not.

13.2.2 CONSTRAINED INEQUALITIES AND IDENTITIES

To check whether an inequality $\mathbf{b}^\top\mathbf{h} \geq 0$ under the constraint Φ is a Shannon-type inequality, we need to check whether $\Gamma_n \cap \Phi$ is a subset of $\{\mathbf{h} : \mathbf{b}^\top\mathbf{h} \geq 0\}$.

THEOREM 13.4 $\mathbf{b}^\top \mathbf{h} \geq 0$ *is a Shannon-type inequality under the constraint* Φ *if and only if the minimum of the problem*

$$\text{Minimize } \mathbf{b}^\top \mathbf{h}, \text{ subject to } \mathbf{G}\mathbf{h} \geq 0 \text{ and } \mathbf{Q}\mathbf{h} = 0 \tag{13.37}$$

is zero. In this case, the minimum occurs at the origin.

The proof of this theorem is similar to that for Theorem 13.3, so it is omitted. By taking advantage of the linear structure of the constraint Φ, we can reformulate the minimization problem in (13.37) as follows. Let r be the rank of $\mathbf{Q}$. Since $\mathbf{h}$ is in the null space of $\mathbf{Q}$, we can write

$$\mathbf{h} = \tilde{\mathbf{Q}}\mathbf{h}', \tag{13.38}$$

where $\tilde{\mathbf{Q}}$ is a $k \times (k-r)$ matrix such that the rows of $\tilde{\mathbf{Q}}^\top$ form a basis of the orthogonal complement of the row space of $\mathbf{Q}$, and $\mathbf{h}'$ is a column $(k-r)$-vector. Then the elemental inequalities can be expressed as

$$\mathbf{G}\tilde{\mathbf{Q}}\mathbf{h}' \geq 0, \tag{13.39}$$

and in terms of $\mathbf{h}'$, Γ_n becomes

$$\Gamma'_n = \{\mathbf{h}' : \mathbf{G}\tilde{\mathbf{Q}}\mathbf{h}' \geq 0\}, \tag{13.40}$$

which is a pyramid in $\Re^{k-r}$ (but not necessarily in the nonnegative orthant). Likewise, $\mathbf{b}^\top \mathbf{h}$ can be expressed as $\mathbf{b}^\top \tilde{\mathbf{Q}}\mathbf{h}'$.

With all the information expressions in terms of $\mathbf{h}'$, the problem in (13.37) becomes

$$\text{Minimize } \mathbf{b}^\top \tilde{\mathbf{Q}}\mathbf{h}', \text{ subject to } \mathbf{G}\tilde{\mathbf{Q}}\mathbf{h}' \geq 0. \tag{13.41}$$

Therefore, to check whether $\mathbf{b}^\top \mathbf{h} \geq 0$ is a Shannon-type inequality under the constraint Φ, all we need to do is to apply the optimality test of the simplex method to check whether the point $\mathbf{h}' = 0$ is optimal for the problem in (13.41). If $\mathbf{h}' = 0$ is optimal, then $\mathbf{b}^\top \mathbf{h} \geq 0$ is a Shannon-type inequality under the constraint Φ, otherwise it is not.

By imposing the constraint Φ, the number of elemental inequalities remains the same, while the dimension of the problem decreases from k to $k - r$.

Finally, to verify that $\mathbf{b}^\top \mathbf{h} = 0$ is a Shannon-type identity under the constraint Φ, i.e., $\mathbf{b}^\top \mathbf{h} = 0$ is implied by the basic inequalities, all we need to do is to verify that both $\mathbf{b}^\top \mathbf{h} \geq 0$ and $\mathbf{b}^\top \mathbf{h} \leq 0$ are Shannon-type inequalities under the constraint Φ.

13.3 A DUALITY

A *nonnegative linear combination* is a linear combination whose coefficients are all nonnegative. It is clear that a nonnegative linear combination of basic inequalities is a Shannon-type inequality. However, it is not clear that all Shannon-type inequalities are of this form. By applying the *duality theorem* in linear programming [194], we will see that this is in fact the case.

The *dual* of the *primal* linear programming problem in (13.33) is

$$\text{Maximize } \mathbf{y}^\top \cdot 0 \text{ subject to } \mathbf{y} \geq 0 \text{ and } \mathbf{y}^\top \mathbf{G} \leq \mathbf{b}^\top, \tag{13.42}$$

where

$$\mathbf{y} = [\ y_1 \quad \cdots \quad y_m\ ]^\top. \tag{13.43}$$

By the duality theorem, if the minimum of the primal problem is zero, which happens when $\mathbf{b}^\top \mathbf{h} \geq 0$ is a Shannon-type inequality, the maximum of the dual problem is also zero. Since the cost function in the dual problem is zero, the maximum of the dual problem is zero if and only if the feasible region

$$\Psi = \{\mathbf{y} : \mathbf{y} \geq 0 \text{ and } \mathbf{y}^\top \mathbf{G} \leq \mathbf{b}^\top\} \tag{13.44}$$

is nonempty.

THEOREM 13.5 $\mathbf{b}^\top \mathbf{h} \geq 0$ *is a Shannon-type inequality if and only if* $\mathbf{b}^\top = \mathbf{x}^\top \mathbf{G}$ *for some* $\mathbf{x} \geq 0$*, where* $\mathbf{x}$ *is a column* m*-vector, i.e.,* $\mathbf{b}^\top$ *is a nonnegative linear combination of the rows of* $\mathbf{G}$.

Proof We have to prove that Ψ is nonempty if and only if $\mathbf{b}^\top = \mathbf{x}^\top \mathbf{G}$ for some $\mathbf{x} \geq 0$. The feasible region Ψ is nonempty if and only if

$$\mathbf{b}^\top \geq \mathbf{z}^\top \mathbf{G} \tag{13.45}$$

for some $\mathbf{z} \geq 0$, where $\mathbf{z}$ is a column m-vector. Consider any $\mathbf{z}$ which satisfies (13.45), and let

$$\mathbf{s}^\top = \mathbf{b}^\top - \mathbf{z}^\top \mathbf{G} \geq 0. \tag{13.46}$$

Denote by $\mathbf{e}_j$ the column k-vector whose jth component is equal to 1 and all the other components are equal to 0, $1 \leq j \leq k$. Then $\mathbf{e}_j^\top \mathbf{h}$ is a joint entropy. Since every joint entropy can be expressed as the sum of elemental forms of Shannon's information measures, $\mathbf{e}_j^\top$ can be expressed as a nonnegative linear combination of the rows of $\mathbf{G}$. Write

$$\mathbf{s} = [\ s_1 \quad s_2 \quad \cdots \quad s_k\ ]^\top, \tag{13.47}$$

where $s_j \geq 0$ for all $1 \leq j \leq k$. Then

$$\mathbf{s}^\top = \sum_{j=1}^{k} s_j \mathbf{e}_j^\top \tag{13.48}$$

can also be expressed as a nonnegative linear combinations of the rows of $\mathbf{G}$, i.e.,

$$\mathbf{s}^\top = \mathbf{w}^\top \mathbf{G} \tag{13.49}$$

for some $\mathbf{w} \geq 0$. From (13.46), we see that

$$\mathbf{b}^\top = (\mathbf{w}^\top + \mathbf{z}^\top)\mathbf{G} = \mathbf{x}^\top \mathbf{G}, \tag{13.50}$$

where $\mathbf{x} \geq 0$. The proof is accomplished. □

From this theorem, we see that all Shannon-type inequalities are actually trivially implied by the basic inequalities! However, the verification of a Shannon-type inequality requires a computational procedure as described in the last section.

13.4 MACHINE PROVING – ITIP

Theorems 13.3 and 13.4 transform the problem of verifying a Shannon-type inequality into a linear programming problem. This enables machine-proving of all Shannon-type inequalities. A software package called ITIP[1], which runs on MATLAB, has been developed for this purpose. Both the PC version and the Unix version of ITIP are included in this book.

Using ITIP is very simple and intuitive. The following examples illustrate the use of ITIP:

1.
```
>> ITIP('H(XYZ) <= H(X) + H(Y) + H(Z)')
True
```

2.
```
>> ITIP('I(X;Z) = 0','I(X;Z|Y) = 0','I(X;Y) = 0')
True
```

3.
```
>> ITIP('I(Z;U) - I(Z;U|X) - I(Z;U|Y) <=
   0.5 I(X;Y) + 0.25 I(X;ZU) + 0.25 I(Y;ZU)')
Not provable by ITIP
```

In the first example, we prove an unconstrained inequality. In the second example, we prove that X and Z are independent if $X \to Y \to Z$ forms a Markov chain and X and Y are independent. The first identity is what we want to prove, while the second and the third expressions specify the Markov chain $X \to Y \to Z$ and the independency of X and Y, respectively. In the third example, ITIP returns the clause "Not provable by ITIP," which means that the inequality is not a Shannon-type inequality. This, however, does not mean that the inequality to be proved cannot always hold. In fact, this inequality is one of the few known non-Shannon-type inequalities which will be discussed in Chapter 14.

We note that most of the results we have previously obtained by using information diagrams can also be proved by ITIP. However, the advantage of

[1]ITIP stands for *Information-Theoretic Inequality Prover.*

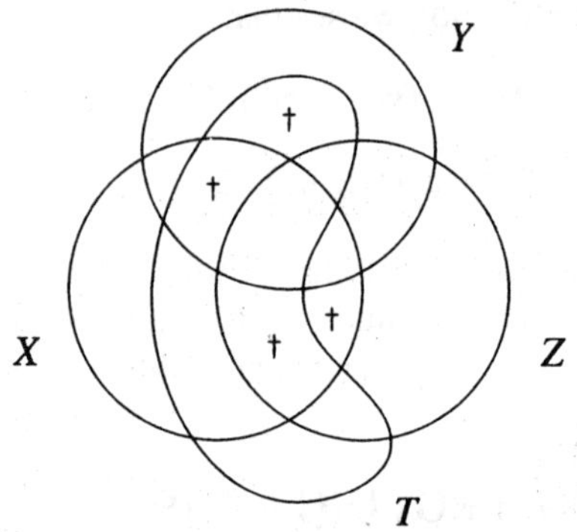

Figure 13.3. The information diagram for X, Y, Z, and T in Example 13.6.

using information diagrams is that one can visualize the structure of the problem. Therefore, the use of information diagrams and ITIP very often complement each other. In the rest of the section, we give a few examples which demonstrate the use of ITIP. The features of ITIP are described in details in the `readme` file.

EXAMPLE 13.6 *By Proposition 2.10, the long Markov chain $X \to Y \to Z \to T$ implies the two short Markov chains $X \to Y \to Z$ and $Y \to Z \to T$. We want to see whether the two short Markov chains also imply the long Markov chain. If so, they are equivalent to each other.*

Using ITIP, we have

```
>> ITIP('X/Y/Z/T', 'X/Y/Z', 'Y/Z/T')
Not provable by ITIP
```

In the above, we have used a macro in ITIP to specify the three Markov chains. The above result from ITIP says that the long Markov chain cannot be proved from the two short Markov chains by means of the basic inequalities. This strongly suggests that the two short Markov chains is weaker than the long Markov chain. However, in order to prove that this is in fact the case, we need an explicit construction of a joint distribution for X, Y, Z, and T which satisfies the two short Markov chains but not the long Markov chain. Toward this end, we resort to the information diagram in Figure 13.3. The Markov chain $X \to Y \to Z$ is equivalent to $I(X;Z|Y) = 0$, i.e.,

$$\mu^*(\tilde{X} \cap \tilde{Y}^c \cap \tilde{Z} \cap \tilde{T}) + \mu^*(\tilde{X} \cap \tilde{Y}^c \cap \tilde{Z} \cap \tilde{T}^c) = 0. \qquad (13.51)$$

Similarly, the Markov chain $Y \to Z \to T$ is equivalent to

$$\mu^*(\tilde{X} \cap \tilde{Y} \cap \tilde{Z}^c \cap \tilde{T}) + \mu^*(\tilde{X}^c \cap \tilde{Y} \cap \tilde{Z}^c \cap \tilde{T}) = 0. \qquad (13.52)$$

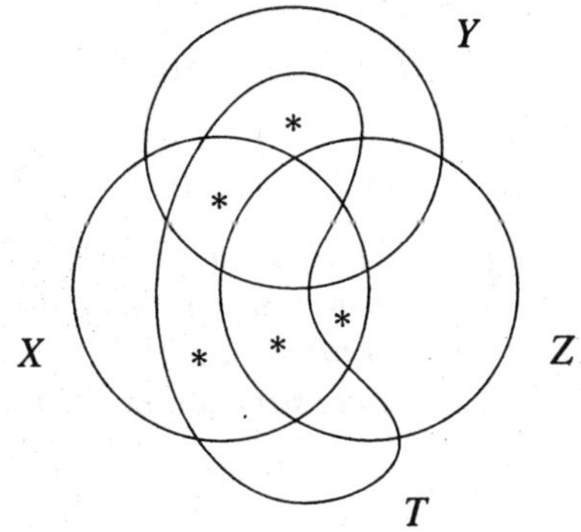

Figure 13.4. The atoms of $\mathcal{F}_4$ on which μ^* vanishes when $X \to Y \to Z \to T$ forms a Markov chain.

The four atoms involved in the constraints (13.51) and (13.52) are marked by a dagger in Figure 13.3. In Section 6.5, we have seen that the Markov chain $X \to Y \to Z \to T$ holds if and only if μ^* takes zero value on the set of atoms in Figure 13.4 which are marked with an asterisk[2]. Comparing Figure 13.3 and Figure 13.4, we see that the only atom marked in Figure 13.4 but not in Figure 13.3 is $\tilde{X} \cap \tilde{Y}^c \cap \tilde{Z}^c \cap \tilde{T}$. Thus if we can construct a μ^* such that it takes zero value on all atoms except for $\tilde{X} \cap \tilde{Y}^c \cap \tilde{Z}^c \cap \tilde{T}$, then the corresponding joint distribution satisfies the two short Markov chains but not the long Markov chain. This would show that the two short Markov chains are in fact weaker than the long Markov chain. Following Theorem 6.11, such a μ^* can be constructed.

In fact, the required joint distribution can be obtained by simply letting $X = T = U$, where U is any random variable such that $H(U) > 0$, and letting Y and Z be degenerate random variables taking constant values. Then it is easy to see that $X \to Y \to Z$ and $Y \to Z \to T$ hold, while $X \to Y \to Z \to T$ does not hold.

EXAMPLE 13.7 *The data processing theorem says that if $X \to Y \to Z \to T$ forms a Markov chain, then*

$$I(Y;Z) \geq I(X;T). \tag{13.53}$$

We want to see whether this inequality holds under the weaker condition that $X \to Y \to Z$ and $Y \to Z \to T$ form two short Markov chains. By using ITIP, we can show that (13.53) is not a Shannon-type inequality under the Markov

[2]This information diagram is essentially a reproduction of Figure 6.8.

conditions

$$I(X;Z|Y) = 0 \quad (13.54)$$

and

$$I(Y;T|Z) = 0. \quad (13.55)$$

This strongly suggests that (13.53) does not always hold under the constraint of the two short Markov chains. However, this has to be proved by an explicit construction of a joint distribution for X, Y, Z, and T which satisfies (13.54) and (13.55) but not (13.53). The construction at the end of the last example serves this purpose.

EXAMPLE 13.8 (SECRET SHARING [183]) *Consider the following secret sharing problem. Let S be a secret to be encoded into three pieces, X, Y, and Z. The scheme has to satisfy the following two secret sharing requirements:*

1. *S can be recovered from any two of the three encoded pieces.*
2. *No information about S can be obtained from any one of the three encoded pieces.*

The first requirement is equivalent to the constraints

$$H(S|X,Y) = H(S|Y,Z) = H(S|X,Z) = 0, \quad (13.56)$$

while the second requirement is equivalent to the constraints

$$I(S;X) = I(S;Y) = I(S;Z) = 0. \quad (13.57)$$

Since the secret S can be recovered if all X, Y, and Z are known,

$$H(X) + H(Y) + H(Z) \geq H(S). \quad (13.58)$$

We are naturally interested in the maximum constant c which satisfies

$$H(X) + H(Y) + H(Z) \geq cH(S). \quad (13.59)$$

We can explore the possible values of c by ITIP. After a few trials, we find that ITIP returns a "True" for all $c \leq 3$, and returns the clause "Not provable by ITIP" for any c slightly larger than 3, say 3.0001. This means that the maximum value of c is lower bounded by 3. This lower bound is in fact tight, as we can see from the following construction. Let S and N be mutually independent ternary random variables uniformly distributed on $\{0,1,2\}$, and define

$$X = N \quad (13.60)$$

$$Y = S + N \bmod 3, \quad (13.61)$$

and

$$Z = S + 2N \bmod 3. \tag{13.62}$$

Then it is easy to verify that

$$\begin{aligned} S &= Y - X \bmod 3 & (13.63)\\ &= 2Y - Z \bmod 3 & (13.64)\\ &= Z - 2X \bmod 3. & (13.65)\end{aligned}$$

Thus the requirements in (13.56) are satisfied. It is also readily verified that the requirements in (13.57) are satisfied. Finally, all S, X, Y, and Z distribute uniformly on $\{0, 1, 2\}$. Therefore,

$$H(X) + H(Y) + H(Z) = 3H(S). \tag{13.66}$$

This proves that the maximum constant c which satisfies (13.59) is 3.

Using the approach in this example, almost all information-theoretic bounds reported in the literature for this class of problems can be obtained when a definite number of random variables are involved.

13.5 TACKLING THE IMPLICATION PROBLEM

We have already mentioned in Section 12.5 that the implication problem of conditional independence is extremely difficult except for the special case that only full conditional mutual independencies are involved. In this section, we employ the tools we have developed in this chapter to tackle this problem.

In Bayesian network (see [160]), the following four axioms are often used for proving implications of conditional independencies:

- *Symmetry:*

$$X \perp Y|Z \Leftrightarrow Y \perp X|Z \tag{13.67}$$

- *Decomposition:*

$$X \perp (Y,T)|Z \Rightarrow (X \perp Y|Z) \,\&\, (X \perp T|Z) \tag{13.68}$$

- *Weak Union:*

$$X \perp (Y,T)|Z \Rightarrow X \perp Y|Z,T \tag{13.69}$$

- *Contraction:*

$$(X \perp Y|Z) \;\&\; (X \perp T|Y,Z) \;\Rightarrow\; X \perp (Y,T)|Z. \tag{13.70}$$

These axioms form a system called *semi-graphoid* and were first proposed by Dawid [61] as heuristic properties of conditional independence.

The axiom of symmetry is trivial in the context of probability[3]. The other three axioms can be summarized by

$$X \perp (Y,T)|Z \;\Leftrightarrow\; (X \perp Y|Z) \;\&\; (X \perp T|Y,Z). \tag{13.71}$$

This can easily be proved as follows. Consider the identity

$$I(X;Y,T|Z) = I(X;Y|Z) + I(X;T|Y,Z). \tag{13.72}$$

Since conditional mutual informations are always nonnegative by the basic inequalities, if $I(X;Y,T|Z)$ vanishes, $I(X;Y|Z)$ and $I(X;T|Y,Z)$ also vanish, and vice versa. This proves (13.71). In other words, (13.71) is the result of a specific application of the basic inequalities. Therefore, any implication which can be proved by invoking these four axioms can also be proved by ITIP.

In fact, ITIP is considerably more powerful than the above four axioms. This will be shown in the next example in which we give an implication which can be proved by ITIP but not by these four axioms[4]. We will see some implications which cannot be proved by ITIP when we discuss non-Shannon-type inequalities in the next chapter.

EXAMPLE 13.9 *We will show that*

$$\left.\begin{array}{l} I(X;Y|Z) = 0 \\ I(X;T|Z) = 0 \\ I(X;T|Y) = 0 \\ I(X;Z|Y) = 0 \\ I(X;Z|T) = 0 \end{array}\right\} \Rightarrow I(X;Y|T) = 0 \tag{13.73}$$

can be proved by invoking the basic inequalities. First, we write

$$I(X;Y|Z) = I(X;Y|Z,T) + I(X;Y;T|Z). \tag{13.74}$$

Since $I(X;Y|Z) = 0$ and $I(X;Y|Z,T) \ge 0$, we let

$$I(X;Y|Z,T) = a \tag{13.75}$$

[3]These four axioms can be used beyond the context of probability.

[4]This example is due to Zhen Zhang, private communication.

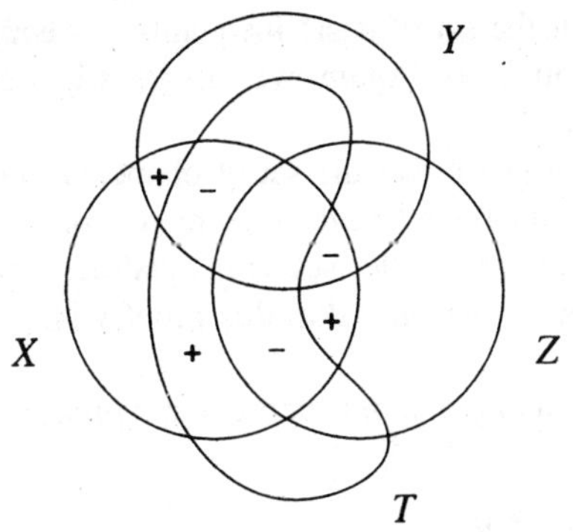

Figure 13.5. The information diagram for X, Y, Z, and T.

for some nonnegative real number a, so that

$$I(X;Y;T|Z) = -a \tag{13.76}$$

from (13.74). In the information diagram in Figure 13.5, we mark the atom $I(X;Y|Z,T)$ by a "+" and the atom $I(X;Y;T|Z)$ by a "−." Then we write

$$I(X;T|Z) = I(X;Y;T|Z) + I(X;T|Y,Z). \tag{13.77}$$

Since $I(X;T|Z) = 0$ and $I(X;Y;T|Z) = -a$, we get

$$I(X;T|Y,Z) = a. \tag{13.78}$$

In the information diagram, we mark the atom $I(X;T|Y,Z)$ with a "+." Continue in this fashion, the five CI's on the left hand side of (13.73) imply that all the atoms marked with a "+" in the information diagram take the value a, while all the atoms marked with a "−" take the value −a. From the information diagram, we see that

$$I(X;Y|T) = I(X;Y;Z|T) + I(X;Y|Z,T) = (-a) + a = 0, \tag{13.79}$$

which proves our claim. Since we base our proof on the basic inequalities, this implication can also be proved by ITIP.

Due to the form of the five given CI's in (13.73), none of the axioms in (13.68) to (13.70) can be applied. Thus we conclude that the implication in (13.73) cannot be proved by the four axioms in (13.67) to (13.70).

13.6 MINIMALITY OF THE ELEMENTAL INEQUALITIES

We have already seen in Section 13.1 that the set of basic inequalities is not minimal in the sense that in the set, some inequalities are implied by the others.

We then showed that the set of basic inequalities is equivalent to the smaller set of elemental inequalities. Again, we can ask whether the set of elemental inequalities is minimal.

In this section, we prove that the set of elemental inequalities is minimal. This result is important for efficient implementation of ITIP because it says that we cannot consider a smaller set of inequalities. The proof, however, is rather technical. The reader may skip this proof without missing the essence of this chapter.

The elemental inequalities in set-theoretic notations have one of the following two forms:

1. $\mu(\tilde{X}_i - \tilde{X}_{\mathcal{N}_n - \{i\}}) \geq 0$,
2. $\mu(\tilde{X}_i \cap \tilde{X}_j - \tilde{X}_K) \geq 0, i \neq j$ and $K \subset \mathcal{N}_n - \{i, j\}$,

where μ denotes a set-additive function defined on $\mathcal{F}_n$. They will be referred to as α-inequalities and β-inequalities, respectively.

We are to show that all the elemental inequalities are nonredundant, i.e., none of them is implied by the others. For an α-inequality

$$\mu(\tilde{X}_i - \tilde{X}_{\mathcal{N}_n - \{i\}}) \geq 0, \tag{13.80}$$

since it is the only elemental inequality which involves the atom $\tilde{X}_i - \tilde{X}_{\mathcal{N}_n - \{i\}}$, it is clearly not implied by the other elemental inequalities. Therefore we only need to show that all β-inequalities are nonredundant. To show that a β-inequality is nonredundant, it suffices to show that there exists a measure $\hat{\mu}$ on $\mathcal{F}_n$ which satisfies all other elemental inequalities except for that β-inequality.

We will show that the β-inequality

$$\mu(\tilde{X}_i \cap \tilde{X}_j - \tilde{X}_K) \geq 0 \tag{13.81}$$

is nonredundant. To facilitate our discussion, we denote $\mathcal{N}_n - K - \{i, j\}$ by $L(i, j, K)$, and we let $C_{ij|K}(S), S \subset L(i, j, K)$ be the atoms in $\tilde{X}_i \cap \tilde{X}_j - \tilde{X}_K$, where

$$C_{ij|K}(S) = \tilde{X}_i \cap \tilde{X}_j \cap \tilde{X}_S \cap \tilde{X}_K^c \cap \tilde{X}_{L(i,j,K)-S}^c. \tag{13.82}$$

We first consider the case when $L(i, j, K) = \emptyset$, i.e., $K = \mathcal{N}_n - \{i, j\}$. We construct a measure $\hat{\mu}$ by

$$\hat{\mu}(A) = \begin{cases} -1 & \text{if } A = \tilde{X}_i \cap \tilde{X}_j - \tilde{X}_K \\ 1 & \text{otherwise,} \end{cases} \tag{13.83}$$

where $A \in \mathcal{A}$. In other words, $\tilde{X}_i \cap \tilde{X}_j - \tilde{X}_K$ is the only atom with measure -1; all other atoms have measure 1. Then $\hat{\mu}(\tilde{X}_i \cap \tilde{X}_j - \tilde{X}_K) < 0$ is trivially true. It is also trivial to check that for any $i' \in \mathcal{N}_n$,

$$\hat{\mu}(\tilde{X}_{i'} - \tilde{X}_{\mathcal{N}_n - \{i'\}}) = 1 \geq 0, \tag{13.84}$$

and for any $(i',j',K') \neq (i,j,K)$ such that $i' \neq j'$ and $K' \subset \mathcal{N}_n - \{i',j'\}$,

$$\hat{\mu}(\tilde{X}_{i'} \cap \tilde{X}_{j'} - \tilde{X}_{K'}) = 1 \geq 0 \tag{13.85}$$

if $K' = \mathcal{N}_n - \{i',j'\}$. On the other hand, if K' is a proper subset of $\mathcal{N}_n - \{i',j'\}$, then $\tilde{X}_{i'} \cap \tilde{X}_{j'} - \tilde{X}_{K'}$ contains at least two atoms, and therefore

$$\hat{\mu}(\tilde{X}_{i'} \cap \tilde{X}_{j'} - \tilde{X}_{K'}) \geq 0. \tag{13.86}$$

This completes the proof for the β-inequality in (13.81) to be nonredundant when $L(i,j,K) = \phi$.

We now consider the case when $L(i,j,K) \neq \phi$, or $|L(i,j,K)| \geq 1$. We construct a measure $\hat{\mu}$ as follows. For the atoms in $\tilde{X}_i \cap \tilde{X}_j - \tilde{X}_K$, let

$$\hat{\mu}(C_{ij|K}(S)) = \begin{cases} (-1)^{|S|} - 1 & S = L(i,j,K) \\ (-1)^{|S|} & S \neq L(i,j,K). \end{cases} \tag{13.87}$$

For $C_{ij|K}(S)$, if $|S|$ is odd, it is referred to as an *odd atom* of $\tilde{X}_i \cap \tilde{X}_j - \tilde{X}_K$, and if $|S|$ is even, it is referred to as an *even atom* of $\tilde{X}_i \cap \tilde{X}_j - \tilde{X}_K$. For any atom $A \notin \tilde{X}_i \cap \tilde{X}_j - \tilde{X}_K$, we let

$$\hat{\mu}(A) = 1. \tag{13.88}$$

This completes the construction of $\hat{\mu}$.

We first prove that

$$\hat{\mu}(\tilde{X}_i \cap \tilde{X}_j - \tilde{X}_K) < 0. \tag{13.89}$$

Consider

$$\begin{aligned} \hat{\mu}(\tilde{X}_i \cap \tilde{X}_j - \tilde{X}_K) &= \sum_{S \subset L(i,j,K)} \hat{\mu}(C_{ij|K}(S)) \\ &= \left(\sum_{r=0}^{|L(i,j,K)|} \binom{|L(i,j,K)|}{r} (-1)^r \right) - 1 \\ &= -1, \end{aligned}$$

where the last equality follows from the binomial formula

$$\sum_{r=0}^{n} \binom{n}{r} (-1)^r = 0 \tag{13.90}$$

for $n \geq 1$. This proves (13.89).

Next we prove that $\hat{\mu}$ satisfies all α-inequalities. We note that for any $i' \in \mathcal{N}_n$, the atom $\tilde{X}_{i'} - \tilde{X}_{\mathcal{N}_n - \{i'\}}$ is not in $\tilde{X}_i \cap \tilde{X}_j - \tilde{X}_K$. Thus

$$\hat{\mu}(\tilde{X}_{i'} - \tilde{X}_{\mathcal{N}_n - \{i\}}) = 1 \geq 0. \tag{13.91}$$

It remains to prove that $\hat{\mu}$ satisfies all β-inequalities except for (13.81), i.e., for any $(i', j', K') \neq (i, j, K)$ such that $i' \neq j'$ and $K' \subset \mathcal{N}_n - \{i', j'\}$,

$$\hat{\mu}(\tilde{X}_{i'} \cap \tilde{X}_{j'} - \tilde{X}_{K'}) \geq 0. \tag{13.92}$$

Consider

$$\begin{aligned} &\hat{\mu}(\tilde{X}_{i'} \cap \tilde{X}_{j'} - \tilde{X}_{K'}) \\ &= \hat{\mu}((\tilde{X}_{i'} \cap \tilde{X}_{j'} - \tilde{X}_{K'}) \cap (\tilde{X}_i \cap \tilde{X}_j - \tilde{X}_K)) \\ &\quad + \hat{\mu}((\tilde{X}_{i'} \cap \tilde{X}_{j'} - \tilde{X}_{K'}) - (\tilde{X}_i \cap \tilde{X}_j - \tilde{X}_K)). \end{aligned} \tag{13.93}$$

The nonnegativity of the second term above follows from (13.88). For the first term,

$$(\tilde{X}_{i'} \cap \tilde{X}_{j'} - \tilde{X}_{K'}) \cap (\tilde{X}_i \cap \tilde{X}_j - \tilde{X}_K) \tag{13.94}$$

is nonempty if and only if

$$\{i', j'\} \cap K = \phi \quad \text{and} \quad \{i, j\} \cap K' = \phi. \tag{13.95}$$

If this condition is not satisfied, then the first term in (13.93) becomes $\hat{\mu}(\phi) = 0$, and (13.92) follows immediately.

Let us assume that the condition in (13.95) is satisfied. Then by simple counting, we see that the number atoms in

$$(\tilde{X}_{i'} \cap \tilde{X}_{j'} - \tilde{X}_{K'}) \cap (\tilde{X}_i \cap \tilde{X}_j - \tilde{X}_K) \tag{13.96}$$

is equal to 2^{φ}, where

$$\varphi = n - |\{i, j\} \cup \{i', j'\} \cup K \cup K'|. \tag{13.97}$$

For example, for $n = 6$, there are $4 = 2^2$ atoms in

$$(\tilde{X}_1 \cap \tilde{X}_2) \cap (\tilde{X}_1 \cap \tilde{X}_3 - \tilde{X}_4), \tag{13.98}$$

namely $\tilde{X}_1 \cap \tilde{X}_2 \cap \tilde{X}_3 \cap \tilde{X}_4^c \cap Y_5 \cap Y_6$, where $Y_i = \tilde{X}_i$ or $\tilde{X}_i^c$ for $i = 5, 6$. We check that

$$\varphi = 6 - |\{1, 2\} \cup \{1, 3\} \cup \phi \cup \{4\}| = 2. \tag{13.99}$$

We first consider the case when $\varphi = 0$, i.e.,

$$\mathcal{N}_n = \{i, j\} \cup \{i', j'\} \cup K \cup K'. \tag{13.100}$$

Then

$$(\tilde{X}_{i'} \cap \tilde{X}_{j'} - \tilde{X}_{K'}) \cap (\tilde{X}_i \cap \tilde{X}_j - \tilde{X}_K) \tag{13.101}$$

contains exactly one atom. If this atom is an even atom of $\tilde{X}_i \cap \tilde{X}_j - \tilde{X}_K$, then the first term in (13.93) is either 0 or 1 (cf., (13.87)), and (13.92) follows immediately. If this atom is an odd atom of $\tilde{X}_i \cap \tilde{X}_j - \tilde{X}_K$, then the first term in (13.93) is equal to -1. This happens if and only if $\{i, j\}$ and $\{i', j'\}$ have one common element, which implies that $(\tilde{X}_{i'} \cap \tilde{X}_{j'} - \tilde{X}_{K'}) - (\tilde{X}_i \cap \tilde{X}_j - \tilde{X}_K)$ is nonempty. Therefore the second term in (13.93) is at least 1, and hence (13.92) follows.

Finally, we consider the case when $\varphi \geq 1$. Using the binomial formula in (13.90), we see that the number of odd atoms and even atoms of $\tilde{X}_i \cap \tilde{X}_j - \tilde{X}_K$ in

$$(\tilde{X}_{i'} \cap \tilde{X}_{j'} - \tilde{X}_{K'}) \cap (\tilde{X}_i \cap \tilde{X}_j - \tilde{X}_K) \tag{13.102}$$

are the same. Therefore the first term in (13.93) is equal to -1 if

$$C_{ij|K}(L(i, j, K)) \in \tilde{X}_{i'} \cap \tilde{X}_{j'} - \tilde{X}_{K'}, \tag{13.103}$$

and is equal to 0 otherwise. The former is true if and only if $K' \subset K$, which implies that $(\tilde{X}_{i'} \cap \tilde{X}_{j'} - \tilde{X}_{K'}) - (\tilde{X}_i \cap \tilde{X}_j - \tilde{X}_K)$ is nonempty, or that the second term is at least 1. Thus in either case (13.92) is true. This completes the proof that (13.81) is nonredundant.

APPENDIX 13.A: THE BASIC INEQUALITIES AND THE POLYMATROIDAL AXIOMS

In this appendix, we show that the basic inequalities for a collection of n random variables $\Theta = \{X_i, i \in \mathcal{N}_n\}$ is equivalent to the following polymatroidal axioms: For all $\alpha, \beta \subset \mathcal{N}_n$,

P1. $H_\Theta(\emptyset) = 0$.

P2. $H_\Theta(\alpha) \leq H_\Theta(\beta)$ if $\alpha \subset \beta$.

P3. $H_\Theta(\alpha) + H_\Theta(\beta) \geq H_\Theta(\alpha \cap \beta) + H_\Theta(\alpha \cup \beta)$.

We first show that the polymatroidal axioms imply the basic inequalities. From P1 and P2, since $\emptyset \subset \alpha$ for any $\alpha \subset \mathcal{N}_n$, we have

$$H_\Theta(\alpha) \geq H_\Theta(\emptyset) = 0, \tag{13.A.1}$$

or

$$H(X_\alpha) \geq 0. \tag{13.A.2}$$

This shows that entropy is nonnegative.

In P2, letting $\gamma = \beta \backslash \alpha$, we have

$$H_\Theta(\alpha) \leq H_\Theta(\alpha \cup \gamma), \tag{13.A.3}$$

or

$$H(X_\gamma | X_\alpha) \geq 0. \tag{13.A.4}$$

Here, γ and α are disjoint subsets of $\mathcal{N}_n$.

In P3, letting $\gamma = \beta \backslash \alpha$, $\delta = \alpha \cap \beta$, and $\sigma = \alpha \backslash \beta$, we have

$$H_\Theta(\sigma \cup \delta) + H_\Theta(\gamma \cup \delta) \geq H_\Theta(\delta) + H_\Theta(\sigma \cup \delta \cup \gamma), \tag{13.A.5}$$

or

$$I(X_\sigma; X_\gamma | X_\delta) \geq 0. \tag{13.A.6}$$

Again, σ, δ, and γ are disjoint subsets of $\mathcal{N}_n$. When $\delta = \emptyset$, from P3, we have

$$I(X_\sigma; X_\gamma) \geq 0. \tag{13.A.7}$$

Thus P1 to P3 imply that entropy is nonnegative, and that conditional entropy, mutual information, and conditional mutual information are nonnegative provided that they are irreducible. However, it has been shown in Section 13.1 that a reducible Shannon's information measure can always be written as the sum of irreducible Shannon's information measures. Therefore, we have shown that the polymatroidal axioms P1 to P3 imply the basic inequalities.

The converse is trivial and the proof is omitted.

PROBLEMS

1. Prove (13.12) for the total number of elemental forms of Shannon's information measures for n random variables.

2. Shannon-type inequalities for n random variables $X_1, X_2, \cdots, X_n$ refer to all information inequalities implied by the basic inequalities for these n random variables. Show that no new information inequality can be generated by considering the basic inequalities for more than n random variables.

3. Show by an example that the decomposition of an information expression into a sum of elemental forms of Shannon's information measures is not unique.

4. *Elemental forms of conditional independencies* Consider random variables $X_1, X_2, \cdots, X_n$. A conditional independency is said to be *elemental* if it corresponds to setting an elemental form of Shannon's information measure to zero. Show that any conditional independency involving $X_1, X_2, \cdots, X_n$ is equivalent to a collection of elemental conditional independencies.

5. *Symmetrical information inequalities*

   a) Show that every symmetrical information expression (cf. Problem 1 in Chapter 12) involving random variable $X_1, X_2, \cdots, X_n$ can be written in the form

   $$E = \sum_{k=0}^{n-1} a_k c_k^{(n)},$$

   where

   $$c_0^{(n)} = \sum_{i=1}^{n} H(X_i|X_{N-i})$$

   and for $1 \leq k \leq n-1$,

   $$c_k^{(n)} = \sum_{\substack{1 \leq i < j \leq n \\ K \subset N - \{i,j\}, |K| = k-1}} I(X_i; X_j | X_K).$$

Note that $c_0^{(n)}$ is the sum of all Shannon's information measures of the first elemental form, and for $1 \le k \le n-1$, $c_k^{(n)}$ is the sum of all Shannon's information measures of the second elemental form conditioning on $k-1$ random variables.

b) Show that $E \ge 0$ always holds if $a_k \ge 0$ for all k.

c) Show that if $E \ge 0$ always holds, then $a_k \ge 0$ for all k. Hint: Construct random variables $X_1, X_2, \cdots, X_n$ for each $0 \le k \le n-1$ such that $c_k^{(n)} > 0$ and $c_{k'}^{(n)} = 0$ for all $0 \le k' \le n-1$ and $k' \ne k$.

(Han [93].)

6. *Strictly positive probability distributions* It was shown in Proposition 2.12 that

$$\left.\begin{array}{l} X_1 \perp X_4|(X_2, X_3) \\ X_1 \perp X_3|(X_2, X_4) \end{array}\right\} \Rightarrow X_1 \perp (X_3, X_4)|X_2$$

if $p(x_1, x_2, x_3, x_4) > 0$ for all x_1, x_2, x_3, and x_4. Show by using ITIP that this implication is not implied by the basic inequalities. This strongly suggests that this implication does not hold in general, which was shown to be the case by the construction following Proposition 2.12.

7. a) Verify by ITIP that

$$I(X_1, X_2; Y_1, Y_2) \le I(X_1; Y_1) + I(X_2; Y_2)$$

under the constraint $H(Y_1, Y_2|X_1, X_2) = H(Y_1|X_1) + H(Y_2|X_2)$. This constrained inequality was used in Problem 10 in Chapter 8 to obtain the capacity of two independent parallel channels.

b) Verify by ITIP that

$$I(X_1, X_2; Y_1, Y_2) \ge I(X_1; Y_1) + I(X_2; Y_2)$$

under the constraint $I(X_1; X_2) = 0$. This constrained inequality was used in Problem 4 in Chapter 9 to obtain the rate-distortion function for a product source.

8. Verify by ITIP the information identity in Example 6.18.

9. Repeat Problem 9 in Chapter 6 with the help of ITIP.

10. Prove the implications in Problem 13 in Chapter 6 by ITIP and show that they cannot be deduced from the semi-graphoidal axioms. (Studený [202].)

HISTORICAL NOTES

For almost half a century, all information inequalities known in the literature are consequences of the basic inequalities due to Shannon [184]. Fujishige [77] showed that the entropy function is a polymatroid (see Appendix 13.A). Yeung [231] showed that verification of all such inequalities, referred to Shannon-type inequalities, can be formulated as a linear programming problem if the number of random variables involved is fixed. ITIP, a software package for this purpose, was developed by Yeung and Yan [234]. Non-Shannon-type inequalities, which have been discovered only recently, will be discussed in the next chapter.

The recent interest in the implication problem of conditional independence has been fueled by Bayesian networks. For a number of years, researchers in Bayesian networks generally believed that the semi-graphoidal axioms form a complete set of axioms for conditional independence until it was refuted by Studený [202].

Chapter 14

BEYOND SHANNON-TYPE INEQUALITIES

In Chapter 12, we introduced the regions Γ_n^* and Γ_n in the entropy space $\mathcal{H}_n$ for n random variables. From Γ_n^*, one in principle can determine whether any information inequality always holds. The region Γ_n, defined by the set of all basic inequalities (equivalently all elemental inequalities) involving n random variables, is an outer bound on Γ_n^*. From Γ_n, one can determine whether any information inequality is implied by the basic inequalities. If so, it is called a Shannon-type inequality. Since the basic inequalities always hold, so do all Shannon-type inequalities. In the last chapter, we have shown how machine-proving of all Shannon-type inequalities can be made possible by taking advantage of the linear structure of Γ_n.

If the two regions Γ_n^* and Γ_n are identical, then all information inequalities which always hold are Shannon-type inequalities, and hence all information inequalities can be completely characterized. However, if Γ_n^* is a proper subset of Γ_n, then there exist constraints on an entropy function which are not implied by the basic inequalities. Such a constraint, if in the form of an inequality, is referred to a non-Shannon-type inequality.

There is a point here which needs further explanation. The fact that $\Gamma_n^* \neq \Gamma_n$ does not necessarily imply the existence of a non-Shannon-type inequality. As an example, suppose Γ_n contains all but an isolated point in Γ_n^*. Then this does not lead to the existence of a non-Shannon-type inequality for n random variables.

In this chapter, we present characterizations of Γ_n^* which are more refined than Γ_n. These characterizations lead to the existence of non-Shannon-type inequalities for $n \geq 4$.

14.1 CHARACTERIZATIONS OF Γ_2^*, Γ_3^*, AND $\overline{\Gamma}_n^*$

Recall from the proof of Theorem 6.6 that the vector $\mathbf{h}$ represents the values of the I-Measure μ^* on the unions in $\mathcal{F}_n$. Moreover, $\mathbf{h}$ is related to the values of μ^* on the atoms of $\mathcal{F}_n$, represented as $\mathbf{u}$, by

$$\mathbf{h} = \mathbf{C}_n \mathbf{u} \tag{14.1}$$

where $\mathbf{C}_n$ is a unique $k \times k$ matrix with $k = 2^n - 1$ (cf. (6.27)).

Let $\mathcal{I}_n$ be the k-dimensional Euclidean space with the coordinates labeled by the components of $\mathbf{u}$. Note that each coordinate in $\mathcal{I}_n$ corresponds to the value of μ^* on a nonempty atom of $\mathcal{F}_n$. Recall from Lemma 12.1 the definition of the region

$$\Psi_n^* = \{\mathbf{u} \in \mathcal{I}_n : \mathbf{C}_n \mathbf{u} \in \Gamma_n^*\}, \tag{14.2}$$

which is obtained from the region Γ_n^* via the linear transformation induced by $\mathbf{C}_n^{-1}$. Analogously, we define the region

$$\Psi_n = \{\mathbf{u} \in \mathcal{I}_n : \mathbf{C}_n \mathbf{u} \in \Gamma_n\}. \tag{14.3}$$

The region Γ_n^*, as we will see, is extremely difficult to characterize for a general n. Therefore, we start our discussion with the simplest case, namely $n = 2$.

THEOREM 14.1 $\Gamma_2^* = \Gamma_2$.

Proof For $n = 2$, the elemental inequalities are

$$H(X_1|X_2) = \mu^*(\tilde{X}_1 - \tilde{X}_2) \geq 0 \tag{14.4}$$
$$H(X_2|X_1) = \mu^*(\tilde{X}_2 - \tilde{X}_1) \geq 0 \tag{14.5}$$
$$I(X_1;X_2) = \mu^*(\tilde{X}_1 \cap \tilde{X}_2) \geq 0. \tag{14.6}$$

Note that the quantities on the left hand sides above are precisely the values of μ^* on the atoms of $\mathcal{F}_2$. Therefore,

$$\Psi_2 = \{\mathbf{u} \in \mathcal{I}_2 : \mathbf{u} \geq 0\}, \tag{14.7}$$

i.e., Ψ_2 is the nonnegative orthant of $\mathcal{I}_2$. Since $\Gamma_2^* \subset \Gamma_2$, $\Psi_2^* \subset \Psi_2$. On the other hand, $\Psi_2 \subset \Psi_2^*$ by Lemma 12.1. Thus $\Psi_2^* = \Psi_2$, which implies $\Gamma_2^* = \Gamma_2$. The proof is accomplished. □

Next, we prove that Theorem 14.1 cannot even be generalized to $n = 3$.

THEOREM 14.2 $\Gamma_3^* \neq \Gamma_3$.

Proof For $n = 3$, the elemental inequalities are

$$H(X_i|X_j, X_k) = \mu^*(\tilde{X}_i - \tilde{X}_j - \tilde{X}_k) \geq 0 \tag{14.8}$$
$$I(X_i; X_j|X_k) = \mu^*(\tilde{X}_i \cap \tilde{X}_j - \tilde{X}_k) \geq 0, \tag{14.9}$$

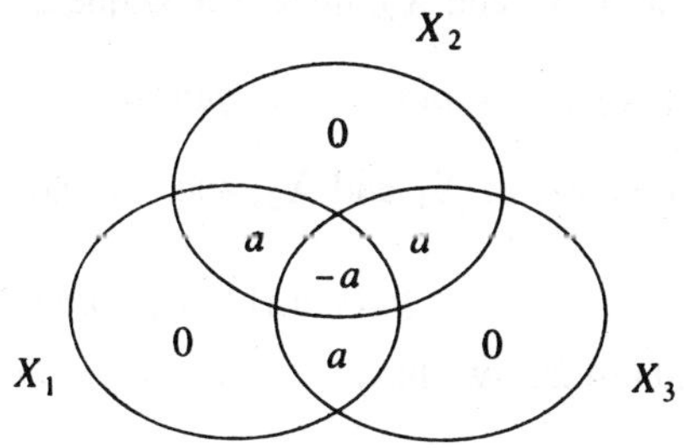

Figure 14.1. The set-theoretic structure of the point $(0, 0, 0, a, a, a, -a)$ in Ψ_3.

and

$$
\begin{aligned}
I(X_i; X_j) &= \mu^*(\tilde{X}_i \cap \tilde{X}_j) && (14.10)\\
&= \mu^*(\tilde{X}_i \cap \tilde{X}_j \cap \tilde{X}_k) + \mu^*(\tilde{X}_i \cap \tilde{X}_j - \tilde{X}_k) && (14.11)\\
&\geq 0 && (14.12)
\end{aligned}
$$

for $1 \leq i < j < k \leq 3$. For $\mathbf{u} \in \mathcal{I}_3$, let

$$\mathbf{u} = (u_1, u_2, u_3, u_4, u_5, u_6, u_7), \tag{14.13}$$

where $u_i, 1 \leq i \leq 7$ correspond to the values

$$
\begin{aligned}
&\mu^*(\tilde{X}_1 - \tilde{X}_2 - \tilde{X}_3),\ \mu^*(\tilde{X}_2 - \tilde{X}_1 - \tilde{X}_3),\ \mu^*(\tilde{X}_3 - \tilde{X}_1 - \tilde{X}_2),\\
&\mu^*(\tilde{X}_1 \cap \tilde{X}_2 - \tilde{X}_3),\ \mu^*(\tilde{X}_1 \cap \tilde{X}_3 - \tilde{X}_2),\ \mu^*(\tilde{X}_2 \cap \tilde{X}_3 - \tilde{X}_1), \qquad (14.14)\\
&\mu^*(\tilde{X}_1 \cap \tilde{X}_2 \cap \tilde{X}_3),
\end{aligned}
$$

respectively. These are the values of μ^* on the nonempty atoms of $\mathcal{F}_3$. Then from (14.8), (14.9), and (14.12), we see that

$$\Psi_3 = \{\mathbf{u} \in \mathcal{I}_3 : u_i \geq 0,\ 1 \leq i \leq 6;\ u_j + u_7 \geq 0,\ 4 \leq j \leq 6\}. \tag{14.15}$$

It is easy to check that the point $(0, 0, 0, a, a, a, -a)$ for any $a \geq 0$ is in Ψ_3. This is illustrated in Figure 14.1, and it is readily seen that the relations

$$H(X_i|X_j, X_k) = 0 \tag{14.16}$$

and

$$I(X_i; X_j) = 0 \tag{14.17}$$

for $1 \leq i < j < k \leq 3$ are satisfied, i.e., each random variable is a function of the other two, and the three random variables are pairwise independent.

Let $\mathcal{S}_{X_i}$ be the support of X_i, $i = 1, 2, 3$. For any $x_1 \in \mathcal{S}_{X_1}$ and $x_2 \in \mathcal{S}_{X_2}$, since X_1 and X_2 are independent, we have

$$p(x_1, x_2) = p(x_1)p(x_2) > 0. \tag{14.18}$$

Since X_3 is a function of X_1 and X_2, there is a unique $x_3 \in \mathcal{S}_{X_3}$ such that

$$p(x_1, x_2, x_3) = p(x_1, x_2) = p(x_1)p(x_2) > 0. \tag{14.19}$$

Now since X_2 is a function of X_1 and X_3, and X_1 and X_3 are independent, we can write

$$p(x_1, x_2, x_3) = p(x_1, x_3) = p(x_1)p(x_3). \tag{14.20}$$

Equating (14.19) and (14.20), we have

$$p(x_2) = p(x_3). \tag{14.21}$$

Now consider any $x_2' \in \mathcal{S}_{X_2}$ such that $x_2' \neq x_2$. Since X_2 and X_3 are independent, we have

$$p(x_2', x_3) = p(x_2')p(x_3) > 0. \tag{14.22}$$

Since X_1 is a function of X_2 and X_3, there is a unique $x_1' \in \mathcal{S}_{X_1}$ such that

$$p(x_1', x_2', x_3) = p(x_2', x_3) = p(x_2')p(x_3) > 0. \tag{14.23}$$

Now since X_2 is a function of X_1 and X_3, and X_1 and X_3 are independent, we can write

$$p(x_1', x_2', x_3) = p(x_1', x_3) = p(x_1')p(x_3). \tag{14.24}$$

Similarly, since X_3 is a function of X_1 and X_2, and X_1 and X_2 are independent, we can write

$$p(x_1', x_2', x_3) = p(x_1', x_2') = p(x_1')p(x_2'). \tag{14.25}$$

Equating (14.24) and (14.25), we have

$$p(x_2') = p(x_3), \tag{14.26}$$

and from (14.21), we have

$$p(x_2') = p(x_2). \tag{14.27}$$

Therefore X_2 must have a uniform distribution on its support. The same can be proved for X_1 and X_3. Now from Figure 14.1,

$$\begin{aligned} H(X_1) &= H(X_1|X_2, X_3) + I(X_1; X_2|X_3) + I(X_1; X_3|X_2) \\ &\quad + I(X_1; X_2; X_3) & (14.28) \\ &= 0 + a + a + (-a) & (14.29) \\ &= a, & (14.30) \end{aligned}$$

and similarly

$$H(X_2) = H(X_3) = a. \tag{14.31}$$

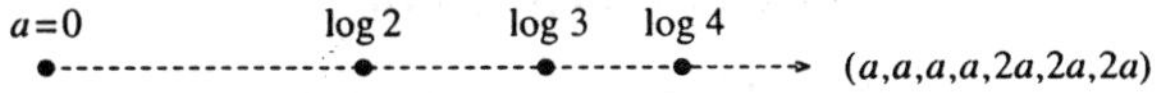

Figure 14.2. The values of a for which $(a, a, a, 2a, 2a, 2a, 2a)$ is in Γ_3.

Then the only values that a can take are $\log M$, where M (a positive integer) is the cardinality of the supports of X_1, X_2, and X_3. In other words, if a is not equal to $\log M$ for some positive integer M, then the point $(0, 0, 0, a, a, a, -a)$ is not in Ψ_3^*. This proves that $\Psi_3^* \neq \Psi_3$, which implies $\Gamma_3^* \neq \Gamma_3$. The theorem is proved. □

The proof above has the following interpretation. For $\mathbf{h} \in \mathcal{H}_3$, let

$$\mathbf{h} = (h_1, h_2, h_3, h_{12}, h_{13}, h_{23}, h_{123}). \tag{14.32}$$

From Figure 14.1, we see that the point $(0, 0, 0, a, a, a, -a)$ in Ψ_3 corresponds to the point $(a, a, a, 2a, 2a, 2a, 2a)$ in Γ_3. Evidently, the point $(a, a, a, 2a, 2a, 2a, 2a)$ in Γ_3 satisfies the 6 elemental inequalities given in (14.8) and (14.12) for $1 \leq i < j < k \leq 3$ with equality. Since Γ_3 is defined by all the elemental inequalities, the set

$$\{(a, a, a, 2a, 2a, 2a, 2a) \in \Gamma_3 : a \geq 0\} \tag{14.33}$$

is in the intersection of 6 hyperplanes in $\mathcal{H}_3$ (i.e., $\Re^7$) defining the boundary of Γ_3, and hence it defines an extreme direction of Γ_3. Then the proof says that along this extreme direction of Γ_3, only certain discrete points, namely those points with a equals $\log M$ for some positive integer M, are entropic. This is illustrated in Figure 14.2. As a consequence, the region Γ_3^* is not convex.

Having proved that $\Gamma_3^* \neq \Gamma_3$, it is natural to conjecture that the gap between Γ_3^* and Γ_3 has Lebesgue measure 0. In other words, $\overline{\Gamma}_3^* = \Gamma_3$, where $\overline{\Gamma}_3^*$ is the closure of Γ_3. This conjecture is indeed true and will be proved at the end of the section.

More generally, we are interested in characterizing $\overline{\Gamma}_n^*$, the closure of Γ_n^*. Although the region $\overline{\Gamma}_n^*$ is not sufficient for characterizing all information inequalities, it is actually sufficient for characterizing all unconstrained information inequalities. This can be seen as follows. Following the discussion in Section 12.3.1, an unconstrained information inequality $f \geq 0$ involving n random variables always hold if and only if

$$\Gamma_n^* \subset \{\mathbf{h} : f(\mathbf{h}) \geq 0\}. \tag{14.34}$$

Since $\{\mathbf{h} : f(\mathbf{h}) \geq 0\}$ is closed, upon taking closure on both sides, we have

$$\overline{\Gamma}_n^* \subset \{\mathbf{h} : f(\mathbf{h}) \geq 0\}. \tag{14.35}$$

On the other hand, if $f \geq 0$ satisfies (14.35), then

$$\Gamma_n^* \subset \overline{\Gamma}_n^* \subset \{\mathbf{h} : f(\mathbf{h}) \geq 0\}. \tag{14.36}$$

Therefore, (14.34) and (14.35) are equivalent, and hence $\overline{\Gamma}_n^*$ is sufficient for characterizing all unconstrained information inequalities.

We will prove in the next theorem an important property of the region $\overline{\Gamma}_n^*$ for all $n \geq 2$. This result will be used in the proof for $\overline{\Gamma}_3^* = \Gamma_3$. Further, this result will be used in Chapter 15 when we use Γ_n^* to characterize the achievable information rate region for multi-source networking coding. It will also be used in Chapter 16 when we establish a fundamental relation between information theory and group theory.

We first prove a simple lemma. In the following, we use $\mathcal{N}_n$ to denote the set $\{1, 2, \cdots, n\}$.

LEMMA 14.3 *If $\mathbf{h}$ and $\mathbf{h}'$ are in Γ_n^*, then $\mathbf{h} + \mathbf{h}'$ is in Γ_n^*.*

Proof Consider $\mathbf{h}$ and $\mathbf{h}'$ in Γ_n^*. Let $\mathbf{h}$ represents the entropy function for random variables $X_1, X_2, \cdots, X_n$, and let $\mathbf{h}'$ represents the entropy function for random variables $X_1', X_2', \cdots, X_n'$. Let $(X_1, X_2, \cdots, X_n)$ and $(X_1', X_2', \cdots, X_n')$ be independent, and define random variables $Y_1, Y_2, \cdots, Y_n$ by

$$Y_i = (X_i, X_i') \tag{14.37}$$

for all $i \in \mathcal{N}_n$. Then for any subset α of $\mathcal{N}_n$,

$$H(Y_\alpha) = H(X_\alpha) + H(X_\alpha') = h_\alpha + h_\alpha'. \tag{14.38}$$

Therefore, $\mathbf{h} + \mathbf{h}'$, which represents the entropy function for $Y_1, Y_2, \cdots, Y_n$, is in Γ_n^*. The lemma is proved. □

COROLLARY 14.4 *If $\mathbf{h} \in \Gamma_n^*$, then $k\mathbf{h} \in \Gamma_n^*$ for any positive integer k.*

Proof It suffices to write

$$k\mathbf{h} = \underbrace{\mathbf{h} + \mathbf{h} + \cdots + \mathbf{h}}_{k} \tag{14.39}$$

and apply Lemma 14.3. □

THEOREM 14.5 *$\overline{\Gamma}_n^*$ is a convex cone.*

Proof Consider the entropy function for random variables $X_1, X_2, \cdots, X_n$ all taking constant values with probability 1. Then for all subset α of $\mathcal{N}_n$,

$$H(X_\alpha) = 0. \tag{14.40}$$

Therefore, Γ_n^* contains the origin in $\mathcal{H}_n$.

Let $\mathbf{h}$ and $\mathbf{h}'$ in Γ_n^* be the entropy functions for any two sets of random variables $Y_1, Y_2, \cdots, Y_n$ and $Z_1, Z_2, \cdots, Z_n$, respectively. In view of Corollary 14.4, in order to prove that $\overline{\Gamma}_n^*$ is a convex cone, we only need to show that if $\mathbf{h}$ and $\mathbf{h}'$ are in Γ_n^*, then $b\mathbf{h}+\bar{b}\mathbf{h}'$ is in $\overline{\Gamma}_n^*$ for all $0 < b < 1$, where $\bar{b} = 1-b$.

Let $(\mathbf{Y}_1, \mathbf{Y}_2, \cdots, \mathbf{Y}_n)$ be k independent copies of $(Y_1, Y_2, \cdots, Y_n)$ and $(\mathbf{Z}_1, \mathbf{Z}_2, \cdots, \mathbf{Z}_n)$ be k independent copies of $(Z_1, Z_2, \cdots, Z_n)$. Let U be a ternary random variable independent of all other random variables such that

$$\Pr\{U=0\} = 1-\delta-\mu,\ \Pr\{U=1\} = \delta,\ \Pr\{U=2\} = \mu.$$

Now construct random variables $X_1, X_2, \cdots, X_n$ by letting

$$X_i = \begin{cases} 0 & \text{if } U = 0 \\ \mathbf{Y}_i & \text{if } U = 1 \\ \mathbf{Z}_i & \text{if } U = 2. \end{cases}$$

Note that $H(U) \to 0$ as $\delta, \mu \to 0$. Then for any nonempty subset α of $\mathcal{N}_n$,

$$\begin{aligned} H(X_\alpha) &\le H(X_\alpha, U) && (14.41)\\ &= H(U) + H(X_\alpha|U) && (14.42)\\ &= H(U) + \delta k H(Y_\alpha) + \mu k H(Z_\alpha). && (14.43) \end{aligned}$$

On the other hand,

$$H(X_\alpha) \ge H(X_\alpha|U) = \delta k H(Y_\alpha) + \mu k H(Z_\alpha). \tag{14.44}$$

Combining the above, we have

$$0 \le H(X_\alpha) - (\delta k H(Y_\alpha) + \mu k H(Z_\alpha)) \le H(U). \tag{14.45}$$

Now take

$$\delta = \frac{b}{k} \tag{14.46}$$

and

$$\mu = \frac{\bar{b}}{k} \tag{14.47}$$

to obtain

$$0 \le H(X_\alpha) - (bH(Y_\alpha) + \bar{b}H(Z_\alpha)) \le H(U). \tag{14.48}$$

By letting k be sufficiently large, the upper bound can be made arbitrarily small. This shows that $b\mathbf{h} + \bar{b}\mathbf{h}' \in \overline{\Gamma}_n^*$. The theorem is proved. □

In the next theorem, we prove that Γ_3^* and Γ_3 are almost identical. Analogous to $\overline{\Gamma}_n^*$, we will use $\overline{\Psi}_n^*$ to denote the closure of Ψ_n^*.

THEOREM 14.6 $\overline{\Gamma}_3^* = \Gamma_3$.

Proof We first note that $\overline{\Gamma}_3^* = \Gamma_3$ if and only if

$$\overline{\Psi}_3^* = \Psi_3. \tag{14.49}$$

Since

$$\Gamma_3^* \subset \Gamma_3 \tag{14.50}$$

and Γ_3 is closed, by taking closure on both sides in the above, we obtain $\overline{\Gamma}_3^* \subset \Gamma_3$. This implies that $\overline{\Psi}_3^* \subset \Psi_3$. Therefore, in order to prove the theorem, it suffices to show that $\Psi_3 \subset \overline{\Psi}_3^*$.

We first show that the point $(0, 0, 0, a, a, a, -a)$ is in $\overline{\Psi}_3^*$ for all $a > 0$. Let random variables X_1, X_2, and X_3 be defined as in Example 6.10, i.e., X_1 and X_2 are two independent binary random variables taking values in $\{0, 1\}$ according to the uniform distribution, and

$$X_3 = X_1 + X_2 \bmod 2. \tag{14.51}$$

Let $\mathbf{h} \in \Gamma_3^*$ represents the entropy function for X_1, X_2, and X_3, and let

$$\mathbf{u} = \mathbf{C}_3^{-1}\mathbf{h}. \tag{14.52}$$

As in the proof of Theorem 14.2, we let u_i, $1 \leq i \leq 7$, be the coordinates of $\mathcal{I}_3$ which correspond to the values of the quantities in (14.14), respectively. From Example 6.10, we have

$$u_i = \begin{cases} 0 & \text{for } i = 1, 2, 3 \\ 1 & \text{for } i = 4, 5, 6 \\ -1 & \text{for } i = 7. \end{cases} \tag{14.53}$$

Thus the point $(0, 0, 0, 1, 1, 1, -1)$ is in Ψ_3^*, and the I-Measure μ^* for X_1, X_2, and X_3 is shown in Figure 14.3. Then by Corollary 14.4, $(0, 0, 0, k, k, k, -k)$ is in Ψ_3^* and hence in $\overline{\Psi}_3^*$ for all positive integer k. Since $\overline{\Gamma}_3^*$ contains the origin, $\overline{\Psi}_3^*$ also contains the origin. By Theorem 14.5, $\overline{\Gamma}_3^*$ is convex. This implies $\overline{\Psi}_3^*$ is also convex. Therefore, $(0, 0, 0, a, a, a, -a)$ is in $\overline{\Psi}_3^*$ for all $a > 0$.

Consider any $\mathbf{u} \in \Psi_3$. Referring to (14.15), we have

$$u_i \geq 0 \tag{14.54}$$

for $1 \leq i \leq 6$. Thus u_7 is the only component of $\mathbf{u}$ which can possibly be negative. We first consider the case when $u_7 \geq 0$. Then $\mathbf{u}$ is in the nonnegative orthant of $\mathcal{I}_3$, and by Lemma 12.1, $\mathbf{u}$ is in Ψ_3^*. Next, consider the case when $u_7 < 0$. Let

$$\mathbf{t} = (0, 0, 0, -u_7, -u_7, -u_7, u_7). \tag{14.55}$$

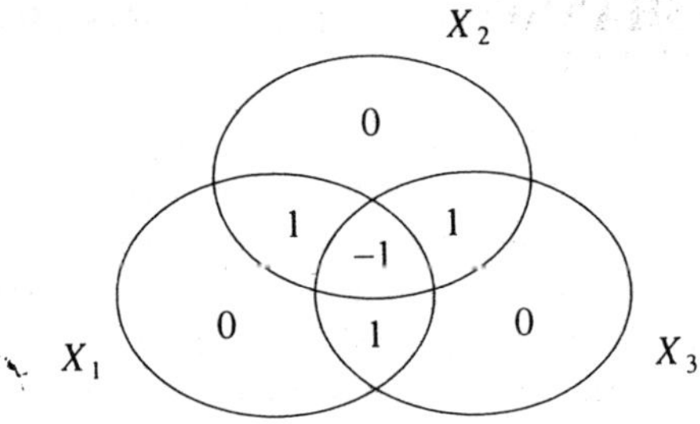

Figure 14.3. The I-Measure μ^* for X_1, X_2, and X_3 in the proof of Theorem 14.6.

Then

$$\mathbf{u} = \mathbf{w} + \mathbf{t}, \tag{14.56}$$

where

$$\mathbf{w} = (u_1, u_2, u_3, u_4 + u_7, u_5 + u_7, u_6 + u_7, 0). \tag{14.57}$$

Since $-u_7 > 0$, we see from the foregoing that $\mathbf{t} \in \overline{\Psi}^*_3$. From (14.15), we have

$$u_i + u_7 \geq 0 \tag{14.58}$$

for $i = 4, 5, 6$. Thus $\mathbf{w}$ is in the nonnegative orthant in $\mathcal{I}_3$ and hence in Ψ^*_3 by Lemma 12.1. Now for any $\epsilon > 0$, let $\mathbf{t}' \in \Psi^*_3$ such that

$$\|\mathbf{t} - \mathbf{t}'\| < \epsilon, \tag{14.59}$$

where $\|\mathbf{t} - \mathbf{t}'\|$ denotes the Euclidean distance between $\mathbf{t}$ and $\mathbf{t}'$, and let

$$\mathbf{u}' = \mathbf{w} + \mathbf{t}'. \tag{14.60}$$

Since both $\mathbf{w}$ and $\mathbf{t}'$ are in Ψ^*_3, by Lemma 14.3, $\mathbf{u}'$ is also in Ψ^*_3, and

$$\|\mathbf{u} - \mathbf{u}'\| = \|\mathbf{t} - \mathbf{t}'\| < \epsilon. \tag{14.61}$$

Therefore, $\mathbf{u} \in \overline{\Psi}^*_3$. Hence, $\Psi_3 \subset \overline{\Psi}^*_3$, and the theorem is proved. □

Remark 1 Han [94] has found that Γ_3 is the smallest cone that contains Γ^*_3. This result together with Theorem 14.5 implies Theorem 14.6. Theorem 14.6 was also obtained by Golić [85], and it is a consequence of the theorem in Matúš [144].

Remark 2 We have shown that the region $\overline{\Gamma}^*_n$ completely characterizes all unconstrained information inequalities involving n random variables. Since $\overline{\Gamma}^*_3 = \Gamma_3$, it follows that there exists no unconstrained information inequalities involving three random variables other than the Shannon-type inequalities. However, whether there exist constrained non-Shannon-type inequalities involving three random variables is still unknown.

14.2 A NON-SHANNON-TYPE UNCONSTRAINED INEQUALITY

We have proved in Theorem 14.6 at the end of the last section that $\overline{\Gamma}_3^* = \Gamma_3$. It is natural to conjecture that this theorem can be generalized to $n \geq 4$. If this conjecture is true, then it follows that all unconstrained information inequalities involving a finite number of random variables are Shannon-type inequalities, and they can all be proved by ITIP running on a sufficiently powerful computer. However, it turns out that this is not the case even for $n = 4$.

We will prove in the next theorem an unconstrained information inequality involving four random variables. Then we will show that this inequality is a non-Shannon-type inequality, and that $\overline{\Gamma}_4^* \neq \Gamma_4$.

THEOREM 14.7 *For any four random variables X_1, X_2, X_3, and X_4,*

$$2I(X_3; X_4) \leq I(X_1; X_2) + I(X_1; X_3, X_4) + 3I(X_3; X_4|X_1) + I(X_3; X_4|X_2). \tag{14.62}$$

Toward proving this theorem, we introduce two auxiliary random variables $\tilde{X}_1$ and $\tilde{X}_2$ jointly distributed with X_1, X_2, X_3, and X_4 such that $\tilde{\mathcal{X}}_1 = \mathcal{X}_1$ and $\tilde{\mathcal{X}}_2 = \mathcal{X}_2$. To simplify notation, we will use $p_{1234\tilde{1}\tilde{2}}(x_1, x_2, x_3, x_4, \tilde{x}_1, \tilde{x}_2)$ to denote $p_{X_1X_2X_3X_4\tilde{X}_1\tilde{X}_2}(x_1, x_2, x_3, x_4, \tilde{x}_1, \tilde{x}_2)$, etc. The joint distribution for the six random variables $X_1, X_2, X_3, X_4, \tilde{X}_1$, and $\tilde{X}_2$ is defined by

$$p_{1234\tilde{1}\tilde{2}}(x_1, x_2, x_3, x_4, \tilde{x}_1, \tilde{x}_2) = \begin{cases} \frac{p_{1234}(x_1,x_2,x_3,x_4)p_{1234}(\tilde{x}_1,\tilde{x}_2,x_3,x_4)}{p_{34}(x_3,x_4)} & \text{if } p_{34}(x_3, x_4) > 0 \\ 0 & \text{if } p_{34}(x_3, x_4) = 0. \end{cases} \tag{14.63}$$

LEMMA 14.8

$$(X_1, X_2) \to (X_3, X_4) \to (\tilde{X}_1, \tilde{X}_2) \tag{14.64}$$

forms a Markov chain. Moreover, (X_1, X_2, X_3, X_4) and $(\tilde{X}_1, \tilde{X}_2, X_3, X_4)$ have the same marginal distribution.

Proof The Markov chain in (14.64) is readily seen by invoking Proposition 2.5. The second part of the lemma is readily seen to be true by noting in (14.63) that $p_{1234\tilde{1}\tilde{2}}$ is symmetrical in X_1 and $\tilde{X}_1$ and in X_2 and $\tilde{X}_2$. □

From the above lemma, we see that the pair of auxiliary random variables $(\tilde{X}_1, \tilde{X}_2)$ corresponds to the pair of random variables (X_1, X_2) in the sense that $(\tilde{X}_1, \tilde{X}_2, X_3, X_4)$ have the same marginal distribution as (X_1, X_2, X_3, X_4). We need to prove two inequalities regarding these six random variables before we prove Theorem 14.7.

LEMMA 14.9 *For any four random variables X_1, X_2, X_3, and X_4 and auxiliary random variables $\tilde{X}_1$ and $\tilde{X}_2$ as defined in (14.63),*

$$I(X_3;X_4) - I(X_3;X_4|X_1) - I(X_3;X_4|X_2) \leq I(X_1;\tilde{X}_2). \tag{14.65}$$

Proof Consider

$$\begin{aligned}
& I(X_3;X_4) - I(X_3;X_4|X_1) - I(X_3;X_4|X_2) \\
& \stackrel{a)}{=} [I(X_3;X_4) - I(X_3;X_4|X_1)] - I(X_3;X_4|\tilde{X}_2) && (14.66) \\
& = I(X_1;X_3;X_4) - I(X_3;X_4|\tilde{X}_2) && (14.67) \\
& = [I(X_1;X_3;X_4;\tilde{X}_2) + I(X_1;X_3;X_4|\tilde{X}_2)] - I(X_3;X_4|\tilde{X}_2) && (14.68) \\
& = I(X_1;X_3;X_4;\tilde{X}_2) - [I(X_3;X_4|\tilde{X}_2) - I(X_1;X_3;X_4|\tilde{X}_2)] && (14.69) \\
& = I(X_1;X_3;X_4;\tilde{X}_2) - I(X_3;X_4|X_1,\tilde{X}_2) && (14.70) \\
& = [I(X_1;X_4;\tilde{X}_2) - I(X_1;X_4;\tilde{X}_2|X_3)] - I(X_3;X_4|X_1,\tilde{X}_2) && (14.71) \\
& = [I(X_1;\tilde{X}_2) - I(X_1;\tilde{X}_2|X_4)] - [I(X_1;\tilde{X}_2|X_3) \\
& \quad -I(X_1;\tilde{X}_2|X_3,X_4)] - I(X_3;X_4|X_1,\tilde{X}_2) && (14.72) \\
& \stackrel{b)}{=} I(X_1;\tilde{X}_2) - I(X_1;\tilde{X}_2|X_4) - I(X_1;\tilde{X}_2|X_3) \\
& \quad -I(X_3;X_4|X_1,\tilde{X}_2) && (14.73) \\
& \leq I(X_1;\tilde{X}_2), && (14.74)
\end{aligned}$$

where a) follows because we see from Lemma 14.8 that (X_2, X_3, X_4) and $(\tilde{X}_2, X_3, X_4)$ have the same marginal distribution, and b) follows because

$$I(X_1;\tilde{X}_2|X_3,X_4) = 0 \tag{14.75}$$

from the Markov chain in (14.64). The lemma is proved. □

LEMMA 14.10 *For any four random variables X_1, X_2, X_3, and X_4 and auxiliary random variables $\tilde{X}_1$ and $\tilde{X}_2$ as defined in (14.63),*

$$I(X_3;X_4) - 2I(X_3;X_4|X_1) \leq I(X_1;\tilde{X}_1). \tag{14.76}$$

Proof Notice that (14.76) can be obtained from (14.65) by replacing X_2 by X_1 and $\tilde{X}_2$ by $\tilde{X}_1$ in (14.65). The inequality (14.76) can be proved by replacing X_2 by X_1 and $\tilde{X}_2$ by $\tilde{X}_1$ in (14.66) through (14.74) in the proof of the last lemma. The details are omitted. □

Proof of Theorem 14.7 By adding (14.65) and (14.76), we have

$$
\begin{aligned}
&2I(X_3;X_4) - 3I(X_3;X_4|X_1) - I(X_3;X_4|X_2) \\
&\leq I(X_1;\tilde{X}_2) + I(X_1;\tilde{X}_1) && (14.77)\\
&= I(X_1;\tilde{X}_2) + [I(X_1;\tilde{X}_1|\tilde{X}_2) + I(X_1;\tilde{X}_1;\tilde{X}_2)] && (14.78)\\
&= [I(X_1;\tilde{X}_2) + I(X_1;\tilde{X}_1|\tilde{X}_2)] + I(X_1;\tilde{X}_1;\tilde{X}_2) && (14.79)\\
&= I(X_1;\tilde{X}_1,\tilde{X}_2) + I(X_1;\tilde{X}_1;\tilde{X}_2) && (14.80)\\
&= I(X_1;\tilde{X}_1,\tilde{X}_2) + [I(\tilde{X}_1;\tilde{X}_2) - I(\tilde{X}_1;\tilde{X}_2|X_1)] && (14.81)\\
&\leq I(X_1;\tilde{X}_1,\tilde{X}_2) + I(\tilde{X}_1;\tilde{X}_2) && (14.82)\\
&\stackrel{a)}{\leq} I(X_1;X_3,X_4) + I(\tilde{X}_1;\tilde{X}_2) && (14.83)\\
&\stackrel{b)}{=} I(X_1;X_3,X_4) + I(X_1;X_2), && (14.84)
\end{aligned}
$$

where a) follows from the Markov chain in (14.64), and b) follows because we see from Lemma 14.8 that $(\tilde{X}_1, \tilde{X}_2)$ and (X_1, X_2) have the same marginal distribution. Note that the auxiliary random variables $\tilde{X}_1$ and $\tilde{X}_2$ disappear in (14.84) after the sequence of manipulations. The theorem is proved. □

THEOREM 14.11 *The inequality (14.62) is a non-Shannon-type inequality, and* $\overline{\Gamma}_4^* \neq \Gamma_4$.

Proof Consider for any $a > 0$ the point $\tilde{\mathbf{h}}(a) \in \mathcal{H}_4$, where

$$
\begin{aligned}
&\tilde{h}_1(a) = \tilde{h}_2(a) = \tilde{h}_3(a) = \tilde{h}_4(a) = 2a,\\
&\tilde{h}_{12}(a) = 4a,\ \tilde{h}_{13}(a) = \tilde{h}_{14}(a) = 3a,\\
&\tilde{h}_{23}(a) = \tilde{h}_{24}(a) = \tilde{h}_{34}(a) = 3a,\\
&\tilde{h}_{123}(a) = \tilde{h}_{124}(a) = \tilde{h}_{134}(a) = \tilde{h}_{234}(a) = \tilde{h}_{1234}(a) = 4a.
\end{aligned}
\qquad (14.85)
$$

The set-theoretic structure of $\tilde{\mathbf{h}}(a)$ is illustrated by the information diagram in Figure 14.4. The reader should check that this information diagram correctly represents $\tilde{\mathbf{h}}(a)$ as defined. It is also easy to check from this diagram that $\tilde{\mathbf{h}}(a)$ satisfies all the elemental inequalities for four random variables, and therefore $\tilde{\mathbf{h}}(a) \in \Gamma_4$. However, upon substituting the corresponding values in (14.62) for $\tilde{\mathbf{h}}(a)$ with the help of Figure 14.4, we have

$$2a \leq 0 + a + 0 + 0 = a, \qquad (14.86)$$

which is a contradiction because $a > 0$. In other words, $\tilde{\mathbf{h}}(a)$ does not satisfy (14.62). Equivalently,

$$\tilde{\mathbf{h}}(a) \notin \{\mathbf{h} \in \mathcal{H}_4 : \mathbf{h} \text{ satisfies } (14.62)\}. \qquad (14.87)$$

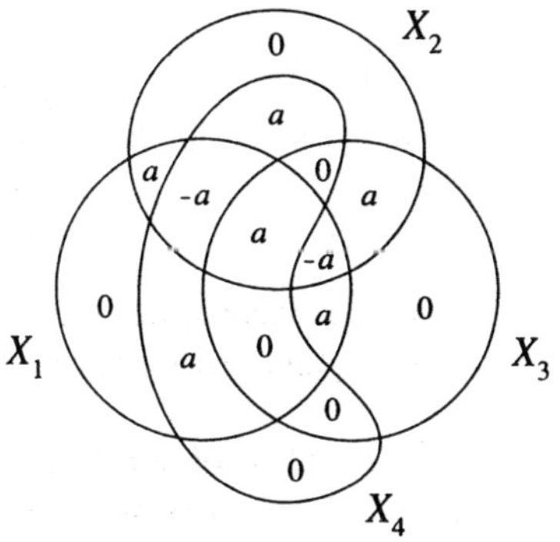

Figure 14.4. The set-theoretic structure of $\tilde{\mathbf{h}}(a)$.

Since $\tilde{\mathbf{h}}(a) \in \Gamma_4$, we conclude that

$$\Gamma_4 \not\subset \{\mathbf{h} \in \mathcal{H}_4 : \mathbf{h} \text{ satisfies (14.62)}\}, \tag{14.88}$$

i.e., (14.62) is not implied by the basic inequalities for four random variables. Hence, (14.62) is a non-Shannon-type inequality.

Since (14.62) is satisfied by all entropy functions for four random variables, we have

$$\Gamma_4^* \subset \{\mathbf{h} \in \mathcal{H}_4 : \mathbf{h} \text{ satisfies (14.62)}\}, \tag{14.89}$$

and upon taking closure on both sides, we have

$$\overline{\Gamma}_4^* \subset \{\mathbf{h} \in \mathcal{H}_4 : \mathbf{h} \text{ satisfies (14.62)}\}. \tag{14.90}$$

Then (14.87) implies $\tilde{\mathbf{h}}(a) \notin \overline{\Gamma}_4^*$. Since $\tilde{\mathbf{h}}(a) \in \Gamma_4$ and $\tilde{\mathbf{h}}(a) \notin \overline{\Gamma}_4^*$, we conclude that $\overline{\Gamma}_4^* \neq \Gamma_4$. The theorem is proved. □

Remark We have shown in the proof of Theorem 14.11 that the inequality (14.62) cannot be proved by invoking the basic inequalities for four random variables. However, (14.62) can be proved by invoking the basic inequalities for the six random variables $X_1, X_2, X_3, X_4, \tilde{X}_1$, and $\tilde{X}_2$ with the joint probability distribution $p_{1234\tilde{1}\tilde{2}}$ as constructed in (14.63).

The inequality (14.62) remains valid when the indices 1, 2, 3, and 4 are permuted. Since (14.62) is symmetrical in X_3 and X_4, $4!/2! = 12$ distinct versions of (14.62) can be obtained by permuting the indices, and all these twelve inequalities are simultaneously satisfied by the entropy function of any set of random variables X_1, X_2, X_3, and X_4. We will denote these twelve inequalities collectively by <14.62>. Now define the region

$$\tilde{\Gamma}_4 = \{\mathbf{h} \in \Gamma_4 : \mathbf{h} \text{ satisfies <14.62>}\}. \tag{14.91}$$

Evidently,

$$\Gamma_4^* \subset \tilde{\Gamma}_4 \subset \Gamma_4. \tag{14.92}$$

Since both $\tilde{\Gamma}_4$ and Γ_4 are closed, upon taking closure, we also have

$$\overline{\Gamma}_4^* \subset \tilde{\Gamma}_4 \subset \Gamma_4. \tag{14.93}$$

Since <14.62> are non-Shannon-type inequalities as we have proved in the last theorem, $\tilde{\Gamma}_4$ is a proper subset of Γ_4 and hence a tighter outer bound on Γ_4^* and $\overline{\Gamma}_4^*$ than Γ_4.

In the course of proving that (14.62) is of non-Shannon-type, it was shown in the proof of Theorem 14.11 that there exists $\tilde{\mathbf{h}}(a) \in \Gamma_4$ as defined in (14.85) which does not satisfy (14.62). By investigating the geometrical relation between $\tilde{\mathbf{h}}(a)$ and Γ_4, we prove in the next theorem that (14.62) in fact induces a class of $2^{14} - 1$ non-Shannon-type constrained inequalities. Applications of some of these inequalities will be discussed in Section 14.4.

THEOREM 14.12 *The inequality (14.62) is a non-Shannon-type inequality conditioning on setting any nonempty subset of the following 14 Shannon's information measures to zero:*

$$\begin{array}{l} I(X_1;X_2), I(X_1;X_2|X_3), I(X_1;X_2|X_4), I(X_1;X_3|X_4), \\ I(X_1;X_4|X_3), I(X_2;X_3|X_4), I(X_2;X_4|X_3), I(X_3;X_4|X_1), \\ I(X_3;X_4|X_2), I(X_3;X_4|X_1,X_2), H(X_1|X_2,X_3,X_4), \\ H(X_2|X_1,X_3,X_4), H(X_3|X_1,X_2,X_4), H(X_4|X_1,X_2,X_3). \end{array} \tag{14.94}$$

Proof It is easy to verify from Figure 14.4 that $\tilde{\mathbf{h}}(a)$ lies in exactly 14 hyperplanes in $\mathcal{H}_4$ (i.e., $\Re^{15}$) defining the boundary of Γ_4 which correspond to setting the 14 Shannon's measures in (14.94) to zero. Therefore, $\tilde{\mathbf{h}}(a)$ for $a \geq 0$ define an extreme direction of Γ_4.

Now for any linear subspace Φ of $\mathcal{H}_4$ containing $\tilde{\mathbf{h}}(a)$, where $a > 0$, we have

$$\tilde{\mathbf{h}}(a) \in \Gamma_4 \cap \Phi \tag{14.95}$$

and $\tilde{\mathbf{h}}(a)$ does not satisfy (14.62). Therefore,

$$(\Gamma_4 \cap \Phi) \not\subset \{\mathbf{h} \in \mathcal{H}_4 : \mathbf{h} \text{ satisfies (14.62)}\}. \tag{14.96}$$

This means that (14.62) is a non-Shannon-type inequality under the constraint Φ. From the above, we see that Φ can be taken to be the intersection of any nonempty subset of the 14 hyperplanes containing $\tilde{\mathbf{h}}(a)$. Thus (14.62) is a non-Shannon-type inequality conditioning on any nonempty subset of the 14 Shannon's measures in (14.94) being equal to zero. Hence, (14.62) induces a class of $2^{14} - 1$ non-Shannon-type constrained inequalities. The theorem is proved. □

Remark It is not true that the inequality (14.62) is of non-Shannon-type under any constraint. Suppose we impose the constraint

$$I(X_3; X_4) = 0. \tag{14.97}$$

Then the left hand side of (14.62) becomes zero, and the inequality is trivially implied by the basic inequalities because only mutual informations with positive coefficients appear on the right hand side. Then (14.62) becomes a Shannon-type inequality under the constraint in (14.97).

14.3 A NON-SHANNON-TYPE CONSTRAINED INEQUALITY

In the last section, we proved a non-Shannon-type unconstrained inequality for four random variables which implies $\overline{\Gamma}_4^* \neq \Gamma_4$. This inequality induces a region $\tilde{\Gamma}_4$ which is a tighter outer bound on Γ_4^* and $\overline{\Gamma}_4^*$ then Γ_4. We further showed that this inequality induces a class of $2^{14} - 1$ non-Shannon-type constrained inequalities for four random variables.

In this section, we prove a non-Shannon-type constrained inequality for four random variables. Unlike the non-Shannon-type unconstrained inequality we proved in the last section, this constrained inequality is not strong enough to imply that $\Gamma_4^* \neq \Gamma_4$. However, the latter is not implied by the former.

LEMMA 14.13 *Let $p(x_1, x_2, x_3, x_4)$ be any probability distribution. Then*

$$\tilde{p}(x_1, x_2, x_3, x_4) = \begin{cases} \frac{p(x_1,x_3,x_4)p(x_2,x_3,x_4)}{p(x_3,x_4)} & \text{if } p(x_3, x_4) > 0 \\ 0 & \text{if } p(x_3, x_4) = 0 \end{cases} \tag{14.98}$$

is also a probability distribution. Moreover,

$$\tilde{p}(x_1, x_3, x_4) = p(x_1, x_3, x_4) \tag{14.99}$$

and

$$\tilde{p}(x_2, x_3, x_4) = p(x_2, x_3, x_4) \tag{14.100}$$

for all x_1, x_2, x_3, and x_4.

Proof The proof for the first part of the lemma is straightforward (see Problem 4 in Chapter 2). The details are omitted here.

To prove the second part of the lemma, it suffices to prove (14.99) for all x_1, x_3, and x_4 because $\tilde{p}(x_1, x_2, x_3, x_4)$ is symmetrical in x_1 and x_2. We first consider x_1, x_3, and x_4 such that $p(x_3, x_4) > 0$. From (14.98), we have

$$\begin{aligned}
\tilde{p}(x_1, x_3, x_4) &= \sum_{x_2} \tilde{p}(x_1, x_2, x_3, x_4) && (14.101)\\
&= \sum_{x_2} \frac{p(x_1, x_3, x_4)p(x_2, x_3, x_4)}{p(x_3, x_4)} && (14.102)\\
&= \frac{p(x_1, x_3, x_4)}{p(x_3, x_4)} \sum_{x_2} p(x_2, x_3, x_4) && (14.103)\\
&= \left[\frac{p(x_1, x_3, x_4)}{p(x_3, x_4)}\right] p(x_3, x_4) && (14.104)\\
&= p(x_1, x_3, x_4). && (14.105)
\end{aligned}$$

For x_1, x_3, and x_4 such that $p(x_3, x_4) = 0$, we have

$$0 \leq p(x_1, x_3, x_4) \leq p(x_3, x_4) = 0, \tag{14.106}$$

which implies

$$p(x_1, x_3, x_4) = 0. \tag{14.107}$$

Therefore, from (14.98), we have

$$\begin{aligned}
\tilde{p}(x_1, x_3, x_4) &= \sum_{x_2} \tilde{p}(x_1, x_2, x_3, x_4) && (14.108)\\
&= \sum_{x_2} 0 && (14.109)\\
&= 0 && (14.110)\\
&= p(x_1, x_3, x_4). && (14.111)
\end{aligned}$$

Thus we have proved (14.99) for all x_1, x_3, and x_4, and the lemma is proved.
□

THEOREM 14.14 *For any four random variables X_1, X_2, X_3, and X_4, if*

$$I(X_1; X_2) = I(X_1; X_2|X_3) = 0, \tag{14.112}$$

then

$$I(X_3; X_4) \leq I(X_3; X_4|X_1) + I(X_3; X_4|X_2). \tag{14.113}$$

Proof Consider

$$I(X_3; X_4) - I(X_3; X_4|X_1) - I(X_3; X_4|X_2)$$
$$= \sum_{\substack{x_1, x_2, x_3, x_4: \\ p(x_1, x_2, x_3, x_4) > 0}} p(x_1, x_2, x_3, x_4) \log \frac{p(x_3, x_4)p(x_1, x_3)p(x_1, x_4)p(x_2, x_3)p(x_2, x_4)}{p(x_3)p(x_4)p(x_1)p(x_2)p(x_1, x_3, x_4)p(x_2, x_3, x_4)}$$

$$= E_p \log \frac{p(X_3,X_4)p(X_1,X_3)p(X_1,X_4)p(X_2,X_3)p(X_2,X_4)}{p(X_3)p(X_4)p(X_1)p(X_2)p(X_1,X_3,X_4)p(X_2,X_3,X_4)}, \tag{14.114}$$

where we have used E_p to denote expectation with respect to $p(x_1,x_2,x_3,x_4)$. We claim that the above expectation is equal to

$$E_{\tilde{p}} \log \frac{p(X_3,X_4)p(X_1,X_3)p(X_1,X_4)p(X_2,X_3)p(X_2,X_4)}{p(X_3)p(X_4)p(X_1)p(X_2)p(X_1,X_3,X_4)p(X_2,X_3,X_4)}, \tag{14.115}$$

where $\tilde{p}(x_1,x_2,x_3,x_4)$ is defined in (14.98).

Toward proving that the claim is correct, we note that (14.115) is the sum of a number of expectations with respect to $\tilde{p}$. Let us consider one of these expectations, say

$$E_{\tilde{p}} \log p(X_1,X_3) = \sum_{\substack{x_1,x_2,x_3,x_4: \\ \tilde{p}(x_1,x_2,x_3,x_4)>0}} \tilde{p}(x_1,x_2,x_3,x_4) \log p(x_1,x_3). \tag{14.116}$$

Note that in the above summation, if $\tilde{p}(x_1,x_2,x_3,x_4) > 0$, then from (14.98), we see that

$$p(x_1,x_3,x_4) > 0, \tag{14.117}$$

and hence

$$p(x_1,x_3) > 0. \tag{14.118}$$

Therefore, the summation in (14.116) is always well-defined. Further, it can be written as

$$\begin{aligned} &\sum_{x_1,x_3,x_4} \log p(x_1,x_3) \sum_{x_2:\tilde{p}(x_1,x_2,x_3,x_4)>0} \tilde{p}(x_1,x_2,x_3,x_4) \\ &= \sum_{x_1,x_3,x_4} \tilde{p}(x_1,x_3,x_4) \log p(x_1,x_3). \end{aligned} \tag{14.119}$$

Thus $E_{\tilde{p}} \log p(X_1,X_3)$ depends on $\tilde{p}(x_1,x_2,x_3,x_4)$ only through $\tilde{p}(x_1,x_3,x_4)$, which by Lemma 14.13 is equal to $p(x_1,x_3,x_4)$. It then follows that

$$\begin{aligned} &E_{\tilde{p}} \log p(X_1,X_3) \\ &= \sum_{x_1,x_3,x_4} \tilde{p}(x_1,x_3,x_4) \log p(x_1,x_3) && (14.120) \\ &= \sum_{x_1,x_3,x_4} p(x_1,x_3,x_4) \log p(x_1,x_3) && (14.121) \\ &= E_p \log p(X_1,X_3). && (14.122) \end{aligned}$$

In other words, the expectation on $\log p(X_1, X_3)$ can be taken with respect to either $\tilde{p}(x_1, x_2, x_3, x_4)$ or $p(x_1, x_2, x_3, x_4)$ without affecting its value. By observing that all the marginals of p in the logarithm in (14.115) involve only subsets of either $\{X_1, X_3, X_4\}$ or $\{X_2, X_3, X_4\}$, we see that similar conclusions can be drawn for all the other expectations in (14.115), and hence the claim is proved.

Thus the claim implies that

$$\begin{aligned} & I(X_3; X_4) - I(X_3; X_4|X_1) - I(X_3; X_4|X_2) \\ &= E_{\tilde{p}} \log \frac{p(X_3, X_4)p(X_1, X_3)p(X_1, X_4)p(X_2, X_3)p(X_2, X_4)}{p(X_3)p(X_4)p(X_1)p(X_2)p(X_1, X_3, X_4)p(X_2, X_3, X_4)} \\ &= \sum_{\substack{x_1, x_2, x_3, x_4: \\ \tilde{p}(x_1, x_2, x_3, x_4) > 0}} \tilde{p}(x_1, x_2, x_3, x_4) \log \frac{p(x_3, x_4)p(x_1, x_3)p(x_1, x_4)p(x_2, x_3)p(x_2, x_4)}{p(x_3)p(x_4)p(x_1)p(x_2)p(x_1, x_3, x_4)p(x_2, x_3, x_4)} \\ &= - \sum_{\substack{x_1, x_2, x_3, x_4: \\ \tilde{p}(x_1, x_2, x_3, x_4) > 0}} \tilde{p}(x_1, x_2, x_3, x_4) \log \frac{\tilde{p}(x_1, x_2, x_3, x_4)}{\hat{p}(x_1, x_2, x_3, x_4)}, \end{aligned} \tag{14.123}$$

where

$$\hat{p}(x_1, x_2, x_3, x_4) = \begin{cases} \frac{p(x_1, x_3)p(x_1, x_4)p(x_2, x_3)p(x_2, x_4)}{p(x_1)p(x_2)p(x_3)p(x_4)} & \text{if } p(x_1), p(x_2), p(x_3), p(x_4) > 0 \\ 0 & \text{otherwise.} \end{cases} \tag{14.124}$$

The equality in (14.123) is justified by observing that if x_1, x_2, x_3, and x_4 are such that $\tilde{p}(x_1, x_2, x_3, x_4) > 0$, then

$$p(x_1, x_3), p(x_1, x_4), p(x_2, x_3), p(x_2, x_4), p(x_1), p(x_2), p(x_3), p(x_4) \tag{14.125}$$

are all strictly positive, and we see from (14.124) that $\hat{p}(x_1, x_2, x_3, x_4) > 0$.

To complete the proof, we only need to show that $\hat{p}(x_1, x_2, x_3, x_4)$ is a probability distribution. Once this is proven, the conclusion of the theorem follows immediately because the summation in (14.123), which is identified as the divergence between $\tilde{p}(x_1, x_2, x_3, x_4)$ and $\hat{p}(x_1, x_2, x_3, x_4)$, is always nonnegative by the divergence inequality (Theorem 2.30). Toward this end, we notice that for x_1, x_2, and x_3 such that $p(x_3) > 0$,

$$p(x_1, x_2, x_3) = \frac{p(x_1, x_3)p(x_2, x_3)}{p(x_3)} \tag{14.126}$$

by the assumption

$$I(X_1; X_2|X_3) = 0, \tag{14.127}$$

and for all x_1 and x_2,

$$p(x_1, x_2) = p(x_1)p(x_2) \tag{14.128}$$

by the assumption

$$I(X_1; X_2) = 0. \tag{14.129}$$

Then

$$\begin{aligned}
&\sum_{x_1,x_2,x_3,x_4} \hat{p}(x_1, x_2, x_3, x_4) \\
&= \sum_{\substack{x_1,x_2,x_3,x_4:\\ \hat{p}(x_1,x_2,x_3,x_4)>0}} \hat{p}(x_1, x_2, x_3, x_4) &(14.130)\\
&= \sum_{\substack{x_1,x_2,x_3,x_4:\\ p(x_1),p(x_2),p(x_3),p(x_4)>0}} \frac{p(x_1, x_3)p(x_1, x_4)p(x_2, x_3)p(x_2, x_4)}{p(x_1)p(x_2)p(x_3)p(x_4)} &(14.131)\\
&\overset{a)}{=} \sum_{\substack{x_1,x_2,x_3,x_4:\\ p(x_1),p(x_2),p(x_3),p(x_4)>0}} \frac{p(x_1, x_2, x_3)p(x_1, x_4)p(x_2, x_4)}{p(x_1)p(x_2)p(x_4)} &(14.132)\\
&\overset{b)}{=} \sum_{\substack{x_1,x_2,x_3,x_4:\\ p(x_1),p(x_2),p(x_3),p(x_4)>0}} \frac{p(x_1, x_2, x_3)p(x_1, x_4)p(x_2, x_4)}{p(x_1, x_2)p(x_4)} &(14.133)\\
&= \sum_{\substack{x_1,x_2,x_4:\\ p(x_1),p(x_2),p(x_4)>0}} \frac{p(x_1, x_4)p(x_2, x_4)}{p(x_4)} \sum_{x_3:p(x_3)>0} p(x_3|x_1, x_2) &(14.134)\\
&= \sum_{\substack{x_1,x_2,x_4:\\ p(x_1),p(x_2),p(x_4)>0}} \frac{p(x_1, x_4)p(x_2, x_4)}{p(x_4)} &(14.135)\\
&= \sum_{\substack{x_2,x_4:\\ p(x_2),p(x_4)>0}} p(x_2, x_4) \sum_{x_1:p(x_1)>0} p(x_1|x_4) &(14.136)\\
&\overset{c)}{=} \sum_{\substack{x_2,x_4:\\ p(x_2),p(x_4)>0}} p(x_2, x_4) &(14.137)\\
&\overset{d)}{=} 1, &(14.138)
\end{aligned}$$

where a) and b) follows from (14.126) and (14.128), respectively. The equality in c) is justified as follows. For x_1 such that $p(x_1) = 0$,

$$p(x_1|x_4) = \frac{p(x_1)p(x_4|x_1)}{p(x_4)} = 0. \tag{14.139}$$

Therefore

$$\sum_{x_1:p(x_1)>0} p(x_1|x_4) = \sum_{x_1} p(x_1|x_4) = 1. \tag{14.140}$$

Finally, the equality in d) is justified as follows. For x_2 and x_4 such that $p(x_2)$ or $p(x_4)$ vanishes, $p(x_2, x_4)$ must vanish because

$$0 \leq p(x_2, x_4) \leq p(x_2) \tag{14.141}$$

and

$$0 \leq p(x_2, x_4) \leq p(x_4). \tag{14.142}$$

Therefore,

$$\sum_{\substack{x_2, x_4: \\ p(x_2), p(x_4) > 0}} p(x_2, x_4) = \sum_{x_2, x_4} p(x_2, x_4) = 1. \tag{14.143}$$

The theorem is proved. □

THEOREM 14.15 *The constrained inequality in Theorem 14.14 is a non-Shannon-type inequality.*

Proof The theorem can be proved by considering the point $\tilde{\mathbf{h}}(a) \in \mathcal{H}_4$ for $a > 0$ as in the proof of Theorem 14.11. The details are left as an exercise. □

The constrained inequality in Theorem 14.14 has the following geometrical interpretation. The constraints in (14.112) correspond to the intersection of two hyperplanes in $\mathcal{H}_4$ which define the boundary of Γ_4. Then the inequality (14.62) says that a certain region on the boundary of Γ_4 is not in Γ_4^*. It can further be proved by computation[1] that the constrained inequality in Theorem 14.14 is not implied by the twelve distinct versions of the unconstrained inequality in Theorem 14.7 (i.e., <14.62>) together with the basic inequalities.

We have proved in the last section that the non-Shannon-type inequality (14.62) implies a class of $2^{14} - 1$ constrained non-Shannon-type inequalities. We end this section by proving a similar result for the non-Shannon-type constrained inequality in Theorem 14.14.

THEOREM 14.16 *The inequality*

$$I(X_3; X_4) \leq I(X_3; X_4|X_1) + I(X_3; X_4|X_2) \tag{14.144}$$

is a non-Shannon-type inequality conditioning on setting both $I(X_1; X_2)$ and $I(X_1; X_2|X_3)$ and any subset of the following 12 Shannon's information measures to zero:

$$\begin{aligned} &I(X_1; X_2|X_4), I(X_1; X_3|X_4), I(X_1; X_4|X_3), \\ &I(X_2; X_3|X_4), I(X_2; X_4|X_3), I(X_3; X_4|X_1), \\ &I(X_3; X_4|X_2), I(X_3; X_4|X_1, X_2), H(X_1|X_2, X_3, X_4), \\ &H(X_2|X_1, X_3, X_4), H(X_3|X_1, X_2, X_4), H(X_4|X_1, X_2, X_3). \end{aligned} \tag{14.145}$$

[1] Ying-On Yan, private communication.

Proof The proof of this theorem is very similar to the proof of Theorem 14.12. We first note that $I(X_1;X_2)$ and $I(X_1;X_2|X_3)$ together with the 12 Shannon's information measures in (14.145) are exactly the 14 Shannon's information measures in (14.94). We have already shown in the proof of Theorem 14.12 that $\tilde{\mathbf{h}}(a)$ (cf. Figure 14.4) lies in exactly 14 hyperplanes defining the boundary of Γ_4 which correspond to setting these 14 Shannon's information measures to zero. We also have shown that $\tilde{\mathbf{h}}(a)$ for $a \geq 0$ define an extreme direction of Γ_4.

Denote by Φ_0 the intersection of the two hyperplanes in $\mathcal{H}_4$ which correspond to setting $I(X_1;X_2)$ and $I(X_1;X_2|X_3)$ to zero. Since $\tilde{\mathbf{h}}(a)$ for any $a > 0$ satisfies

$$I(X_1;X_2) = I(X_1;X_2|X_3) = 0, \tag{14.146}$$

$\tilde{\mathbf{h}}(a)$ is in Φ_0. Now for any linear subspace Φ of $\mathcal{H}_4$ containing $\tilde{\mathbf{h}}(a)$ such that $\Phi \subset \Phi_0$, we have

$$\tilde{\mathbf{h}}(a) \in \Gamma_4 \cap \Phi. \tag{14.147}$$

Upon substituting the corresponding values in (14.113) for $\tilde{\mathbf{h}}(a)$ with the help of Figure 14.4, we have

$$a \leq 0 + 0 = 0, \tag{14.148}$$

which is a contradiction because $a > 0$. Therefore, $\tilde{\mathbf{h}}(a)$ does not satisfy (14.113). Therefore,

$$(\Gamma_4 \cap \Phi) \not\subset \{\mathbf{h} \in \mathcal{H}_4 : \mathbf{h} \text{ satisfies } (14.113)\}. \tag{14.149}$$

This means that (14.113) is a non-Shannon-type inequality under the constraint Φ. From the above, we see that Φ can be taken to be the intersection of Φ_0 and any subset of the 12 hyperplanes which correspond to setting the 12 Shannon's information measures in (14.145) to zero. Hence, (14.113) is a non-Shannon-type inequality conditioning on $I(X_1;X_2)$, $I(X_1;X_2|X_3)$, and any subset of the 12 Shannon's information measures in (14.145) being equal to zero. In other words, the constrained inequality in Theorem 14.14 in fact induces a class of 2^{12} constrained non-Shannon-type inequalities. The theorem is proved. □

14.4 APPLICATIONS

As we have mentioned in Chapter 12, information inequalities are the laws of information theory. In this section, we give several applications of the non-Shannon-type inequalities we have proved in this chapter in probability theory and information theory. An application of the unconstrained inequality proved in Section 14.2 in group theory will be discussed in Chapter 16.

EXAMPLE 14.17 *For the constrained inequality in Theorem 14.14, if we further impose the constraints*

$$I(X_3;X_4|X_1) = I(X_3;X_4|X_2) = 0, \tag{14.150}$$

then the right hand side of (14.113) becomes zero. This implies

$$I(X_3; X_4) = 0 \tag{14.151}$$

because $I(X_3; X_4)$ is nonnegative. This means that

$$\left.\begin{array}{l} X_1 \perp X_2 \\ X_1 \perp X_2 | X_3 \\ X_3 \perp X_4 | X_1 \\ X_3 \perp X_4 | X_2 \end{array}\right\} \Rightarrow X_3 \perp X_4. \tag{14.152}$$

We leave it as an exercise for the reader to show that this implication cannot be deduced from the basic inequalities.

EXAMPLE 14.18 *If we impose the constraints*

$$I(X_1; X_2) = I(X_1; X_3, X_4) = I(X_3; X_4 | X_1) = I(X_3; X_4 | X_2) = 0, \tag{14.153}$$

then the right hand side of (14.62) becomes zero, which implies

$$I(X_3; X_4) = 0. \tag{14.154}$$

This means that

$$\left.\begin{array}{l} X_1 \perp X_2 \\ X_1 \perp (X_3, X_4) \\ X_3 \perp X_4 | X_1 \\ X_3 \perp X_4 | X_2 \end{array}\right\} \Rightarrow X_3 \perp X_4. \tag{14.155}$$

Note that (14.152) and (14.155) differ only in the second constraint. Again, we leave it as an exercise for the reader to show that this implication cannot be deduced from the basic inequalities.

EXAMPLE 14.19 *Consider a fault-tolerant data storage system consisting of random variables X_1, X_2, X_3, X_4 such that any three random variables can recover the remaining one, i.e.,*

$$H(X_i | X_j, j \neq i) = 0, \quad 1 \leq i, j \leq 4. \tag{14.156}$$

We are interested in the set of all entropy functions subject to these constraints, denoted by Υ, which characterizes the amount of joint information which can possibly be stored in such a data storage system. Let

$$\Phi = \{\mathbf{h} \in \mathcal{H}_4 : \mathbf{h} \text{ satisfies (14.156)}\}. \tag{14.157}$$

Then the set Υ is equal to the intersection between Γ_4^* and Φ, i.e., $\Gamma_4^* \cap \Phi$.

Since each constraint in (14.156) is one of the 14 constraints specified in Theorem 14.12, we see that (14.62) is a non-Shannon-type inequality under the constraints in (14.156). Then $\tilde{\Gamma}_4 \cap \Phi$ *(cf. (14.91)) is a tighter outer bound on* Υ *than* $\Gamma_4 \cap \Phi$.

EXAMPLE 14.20 *Consider four random variables* X_1, X_2, X_3, *and* X_4 *such that* $X_3 \to (X_1, X_2) \to X_4$ *forms a Markov chain. This Markov condition is equivalent to*

$$I(X_3; X_4|X_1, X_2) = 0. \tag{14.158}$$

It can be proved by invoking the basic inequalities (using ITIP) that

$$I(X_3; X_4) \leq I(X_3; X_4|X_1) + I(X_3; X_4|X_2) + 0.5I(X_1; X_2)$$

$$+cI(X_1; X_3, X_4) + (1-c)I(X_2; X_3, X_4), \tag{14.159}$$

where $0.25 \leq c \leq 0.75$, *and this is the best possible.*

Now observe that the Markov condition (14.158) is one of the 14 constraints specified in Theorem 14.12. Therefore, (14.62) is a non-Shannon-type inequality under this Markov condition. By replacing X_1 *and* X_2 *by each other in (14.62), we obtain*

$$2I(X_3; X_4) \leq I(X_1; X_2) + I(X_2; X_3, X_4)$$

$$+3I(X_3; X_4|X_2) + I(X_3; X_4|X_1). \tag{14.160}$$

Upon adding (14.62) and (14.160) and dividing by 4, we obtain

$$I(X_3; X_4) \leq I(X_3; X_4|X_1) + I(X_3; X_4|X_2) + 0.5I(X_1; X_2)$$

$$+0.25I(X_1; X_3, X_4) + 0.25I(X_2; X_3, X_4). \tag{14.161}$$

Comparing the last two terms in (14.159) and the last two terms in (14.161), we see that (14.161) is a sharper upper bound than (14.159).

The Markov chain $X_3 \to (X_1, X_2) \to X_4$ *arises in many communication situations. As an example, consider a person listening to an audio source. Then the situation can be modeled by this Markov chain with* X_3 *being the sound wave generated at the source,* X_1 *and* X_2 *being the sound waves received at the two ear drums, and* X_4 *being the nerve impulses which eventually arrive at the brain. The inequality (14.161) gives an upper bound on* $I(X_3; X_4)$ *which is tighter than what can be implied by the basic inequalities.*

There is some resemblance between the constrained inequality (14.161) and the data processing theorem, but there does not seem to be any direct relation between them.

PROBLEMS

1. Verify by ITIP that the unconstrained information inequality in Theorem 14.7 is of non-Shannon-type.

2. Verify by ITIP and prove analytically that the constrained information inequality in Theorem 14.14 is of non-Shannon-type.

3. Use ITIP to verify the unconstrained information inequality in Theorem 14.7. Hint: Create two auxiliary random variables as in the proof of Theorem 14.7 and impose appropriate constraints on the random variables.

4. Verify by ITIP that the implications in Examples 14.17 and 14.18 cannot be deduced from the basic inequalities.

5. Can you show that the sets of constraints in Examples 14.17 and 14.18 are in fact different?

6. Let $X_i, i = 1, 2, \cdots, n$, Z, and T be discrete random variables.

   a) Prove that

$$\begin{aligned} & nI(Z;T) - \sum_{j=1}^{n} I(Z;T|X_j) - nI(Z;T|X_i) \\ \leq \; & I(X_i; Z, T) + \sum_{j=1}^{n} H(X_j) - H(X_1, X_2, \cdots, X_n). \end{aligned}$$

   Hint: When $n = 2$, this inequality reduces to the unconstrained non-Shannon-type inequality in Theorem 14.7.

   b) Prove that

$$\begin{aligned} & nI(Z;T) - 2\sum_{j=1}^{n} I(Z;T|X_j) \\ \leq \; & \frac{1}{n}\sum_{i=1}^{n} I(X_i; Z, T) + \sum_{j=1}^{n} H(X_j) - H(X_1, X_2, \cdots, X_n). \end{aligned}$$

   (Zhang and Yeung [239].)

7. Let $p(x_1, x_2, x_3, x_4)$ be the joint distribution for random variables X_1, X_2, X_3, and X_4 such that $I(X_1; X_2|X_3) = I(X_2; X_4|X_3) = 0$, and let $\tilde{p}$ be defined in (14.98).

   a) Show that

$$\begin{aligned} & \check{p}(x_1, x_2, x_3, x_4) \\ = & \begin{cases} c \cdot \frac{p(x_1,x_2,x_3)p(x_1,x_4)p(x_2,x_4)}{p(x_1,x_2)p(x_4)} & \text{if } p(x_1, x_2), p(x_4) > 0 \\ 0 & \text{otherwise} \end{cases} \end{aligned}$$

defines a probability distribution for an appropriate $c \geq 1$.

b) Prove that $\tilde{p}(x_1, x_2, x_3) = p(x_1, x_2, x_3)$ for all x_1, x_2, and x_3.

c) By considering $D(\tilde{p}\|\check{p}) \geq 0$, prove that

$$\begin{aligned} & H(X_{13}) + H(X_{14}) + H(X_{23}) + H(X_{24}) + H(X_{34}) \\ \geq \; & H(X_3) + H(X_4) + H(X_{12}) + H(X_{134}) + H(X_{234}), \end{aligned}$$

where $H(X_{134})$ denotes $H(X_1, X_3, X_4)$, etc.

d) Prove that under the constraints in (14.112), the inequality in (14.113) is equivalent to the inequality in c).

The inequality in c) is referred to as the *Ingleton inequality* for entropy in the literature. For the origin of the Ingleton inequality, see Problem 9 in Chapter 16. (Matúš [146].)

HISTORICAL NOTES

In 1986, Pippenger [165] asked whether there exist constraints on the entropy function other than the polymatroidal axioms, which are equivalent to the basic inequalities. He called the constraints on the entropy function the *laws of information theory.* The problem had been open since then until Zhang and Yeung discovered for four random variables first the constrained non-Shannon-type inequality in Theorem 14.14 [238] and then the unconstrained non-Shannon-type inequality in Theorem 14.7 [239].

Yeung and Zhang [237] have subsequently shown that each of the inequalities reported in [238] and [239] implies a class of non-Shannon-type inequalities, and they have applied some of these inequalities in information theory problems. The existence of these inequalities implies that there are laws in information theory beyond those laid down by Shannon [184].

Meanwhile, Matúš and Studený [144][147][145] had been studying the structure of conditional independence (which subsumes the implication problem) of random variables. Matúš [146] finally settled the problem for four random variables by means of a constrained non-Shannon-type inequality which is a variation of the inequality reported in [238].

The non-Shannon-type inequalities that have been discovered induce outer bounds on the region Γ_4^* which are tighter than Γ_4. Matúš and Studený [147] showed that an entropy function in Γ_4 is entropic if it satisfies the Ingleton inequality (see Problem 9 in Chapter 16). This gives an inner bound on Γ_4^*. A more explicit proof of this inner bound can be found in Zhang and Yeung [239], where they showed that this bound is not tight.

Along a related direction, Hammer *et al.* [90] have shown that all linear inequalities which always hold for Kolmogorov complexity also always hold for entropy, and vice versa.

Chapter 15

MULTI-SOURCE NETWORK CODING

In Chapter 11, we have discussed the single-source network coding problem in which an information source is multicast in a point-to-point communication network. The maximum rate at which information can be multicast has a simple characterization in terms of the maximum flows in the graph representing the network. In this chapter, we consider the more general multi-source network coding problem in which more than one *mutually independent* information sources are generated at possibly different nodes, and each of the information sources is multicast to a specific set of nodes. We continue to assume that the point-to-point communication channels in the network are free of error.

The *achievable information rate region* for a multi-source network coding problem, which will be formally defined in Section 15.3, refers to the set of all possible rates at which multiple information sources can be multicast simultaneously on a network. In a single-source network coding problem, we are interested in characterizing the maximum rate at which information can be multicast from the source node to all the sink nodes. In a multi-source network coding problem, we are interested in characterizing the achievable information rate region.

Multi-source network coding turns out *not* to be a simple extension of single-source network coding. This will become clear after we have discussed two characteristics of multi-source network coding in the next section. Unlike the single-source network coding problem which has an explicit solution, the multi-source network coding problem has not been completely solved. The best characterizations of the achievable information rate region of the latter problem (for acyclic networks) which have been obtained so far make use of the tools we have developed for information inequalities in Chapter 12 to Chapter 14. This will be explained in the subsequent sections.

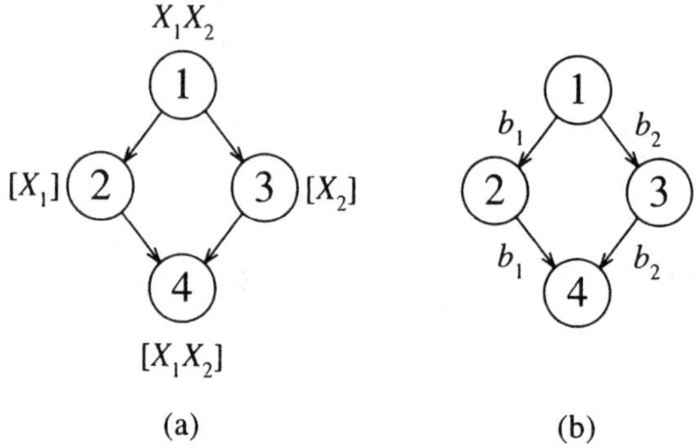

Figure 15.1. A network which achieves the max-flow bound.

15.1 TWO CHARACTERISTICS

In this section, we discuss two characteristics of multi-source networking coding which differentiate it from single-source network coding. In the following discussion, the unit of information is the bit.

15.1.1 THE MAX-FLOW BOUNDS

The max-flow bound, which fully characterizes the maximum rate at which information can be multicast, plays a central role in single-source network coding. We now revisit this bound in the context of multi-source network coding.

Consider the graph in Figure 15.1(a). The capacity of each edge is equal to 1. Two independent information sources X_1 and X_2 with rates ω_1 and ω_2, respectively are generated at node 1. Suppose we want to multicast X_1 to nodes 2 and 4 and multicast X_2 to nodes 3 and 4. In the figure, an information source in square brackets is one which is to be received at that node.

It is easy to see that the values of a max-flow from node 1 to node 2, from node 1 to node 3, and from node 1 to node 4 are respectively 1, 1, and 2. At node 2 and node 3, information is received at rates ω_1 and ω_2, respectively. At node 4, information is received at rate $\omega_1 + \omega_2$ because X_1 and X_2 are independent. Applying the max-flow bound at nodes 2, 3, and 4, we have

$$\omega_1 \leq 1 \tag{15.1}$$

$$\omega_2 \leq 1 \tag{15.2}$$

and

$$\omega_1 + \omega_2 \leq 2, \tag{15.3}$$

respectively. We refer to (15.1) to (15.3) as the max-flow bounds. Figure 15.2 is an illustration of all (ω_1, ω_2) which satisfy these bounds, where ω_1 and ω_2 are obviously nonnegative.

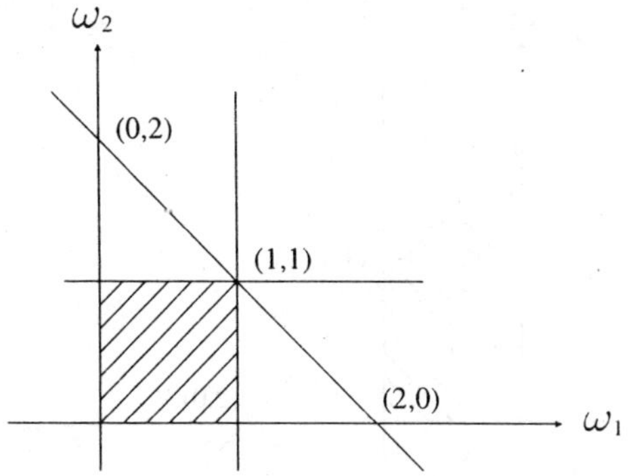

Figure 15.2. The max-flow bounds for the network in Figure 15.1.

We now show that the rate pair $(1, 1)$ is achievable. Let b_1 be a bit generated by X_1 and b_2 be a bit generated by X_2. In the scheme in Figure 15.1(b), b_1 is received at node 2, b_2 is received at node 3, and both b_1 and b_2 are received at node 4. Thus the multicast requirements are satisfied, and the information rate pair $(1, 1)$ is achievable. This implies that all (ω_1, ω_2) which satisfy the max-flow bounds are achievable because they are all inferior to $(1, 1)$ (see Figure. 15.2). In this sense, we say that the max-flow bounds are achievable.

Suppose we now want to multicast X_1 to nodes 2, 3, and 4 and multicast X_2 to node 4 as illustrated in Figure 15.3. Applying the max-flow bound at either node 2 or node 3 gives

$$\omega_1 \leq 1, \tag{15.4}$$

and applying the max-flow bound at node 4 gives

$$\omega_1 + \omega_2 \leq 2. \tag{15.5}$$

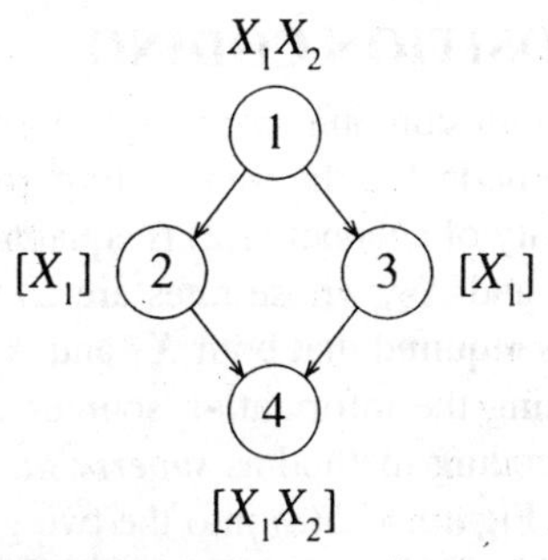

Figure 15.3. A network which does not achieve the max-flow bounds.

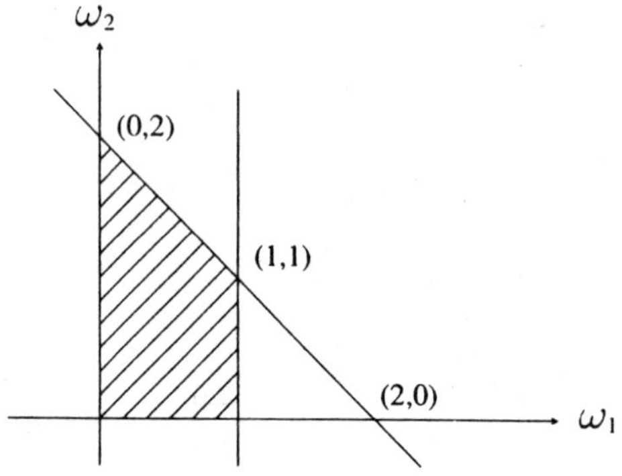

Figure 15.4. The max-flow bounds for the network in Figure 15.3.

Figure 15.4 is an illustration of all (ω_1, ω_2) which satisfy these bounds.

We now show that the information rate pair $(1, 1)$ is not achievable. Suppose we need to send a bit b_1 generated by X_1 to nodes 2, 3, and 4 and send a bit b_2 generated by X_2 to node 4. Since b_1 has to be recovered at node 2, the bit sent to node 2 must be an invertible transformation of b_1. This implies that the bit sent to node 2 cannot not depend on b_2. Similarly, the bit sent to node 3 also cannot depend on b_2. Therefore, it is impossible for node 4 to recover b_2 because both the bits received at nodes 2 and 3 do not depend on b_2. Thus the information rate pair $(1, 1)$ is not achievable, which implies that the max-flow bounds (15.4) and (15.5) are not achievable.

From the last example, we see that the max-flow bounds do not always fully characterize the achievable information rate region. Nevertheless, the max-flow bounds always give an outer bound on the achievable information rate region.

15.1.2 SUPERPOSITION CODING

Consider a point-to-point communication system represented by the graph in Figure 15.5(a), where node 1 is the transmitting point and node 2 is the receiving point. The capacity of channel (1,2) is equal to 1. Let two independent information sources X_1 and X_2, whose rates are ω_1 and ω_2, respectively, be generated at node 1. It is required that both X_1 and X_2 are received at node 2.

We first consider coding the information sources X_1 and X_2 individually. We will refer to such a coding method as *superposition coding*. To do so, we decompose the graph in Figure 15.5(a) into the two graphs in Figures 15.5(b) and (c). In Figure 15.5(b), X_1 is generated at node 1 and received at node 2. In Figure 15.5(c), X_2 is generated at node 1 and received at node 2. Let $r_{12}^{(l)}$ be

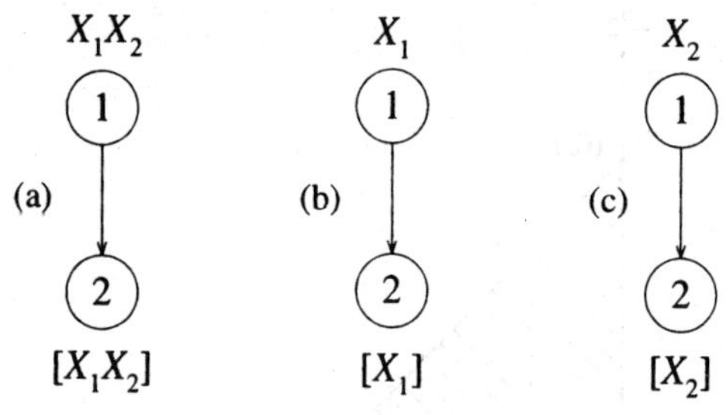

Figure 15.5. A network for which superposition coding is optimal.

the bit rate on channel (1,2) for transmitting X_l, $l = 1, 2$. Then

$$r_{12}^{(1)} \geq \omega_1 \tag{15.6}$$

and

$$r_{12}^{(2)} \geq \omega_2, \tag{15.7}$$

which imply

$$r_{12}^{(1)} + r_{12}^{(2)} \geq \omega_1 + \omega_2. \tag{15.8}$$

On the other hand, from the rate constraint for edge (1,2), we have

$$r_{12}^{(1)} + r_{12}^{(2)} \leq 1. \tag{15.9}$$

From (15.8) and (15.9), we obtain

$$\omega_1 + \omega_2 \leq 1. \tag{15.10}$$

Next, we consider coding the information sources X_1 and X_2 jointly. To this end, since X_1 and X_2 are independent, the total rate at which information is received at node 2 is equal to $\omega_1 + \omega_2$. From the rate constraint for channel (1,2), we again obtain (15.10). Therefore, we conclude that when X_1 and X_2 are independent, superposition coding is always optimal. The achievable information rate region is illustrated in Figure 15.6.

The above example is a degenerate example of multi-source networking coding because there are only two nodes in the network. Nevertheless, one may hope that superposition coding is optimal for all multi-source networking coding problems, so that any multi-source network coding problem can be decomposed into single-source network coding problems which can be solved individually. Unfortunately, this is not the case, as we now explain.

Consider the graph in Figure 15.7. The capacities of all the edges are equal

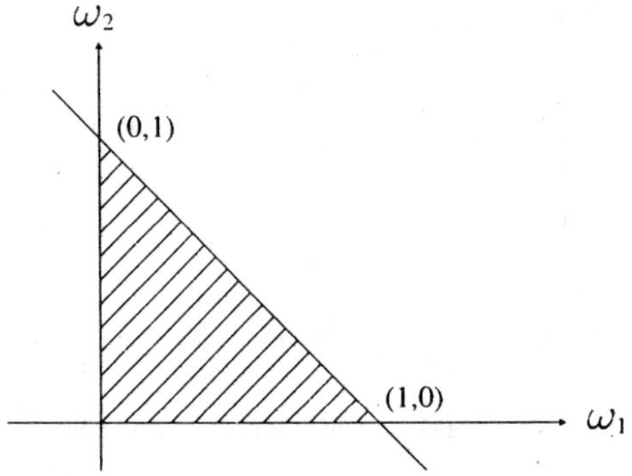

Figure 15.6. The achievable information rate region for the network in Figure 15.5(a).

to 1. We want to multicast X_1 to nodes 2, 5, 6, and 7, and multicast X_2 to nodes 5, 6, and 7.

We first consider coding X_1 and X_2 individually. Let

$$r_{1j}^{(l)} \geq 0 \tag{15.11}$$

be the bit rate on edge $(1, j)$ for the transmission of X_l, where $j = 2, 3, 4$ and $l = 1, 2$. Then the rate constraints on edges $(1, 2)$, $(1, 3)$, and $(1, 4)$ imply

$$r_{12}^{(1)} + r_{12}^{(2)} \leq 1 \tag{15.12}$$
$$r_{13}^{(1)} + r_{13}^{(2)} \leq 1 \tag{15.13}$$
$$r_{14}^{(1)} + r_{14}^{(2)} \leq 1. \tag{15.14}$$

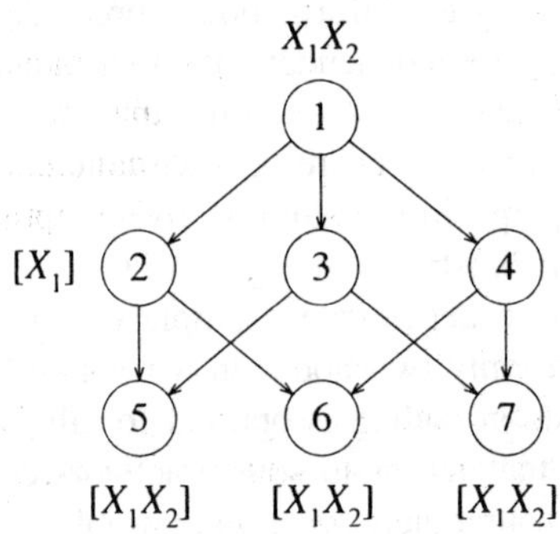

Figure 15.7. A network for which superposition coding is not optimal.

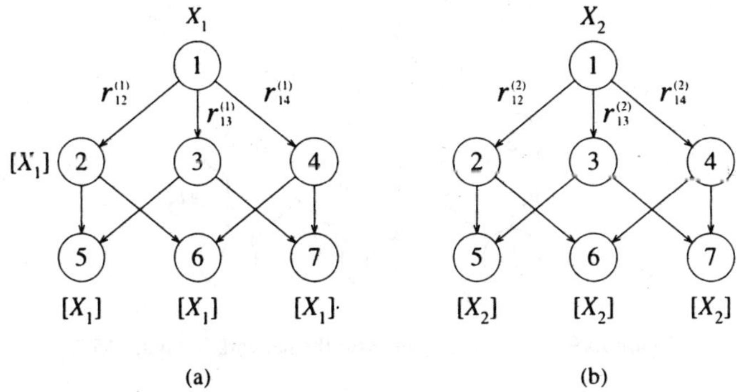

Figure 15.8. The individual multicast requirements for the network in Figure 15.7.

Together with (15.11), we see that

$$0 \leq r_{1j}^{(l)} \leq 1 \tag{15.15}$$

for all $j = 2, 3, 4$ and $l = 1, 2$.

We now decompose the graph in Figure 15.7 into the two graphs in Figures 15.8(a) and (b) which show the individual multicast requirements for X_1 and X_2, respectively. In these two graphs, the edges (1,2), (1,3), and (1,4) are labeled by the bit rates for transmitting the corresponding information source.

We now show by contradiction that the information rate pair (1,1) is not achievable by coding X_1 and X_2 individually. Assume that the contrary is true, i.e., the information rate pair (1,1) is achievable. Referring to Figure 15.8(a), since node 2 receives X_1 at rate 1,

$$r_{12}^{(1)} \geq 1. \tag{15.16}$$

At node 7, since X_1 is received via nodes 3 and 4,

$$r_{13}^{(1)} + r_{14}^{(1)} \geq 1. \tag{15.17}$$

Similar considerations at nodes 5 and 6 in Figure 15.8(b) give

$$r_{12}^{(2)} + r_{13}^{(2)} \geq 1 \tag{15.18}$$

and

$$r_{12}^{(2)} + r_{14}^{(2)} \geq 1. \tag{15.19}$$

Now (15.16) and (15.15) implies

$$r_{12}^{(1)} = 1. \tag{15.20}$$

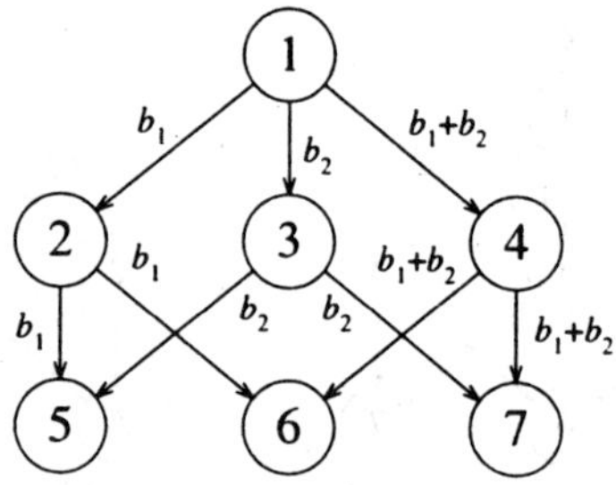

Figure 15.9. A coding scheme for the network in Figure 15.7.

With (15.12) and (15.15), this implies

$$r_{12}^{(2)} = 0. \tag{15.21}$$

From (15.21), (15.18), and (15.15), we have

$$r_{13}^{(2)} = 1. \tag{15.22}$$

With (15.13) and (15.15), this implies

$$r_{13}^{(1)} = 0. \tag{15.23}$$

It then follows from (15.23), (15.17), and (15.15) that

$$r_{14}^{(1)} = 1. \tag{15.24}$$

Together with (15.14) and (15.15), this implies

$$r_{14}^{(2)} = 0. \tag{15.25}$$

However, upon adding (15.21) and (15.25), we obtain

$$r_{12}^{(2)} + r_{14}^{(2)} = 0, \tag{15.26}$$

which is a contradiction to (15.19). Thus we have shown that the information rate pair (1,1) cannot be achieved by coding X_1 and X_2 individually.

However, the information rate pair (1,1) can actually be achieved by coding X_1 and X_2 jointly. Figure 15.9 shows such a scheme, where b_1 is a bit generated by X_1 and b_2 is a bit generated by X_2. Note that at node 4, the bits b_1 and b_2, which are generated by two different information sources, are jointly coded.

Hence, we conclude that superposition coding is not necessarily optimal in multi-source network coding. In other words, in certain problems, optimality can be achieved only by coding the information sources jointly.

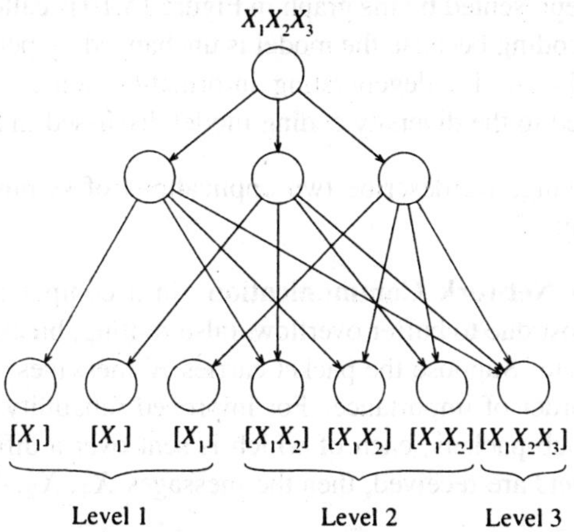

Figure 15.10. A 3-level diversity coding system.

15.2 EXAMPLES OF APPLICATION

Multi-source network coding is a very rich model which encompasses many communication situations arising from fault-tolerant network communication, disk array, satellite communication, etc. In this section, we discuss some applications of the model.

15.2.1 MULTILEVEL DIVERSITY CODING

Let $X_1, X_2, \cdots, X_K$ be K information sources in decreasing order of importance. These information sources are encoded into pieces of information. There are a number of users, each of them having access to a certain subset of the information pieces. Each user belongs to a *level* between 1 and K, where a Level k user can decode $X_1, X_2, \cdots, X_k$. This model, called multilevel diversity coding, finds applications in fault-tolerant network communication, disk array, and distributed data retrieval.

Figure 15.10 shows a graph which represents a 3-level diversity coding system. The graph consists of three layers of nodes. The top layer consists of a node at which information sources X_1, X_2, and X_3 are generated. These information sources are encoded into three pieces, each of which is stored in a distinct node in the middle layer. The nodes in the bottom layer represent the users, each of them belonging to one of the three levels. Each of the three Level 1 users has access to a distinct node in the middle layer and decodes X_1. Each of the three Level 2 users has access to a distinct set of two nodes in the middle layer and decodes X_1 and X_2. There is only one Level 3 user, who has access to all the three nodes in the middle layer and decodes X_1, X_2, and X_3.

The model represented by the graph in Figure 15.10 is called *symmetrical* 3-level diversity coding because the model is unchanged by permuting the nodes in the middle layer. By degenerating information sources X_1 and X_3, the model is reduced to the diversity coding model discussed in Section 11.2.

In the following, we describe two applications of symmetrical multilevel diversity coding:

Fault-Tolerant Network Communication In a computer network, a data packet can be lost due to buffer overflow, false routing, breakdown of communication links, etc. Suppose the packet carries K messages, $X_1, X_2, \cdots, X_K$, in decreasing order of importance. For improved reliability, the packet is encoded into K sub-packets, each of which is sent over a different channel. If any k sub-packets are received, then the messages $X_1, X_2, \cdots, X_k$ can be recovered.

Disk Array Consider a disk array which consists of K disks. The data to be stored in the disk array are segmented into K pieces, $X_1, X_2, \cdots, X_K$, in decreasing order of importance. Then $X_1, X_2, \cdots, X_K$ are encoded into K pieces, each of which is stored on a separate disk. When any k out of the K disks are functioning, the data $X_1, X_2, \cdots, X_k$ can be recovered.

15.2.2 SATELLITE COMMUNICATION NETWORK

In a satellite communication network, a user is at any time covered by one or more satellites. A user can be a transmitter, a receiver, or both. Through the satellite network, each information source generated at a transmitter is multicast to a certain set of receivers. A transmitter can transmit to all the satellites within the line of sight, while a receiver can receive from all the satellites within the line of sight. Neighboring satellites may also communicate with each other. Figure 15.11 is an illustration of a satellite communication network.

The satellite communication network in Figure 15.11 can be represented by the graph in Figure 15.12 which consists of three layers of nodes. The top layer represents the transmitters, the middle layer represents the satellites, and the bottom layer represents the receivers. If a satellite is within the line-of-sight of a transmitter, then the corresponding pair of nodes are connected by a directed edge. Likewise, if a receiver is within the line-of-sight of a satellite, then the corresponding pair nodes are connected by a directed edge. The edges between two nodes in the middle layer represent the communication links between the two neighboring satellites corresponding to the two nodes. Each information source is multicast to a specified set of receiving nodes as shown.

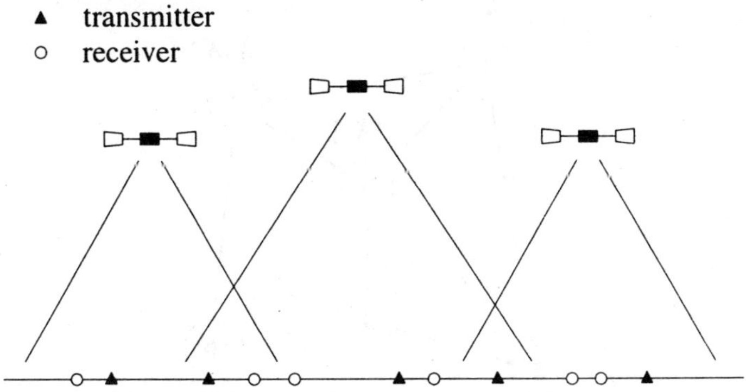

Figure 15.11. A satellite communication network.

15.3 A NETWORK CODE FOR ACYCLIC NETWORKS

Consider a network represented by an *acyclic* directed graph $G = (V, E)$, where $|V| < \infty$. Without loss of generality, assume that

$$V = \{1, 2, \cdots, |V|\}, \tag{15.27}$$

and the nodes are indexed such that if $(i, j) \in E$, then $i < j$. Such an indexing of the nodes is possible by Theorem 11.5. Let R_{ij} be the rate constraint on channel (i, j), and let

$$\mathbf{R} = [R_{ij} : (i, j) \in E] \tag{15.28}$$

be the rate constraints for graph G.

A set of multicast requirements on G consists of the following elements:

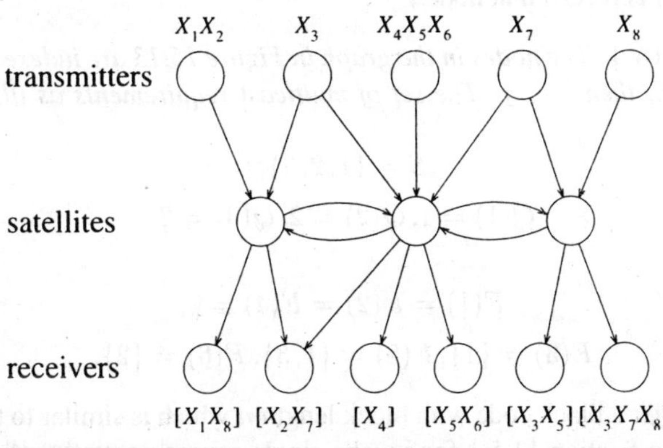

Figure 15.12. A graph representing a satellite communication network.

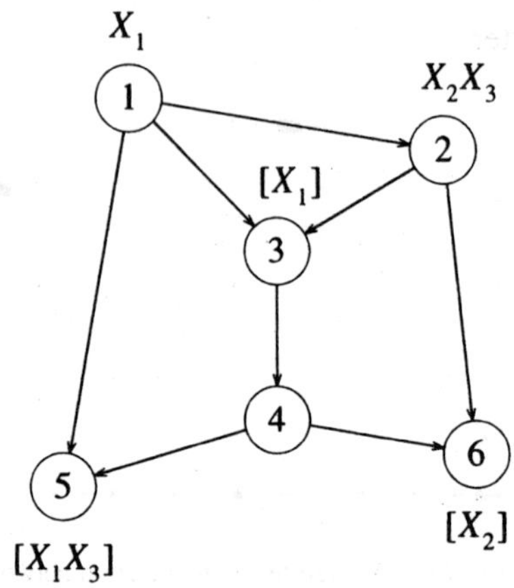

Figure 15.13. The acyclic graph for Example 15.1.

1) S, the set of information sources;

2) $Q : S \to V$, which specifies the node at which an information source is generated;

3) $F : V \to 2^S$, which specifies the set of information sources received at each node.

The information source s is generated at node $Q(s)$. For all $i \in V$, let

$$J(i) = \{s \in S : Q(s) = i\} \tag{15.29}$$

be the set of information sources generated at node i. The set of information sources $F(i)$ is received at node i.

EXAMPLE 15.1 *The nodes in the graph in Figure 15.13 are indexed such that if $(i, j) \in E$, then $i < j$. The set of multicast requirements as illustrated is specified by*

$$S = \{1, 2, 3\}, \tag{15.30}$$

$$Q(1) = 1, Q(2) = 2, Q(3) = 2, \tag{15.31}$$

and

$$F(1) = F(2) = F(4) = \emptyset, \tag{15.32}$$

$$F(3) = \{1\}, F(5) = \{1, 3\}, F(6) = \{2\}. \tag{15.33}$$

We consider a block code with block length n which is similar to the β-code we defined in Section 11.5.1 for acyclic single-source networks. An information source s is represented by a random variable X_s which takes values in the set

$$\mathcal{X}_s = \{1, 2, \cdots, \lceil 2^{n\tau_s} \rceil\} \tag{15.34}$$

according to the uniform distribution. The rate of information source s is τ_s. It is assumed that $X_s, s \in S$ are mutually independently.

Unlike the β-code we defined for acyclic single-source networks which is zero-error, the code we now define allows an arbitrarily small probability of error. Let

$$\Xi = \{i \in V : F(i) \neq \emptyset\} \tag{15.35}$$

be the set of nodes which receive at least one information source. An

$$(n, (\eta_{ij} : (i,j) \in E), (\tau_s : s \in S)) \tag{15.36}$$

code on graph G (with respect to a set of multicast requirements) is defined by

1) for all $(i,j) \in E$, an encoding function

$$f_{ij} : \prod_{s \in J(i)} \mathcal{X}_s \times \prod_{i':(i',i) \in E} \{0, 1, \cdots, \eta_{i'i}\} \to \{0, 1, \cdots, \eta_{ij}\} \tag{15.37}$$

(if both $J(i)$ and $\{i' : (i',i) \in E\}$ are empty, we adopt the convention that f_{ij} is an arbitrary constant taken from $\{0, 1, \cdots, \eta_{ij}\}$);

2) for all $i \in \Xi$, a decoding function

$$g_i : \prod_{i':(i',i) \in E} \{0, 1 \cdots, \eta_{i'i}\} \to \prod_{s \in F(i)} \mathcal{X}_s. \tag{15.38}$$

In the above, f_{ij} is the encoding function for edge (i,j). For $i \in \Xi$, g_i is the decoding function for node i. In a coding session, f_{ij} is applied before $f_{i'j'}$ if $i < i'$, and f_{ij} is applied before $f_{ij'}$ if $j < j'$. This defines the order in which the encoding functions are applied. Since $i' < i$ if $(i', i) \in E$, a node does not encode until all the necessary information is received on the input channels.

For all $i \in \Xi$, define

$$\Delta_i = \Pr\{\tilde{g}_i(X_s : s \in S) \neq (X_s : s \in F(i))\}, \tag{15.39}$$

where $\tilde{g}_i(X_s : s \in S)$ denotes the value of g_i as a function of $(X_s : s \in S)$. Δ_i is the probability that the set of information sources $F(i)$ is decoded incorrectly at node i.

Throughout this chapter, all the logarithms are in the base 2 unless otherwise specified.

DEFINITION 15.2 *For a graph G with rate constraints $\mathbf{R}$, an information rate tuple*

$$\boldsymbol{\omega} = (\omega_s : s \in S), \tag{15.40}$$

where $\boldsymbol{\omega} \geq 0$ (componentwise), is asymptotically achievable if for any $\epsilon > 0$, there exists for sufficiently large n an

$$(n, (\eta_{ij} : (i,j) \in E), (\tau_s : s \in S)) \tag{15.41}$$

code on G such that

$$n^{-1} \log \eta_{ij} \leq R_{ij} + \epsilon \tag{15.42}$$

for all $(i,j) \in E$, where $n^{-1} \log \eta_{ij}$ is the average bit rate of the code on channel (i,j),

$$\tau_s \geq \omega_s - \epsilon \tag{15.43}$$

for all $s \in S$, and

$$\Delta_i \leq \epsilon \tag{15.44}$$

for all $i \in \Xi$. For brevity, an asymptotically achievable information rate tuple will be referred to as an achievable information rate tuple.

DEFINITION 15.3 *The achievable information rate region, denoted by $\mathcal{R}$, is is the set of all achievable information rate tuples $\boldsymbol{\omega}$.*

Remark It follows from the definition of the achievability of an information rate vector that if $\boldsymbol{\omega}$ is achievable, then $\boldsymbol{\omega}'$ is achievable for all $0 \leq \boldsymbol{\omega}' \leq \boldsymbol{\omega}$. Also, if $\boldsymbol{\omega}^{(k)}$, $k \geq 1$ are achievable, then it can be proved by techniques similar to those in the proof of Theorem 9.12 that

$$\boldsymbol{\omega} = \lim_{k \to \infty} \boldsymbol{\omega}^{(k)} \tag{15.45}$$

is also achievable, i.e., $\mathcal{R}$ is closed. The details are omitted here.

In the rest of the chapter, we will prove inner and outer bounds on the achievable information rate region $\mathcal{R}$.

15.4 AN INNER BOUND

In this section, we first state an inner bound $\mathcal{R}_{in}$ on $\mathcal{R}$ in terms of a set of auxiliary random variables. Subsequently, we will cast this inner bound in the framework of information inequalities developed in Chapter 12.

DEFINITION 15.4 *Let $\mathcal{R}'$ be the set of all information rate tuples $\boldsymbol{\omega}$ such that there exist auxiliary random variables $Y_s, s \in S$ and $U_{ij}, (i,j) \in E$ which satisfy the following conditions:*

$$\begin{aligned}
H(Y_s : s \in S) &= \sum_{s \in S} H(Y_s) &(15.46)\\
H(U_{ij} | (Y_s : s \in J(i)), (U_{i'i} : (i',i) \in E)) &= 0 &(15.47)\\
H(Y_s : s \in F(i) | U_{i'i} : (i',i) \in E) &= 0 &(15.48)\\
R_{ij} &> H(U_{ij}) &(15.49)\\
H(Y_s) &> \omega_s. &(15.50)
\end{aligned}$$

Note that for $i \notin \Xi$, the constraint (15.48) in the above definition is degenerate because $F(i)$ is empty.

DEFINITION 15.5 *Let* $\mathcal{R}_{in} = \overline{\text{con}}(\mathcal{R}')$, *the convex closure of* $\mathcal{R}'$.

THEOREM 15.6 $\mathcal{R}_{in} \subset \mathcal{R}$.

In the definition of $\mathcal{R}'$, Y_s is an auxiliary random variable associated with information source s, and U_{ij} is an auxiliary random variable associated with the codeword sent on channel (i,j). The interpretations of (15.46) to (15.50) are as follows. The equality in (15.46) says that $Y_s, s \in S$ are mutually independent, which corresponds to the assumption that all the information sources are mutually independent. The equality in (15.47) says that U_{ij} is a function of $(Y_s : s \in J(i))$ and $(U_{i'i} : (i',i) \in E)$, which corresponds to the requirement that the codeword received from channel (i,j) depends only on the information sources generated at node i and the codewords received from the input channels at node i. The equality in (15.48) says that $(Y_s : s \in F(i))$ is a function of $(U_{i'i} : (i',i) \in E)$, which corresponds to the requirement that the information sources to be received at node i can be decoded from the codewords received from the input channels at node i. The inequality in (15.49) says that the entropy of U_{ij} is strictly less than R_{ij}, the rate constraint for channel (i,j). The inequality in (15.50) says that the entropy of Y_s is strictly greater than ω_s, the rate of information source s.

The proof of Theorem 15.6 is very tedious and will be postponed to Section 15.7. We note that the same inner bound has been proved in [198] for *variable length zero-error* network codes.

In the rest of the section, we cast the region $\mathcal{R}_{in}$ in the framework of information inequalities we developed in Chapter 12. The notation we will use is a slight modification of the notation in Chapter 12. Let

$$N = \{Y_s : s \in S; U_{ij} : (i,j) \in E\} \tag{15.51}$$

be a collection of discrete random variables whose joint distribution is unspecified, and let

$$Q(N) = 2^N \backslash \{\emptyset\}. \tag{15.52}$$

Then

$$|Q(N)| = 2^{|N|} - 1. \tag{15.53}$$

Let $\mathcal{H}_N$ be the $|Q(N)|$-dimensional Euclidean space with the coordinates labeled by $h_A, A \in Q(N)$. We will refer to $\mathcal{H}_N$ as the entropy space for the set of random variables N. A vector

$$\mathbf{h} = (h_A : A \in Q(N)) \tag{15.54}$$

is said to be *entropic* if there exists a joint distribution for $(Z : Z \in N)$ such that

$$H(X : X \in A) = h_A \tag{15.55}$$

for all $A \in Q(N)$. We then define

$$\Gamma_N^* = \{\mathbf{h} \in \mathcal{H}_N : \mathbf{h} \text{ is entropic}\}. \tag{15.56}$$

To simplify notation in the sequel, for any nonempty $A, A' \in Q(N)$, we define

$$h_{A|A'} = h_{AA'} - h_{A'}, \tag{15.57}$$

where we use juxtaposition to denote the union of two sets. In using the above notation, we do not distinguish elements and singletons of N, i.e., for a random variable $Z \in N$, h_Z is the same as $h_{\{Z\}}$.

To describe $\mathcal{R}_{in}$ in terms of Γ_N^*, we observe that the constraints (15.46) to (15.50) in the definition of $\mathcal{R}'$ correspond to the following constraints in $\mathcal{H}_N$, respectively:

$$\begin{aligned}
h_{(Y_s:s\in S)} &= \sum_{s\in S} h_{Y_s} && (15.58)\\
h_{U_{ij}|(Y_s:s\in J(i)),(U_{i'i}:(i',i)\in E)} &= 0 && (15.59)\\
h_{(Y_s:s\in F(i))|(U_{i'i}:(i',i)\in E)} &= 0 && (15.60)\\
R_{ij} &> h_{U_{ij}} && (15.61)\\
h_{Y_s} &> \omega_s. && (15.62)
\end{aligned}$$

Then we have the following alternative definition of $\mathcal{R}'$.

DEFINITION 15.7 *Let $\mathcal{R}'$ be the set of all information rate tuples $\boldsymbol{\omega}$ such that there exists $\mathbf{h} \in \Gamma_N^*$ which satisfies (15.58) to (15.62) for all $s \in S$ and $(i, j) \in E$.*

Although the original definition of $\mathcal{R}'$ as given in Definition 15.4 is more intuitive, the region so defined appears to be totally different from case to case. On the other hand, the alternative definition of $\mathcal{R}'$ above enables the region to be described on the same footing for all cases. Moreover, if $\tilde{\Gamma}_N$ is an explicit inner bound on Γ_N^*, upon replacing Γ_N^* by $\tilde{\Gamma}_N$ in the above definition of $\mathcal{R}'$, we immediately obtain an explicit inner bound on $\mathcal{R}_{in}$ for all cases. We will see further advantage of this alternative definition when we discuss an explicit outer bound on $\mathcal{R}$ in Section 15.6.

15.5 AN OUTER BOUND

In this section, we prove an outer bound $\mathcal{R}_{out}$ on $\mathcal{R}$. This outer bound is in terms of $\overline{\Gamma}_N^*$, the closure of Γ_N^*.

DEFINITION 15.8 *Let $\mathcal{R}_{out}$ be the set of all information rate tuples ω such that there exists $\mathbf{h} \in \overline{\Gamma}^*_N$ which satisfies the following constraints for all $s \in S$ and $(i, j) \in E$:*

$$h_{(Y_s:s\in S)} = \sum_{s\in S} h_{Y_s} \quad (15.63)$$

$$h_{U_{ij}|(Y_s:s\in J(i)),(U_{i'i}:(i',i)\in E)} = 0 \quad (15.64)$$

$$h_{(Y_s:s\in F(i))|(U_{i'i}:(i',i)\in E)} = 0 \quad (15.65)$$

$$R_{ij} \geq h_{U_{ij}} \quad (15.66)$$

$$h_{Y_s} \geq \omega_s. \quad (15.67)$$

The definition of $\mathcal{R}_{out}$ is the same as the alternative definition of $\mathcal{R}'$ (Definition 15.7) except that

1. Γ^*_N is replaced by $\overline{\Gamma}^*_N$.
2. The inequalities in (15.61) and (15.62) are strict, while the inequalities in (15.66) and (15.67) are nonstrict.

From the definition of $\mathcal{R}'$ and $\mathcal{R}_{out}$, it is clear that

$$\mathcal{R}' \subset \mathcal{R}_{out}. \quad (15.68)$$

It is easy to verify that the convexity of $\overline{\Gamma}^*_N$ (Theorem 14.5) implies the convexity of $\mathcal{R}_{out}$. Also, it is readily seen that $\mathcal{R}_{out}$ is closed. Then upon taking convex closure in (15.68), we see that

$$\mathcal{R}_{in} = \overline{\text{con}}(\mathcal{R}') \subset \overline{\text{con}}(\mathcal{R}_{out}) = \mathcal{R}_{out}. \quad (15.69)$$

However, it is not apparent that the two regions coincide in general. This will be further discussed in the next section.

THEOREM 15.9 $\mathcal{R} \subset \mathcal{R}_{out}$.

Proof Let ω be an achievable information rate tuple and n be a sufficiently large integer. Then for any $\epsilon > 0$, there exists an

$$(n, (\eta_{ij} : (i, j) \in E), (\tau_s : s \in S)) \quad (15.70)$$

code on G such that

$$n^{-1} \log \eta_{ij} \leq R_{ij} + \epsilon \quad (15.71)$$

for all $(i, j) \in E$,

$$\tau_s \geq \omega_s - \epsilon \quad (15.72)$$

for all $s \in S$, and

$$\Delta_i \leq \epsilon \quad (15.73)$$

for all $i \in \Xi$.

We consider such a code for a fixed ϵ and a sufficiently large n. For $(i,j) \in E$, let

$$U_{ij} = \tilde{f}_{ij}(X_s : s \in S) \tag{15.74}$$

be the codeword sent on channel (i,j). Since U_{ij} is a function of the information sources generated at node i and the codewords received on the input channels at node i,

$$H(U_{ij}|(X_s : s \in J(i)), (U_{i'i} : (i',i) \in E)) = 0. \tag{15.75}$$

For $i \in \Xi$, by Fano's inequality (Corollary 2.48), we have

$$\begin{aligned}
& H(X_s : s \in F(i)|U_{i'i} : (i',i) \in E) \\
&\leq 1 + \Delta_i \log\left(\prod_{s\in F(i)} |\mathcal{X}_s|\right) && (15.76)\\
&= 1 + \Delta_i H(X_s : s \in F(i)) && (15.77)\\
&\leq 1 + \epsilon H(X_s : s \in F(i)), && (15.78)
\end{aligned}$$

where (15.77) follows because X_s distributes uniformly on $\mathcal{X}_s$ and X_s, $s \in S$ are mutually independent, and (15.78) follows from (15.73). Then

$$\begin{aligned}
& H(X_s : s \in F(i)) \\
&= I((X_s : s \in F(i)); (U_{i'i} : (i',i) \in E)) \\
&\quad + H(X_s : s \in F(i)|U_{i'i} : (i',i) \in E) && (15.79)\\
&\overset{a)}{\leq} I((X_s : s \in F(i)); (U_{i'i} : (i',i) \in E)) \\
&\quad + 1 + \epsilon H(X_s : s \in F(i)) && (15.80)\\
&\leq H(U_{i'i} : (i',i) \in E) + 1 + \epsilon H(X_s : s \in F(i)) && (15.81)\\
&\overset{b)}{\leq} \left(\sum_{(i',i)\in E} \log \eta_{i'i}\right) + 1 + \epsilon H(X_s : s \in F(i)) && (15.82)\\
&\overset{c)}{\leq} \left(\sum_{(i',i)\in E} n(R_{i'i} + \epsilon)\right) + 1 + \epsilon H(X_s : s \in F(i)), && (15.83)
\end{aligned}$$

where

a) follows from (15.78);

b) follows from Theorem 2.43;

c) follows from (15.71).

Rearranging the terms in (15.83), we obtain

$$
\begin{aligned}
& H(X_s : s \in F(i)) \\
& \leq \frac{n}{1-\epsilon}\left(\sum_{(i',i)\in E} (R_{i'i}+\epsilon) + \frac{1}{n} \right) && (15.84) \\
& < 2n \sum_{(i',i)\in E} (R_{i'i}+\epsilon) && (15.85)
\end{aligned}
$$

for sufficiently small ϵ and sufficiently large n. Substituting (15.85) into (15.78), we have

$$
\begin{aligned}
& H(X_s : s \in F(i) | U_{i'i} : (i',i) \in E) \\
& < n\left(\frac{1}{n} + 2\epsilon \sum_{(i',i)\in E} (R_{i'i}+\epsilon) \right) && (15.86) \\
& = n\varrho_i(n,\epsilon), && (15.87)
\end{aligned}
$$

where

$$
\varrho_i(n,\epsilon) = \left(\frac{1}{n} + 2\epsilon \sum_{(i',i)\in E} (R_{i'i}+\epsilon) \right) \to 0 \tag{15.88}
$$

as $n \to \infty$ and then $\epsilon \to 0$. From (15.71), for all $(i,j) \in E$,

$$
n(R_{ij}+2\epsilon) \geq \log(\eta_{ij}+1) = \log|U_{ij}| \geq H(U_{ij}). \tag{15.89}
$$

For all $s \in S$, from (15.72),

$$
H(X_s) = \log|\mathcal{X}_s| = \log\lceil 2^{n\tau_s}\rceil \geq n\tau_s \geq n(\omega_s - \epsilon). \tag{15.90}
$$

Thus for this code, we have

$$
\begin{aligned}
H(X_s : s \in S) &= \sum_{s\in S} H(X_s) && (15.91) \\
H(U_{ij} | (X_s : s \in J(i)), (U_{i'i} : (i',i) \in E)) &= 0 && (15.92) \\
H(X_s : s \in F(i) | U_{i'i} : (i',i) \in E) &\leq n\varrho_i(n,\epsilon) && (15.93) \\
n(R_{ij}+2\epsilon) &\geq H(U_{ij}) && (15.94) \\
H(X_s) &\geq n(\omega_s - \epsilon). && (15.95)
\end{aligned}
$$

We note the one-to-one correspondence between (15.91) to (15.95) and (15.63) to (15.67). By letting $Y_s = X_s$ for all $s \in S$, we see that there exists $\mathbf{h} \in \Gamma_N^*$ such that

$$
h_{(Y_s : s\in S)} = \sum_{s\in S} h_{Y_s} \tag{15.96}
$$

$$\begin{aligned} h_{U_{ij}|(Y_s:s\in J(i)),(U_{i'i}:(i',i)\in E)} &= 0 & (15.97)\\ h_{(Y_s:s\in F(i))|(U_{i'i}:(i',i)\in E)} &\leq n\varrho_i(n,\epsilon) & (15.98)\\ n(R_{ij}+2\epsilon) &\geq h_{U_{ij}} & (15.99)\\ h_{Y_s} &\geq n(\omega_s-\epsilon). & (15.100) \end{aligned}$$

By Theorem 14.5, $\overline{\Gamma}^*_N$ is a convex cone. Therefore, if $\mathbf{h} \in \Gamma^*_N$, then $n^{-1}\mathbf{h} \in \overline{\Gamma}^*_N$. Dividing (15.96) through (15.100) by n and replacing $n^{-1}\mathbf{h}$ by $\mathbf{h}$, we see that there exists $\mathbf{h} \in \overline{\Gamma}^*_N$ such that

$$\begin{aligned} h_{(Y_s:s\in S)} &= \sum_{s\in S} h_{Y_s} & (15.101)\\ h_{U_{ij}|(Y_s:s\in J(i)),(U_{i'i}:(i',i)\in E)} &= 0 & (15.102)\\ h_{(Y_s:s\in F(i))|(U_{i'i}:(i',i)\in E)} &\leq \varrho_i(n,\epsilon) & (15.103)\\ R_{ij}+2\epsilon &\geq h_{U_{ij}} & (15.104)\\ h_{Y_s} &\geq \omega_s-\epsilon. & (15.105) \end{aligned}$$

We then let $n \to \infty$ and then $\epsilon \to 0$ to conclude that there exists $\mathbf{h} \in \overline{\Gamma}^*_N$ which satisfies (15.63) to (15.67). Hence, $\mathcal{R} \subset \mathcal{R}_{out}$, and the theorem is proved. □

15.6 THE LP BOUND AND ITS TIGHTNESS

In Section 15.4, we stated the inner bound $\mathcal{R}_{in}$ on $\mathcal{R}$ in terms of Γ^*_N (the proof is deferred to Section 15.7), and in Section 15.5, we proved the outer bound $\mathcal{R}_{out}$ on $\mathcal{R}$ in terms of $\overline{\Gamma}^*_N$. So far, there exists no full characterization on either Γ^*_N or $\overline{\Gamma}^*_N$. Therefore, these bounds cannot be evaluated explicitly. In this section, we give a geometrical interpretation of these bounds which leads to an explicit outer bound on $\mathcal{R}$ called the LP bound (LP for *linear programming*).

Let A be a subset of $Q(N)$. For a vector $\mathbf{h} \in \mathcal{H}_N$, let

$$\mathbf{h}_A = (h_Z : Z \in A). \qquad (15.106)$$

For a subset $\mathcal{B}$ of $\mathcal{H}_N$, let

$$\text{proj}_A(\mathcal{B}) = \{\mathbf{h}_A : \mathbf{h} \in \mathcal{B}\} \qquad (15.107)$$

be the projection of the set $\mathcal{B}$ on the coordinates h_Z, $Z \in A$. For a subset $\mathcal{B}$ of $\mathcal{H}_N$, define

$$\Lambda(\mathcal{B}) = \{\mathbf{h} \in \mathcal{H}_N : 0 \leq \mathbf{h} < \mathbf{h}' \text{ for some } \mathbf{h}' \in \mathcal{B}\} \qquad (15.108)$$

and

$$\bar{\Lambda}(\mathcal{B}) = \{\mathbf{h} \in \mathcal{H}_N : 0 \leq \mathbf{h} \leq \mathbf{h}' \text{ for some } \mathbf{h}' \in \mathcal{B}\}. \qquad (15.109)$$

A vector $\mathbf{h} \geq 0$ is in $\Lambda(\mathcal{B})$ if and only if it is *strictly* inferior to some vector $\mathbf{h}'$ in $\mathcal{B}$, and is in $\bar{\Lambda}(\mathcal{B})$ if and only if it is inferior to some vector $\mathbf{h}'$ in $\mathcal{B}$.

Define the following subsets of $\mathcal{H}_N$:

$$\mathcal{C}_1 = \left\{ \mathbf{h} \in \mathcal{H}_N : h_{(Y_s:s\in S)} = \sum_{s\in S} h_{Y_s} \right\} \tag{15.110}$$

$$\mathcal{C}_2 = \left\{ \mathbf{h} \in \mathcal{H}_N : h_{U_{ij}|(Y_s:s\in J(i)),(U_{i'i}:(i',i)\in E)} = 0 \text{ for all } (i,j) \in E \right\} \tag{15.111}$$

$$\mathcal{C}_3 = \left\{ \mathbf{h} \in \mathcal{H}_N : h_{(Y_s:s\in F(i))|(U_{i'i}:(i',i)\in E)} = 0 \text{ for all } i \in \Xi \right\} \tag{15.112}$$

$$\mathcal{C}_4 = \left\{ \mathbf{h} \in \mathcal{H}_N : R_{ij} > h_{U_{ij}} \text{ for all } (i,j) \in E \right\}. \tag{15.113}$$

The set $\mathcal{C}_1$ is a hyperplane in $\mathcal{H}_N$. Each of the sets $\mathcal{C}_2$ and $\mathcal{C}_3$ is the intersection of a collection of hyperplanes in $\mathcal{H}_N$. The set $\mathcal{C}_4$ is the intersection of a collection of open half-spaces in $\mathcal{H}_N$. Then from the alternative definition of $\mathcal{R}'$ (Definition 15.7), we see that

$$\mathcal{R}' = \Lambda(\text{proj}_{(Y_s:s\in S)}(\Gamma_N^* \cap \mathcal{C}_1 \cap \mathcal{C}_2 \cap \mathcal{C}_3 \cap \mathcal{C}_4)). \tag{15.114}$$

and

$$\mathcal{R}_{in} = \overline{\text{con}}(\Lambda(\text{proj}_{(Y_s:s\in S)}(\Gamma_N^* \cap \mathcal{C}_1 \cap \mathcal{C}_2 \cap \mathcal{C}_3 \cap \mathcal{C}_4))). \tag{15.115}$$

Similarly, we see that

$$\mathcal{R}_{out} = \bar{\Lambda}(\text{proj}_{(Y_s:s\in S)}(\overline{\Gamma}_N^* \cap \mathcal{C}_1 \cap \mathcal{C}_2 \cap \mathcal{C}_3 \cap \overline{\mathcal{C}_4})). \tag{15.116}$$

It can be shown that if $\Gamma_N^* \cap (\mathcal{C}_1 \cap \mathcal{C}_2 \cap \mathcal{C}_3)$ is dense in $\overline{\Gamma}_N^* \cap (\mathcal{C}_1 \cap \mathcal{C}_2 \cap \mathcal{C}_3)$, i.e.,

$$\overline{\Gamma_N^* \cap (\mathcal{C}_1 \cap \mathcal{C}_2 \cap \mathcal{C}_3)} = \overline{\Gamma}_N^* \cap (\mathcal{C}_1 \cap \mathcal{C}_2 \cap \mathcal{C}_3), \tag{15.117}$$

then

$$\mathcal{R}_{out} = \overline{\mathcal{R}'} \subset \overline{\text{con}}(\mathcal{R}') = \mathcal{R}_{in}, \tag{15.118}$$

which implies

$$\mathcal{R}_{in} = \mathcal{R}_{out}. \tag{15.119}$$

Note that $(\mathcal{C}_1 \cap \mathcal{C}_2 \cap \mathcal{C}_3)$ is a closed subset of $\mathcal{H}_N$. However, while

$$\overline{\Gamma_N^* \cap \mathcal{C}} \subset \overline{\Gamma}_N^* \cap \mathcal{C} \tag{15.120}$$

for any closed subset $\mathcal{C}$ of $\mathcal{H}_N$, it is not in general true that

$$\overline{\Gamma_N^* \cap \mathcal{C}} = \overline{\Gamma}_N^* \cap \mathcal{C}. \tag{15.121}$$

As a counterexample, as we have shown in the proof of Theorem 14.2, $\overline{\Gamma_3^* \cap \tilde{\mathcal{C}}}$ is a proper subset of $\overline{\Gamma}_3^* \cap \tilde{\mathcal{C}}$, where

$$\tilde{\mathcal{C}} = \{\mathbf{h} \in \Gamma_3^* : i_{1;2} = i_{2;3} = i_{1;3} = 0\}, \tag{15.122}$$

and we have used $i_{j;k}$ to denote $h_j - h_{j|k}$.

To facilitate our discussion, we further define

$$i_{A;A'} = h_A - h_{A|A'} \tag{15.123}$$

and

$$i_{A;A'|A''} = h_{A|A''} - h_{A|A'A''} \tag{15.124}$$

for $A, A', A'' \in Q(N)$. Let Γ_N be the set of $\mathbf{h} \in \mathcal{H}_N$ such that $\mathbf{h}$ satisfies all the basic inequalities involving some or all of the random variables in N, i.e., for all $A, A', A'' \in Q(N)$,

$$h_A \geq 0 \tag{15.125}$$

$$h_{A|A'} \geq 0 \tag{15.126}$$

$$i_{A;A'} \geq 0 \tag{15.127}$$

$$i_{A;A'|A''} \geq 0. \tag{15.128}$$

We know from Section 13.2 that $\overline{\Gamma}_N^* \subset \Gamma_N$. Then upon replacing $\overline{\Gamma}_N^*$ by Γ_N in the definition of $\mathcal{R}_{out}$, we immediately obtain an outer bound on $\mathcal{R}_{out}$. This is called the LP bound, which is denoted by $\mathcal{R}_{LP}$. In other words, $\mathcal{R}_{LP}$ is obtained by replacing $\overline{\Gamma}_N^*$ by Γ_N on the right hand side of (15.116), i.e.,

$$\mathcal{R}_{LP} = \bar{\Lambda}(\mathrm{proj}_{(Y_s:s\in S)}(\Gamma_N \cap \mathcal{C}_1 \cap \mathcal{C}_2 \cap \mathcal{C}_3 \cap \overline{\mathcal{C}_4})). \tag{15.129}$$

Since all the constraints defining $\mathcal{R}_{LP}$ are linear, $\mathcal{R}_{LP}$ can be evaluated explicitly.

We now show that the outer bound $\mathcal{R}_{LP}$ is tight for the special case of single-source network coding. Without loss of generality, let

$$S = \{1\}. \tag{15.130}$$

Then the information source represented by X_1 is generated at node $Q(1)$, and for all $i \in \Xi$,

$$F(i) = S = \{1\}. \tag{15.131}$$

Consider any $\omega_1 \in \mathcal{R}_{out}$. Following the proof that $\mathcal{R}_{out}$ is an outer bound on $\mathcal{R}$ in the last section (Theorem 15.9), we see that for any $\epsilon > 0$ and a sufficiently large n, there exist random variables $U_{ij}, (i,j) \in E$ which satisfy the constraints in (15.92) to (15.94). Upon specializing for a single source, these constraints become

$$\begin{aligned} H(U_{Q(1),j}|X_1) &= 0 && \text{for } j : (Q(1), j) \in E && (15.132)\\ H(U_{ij}|U_{i'i} : (i', i) \in E) &= 0 && \text{for } (i, j) \in E && (15.133)\\ H(X_1|U_{it} : (i, t) \in E) &\leq n\varrho_t(n, \epsilon) && \text{for } t \in \Xi && (15.134)\\ n(R_{ij} + 2\epsilon) &\geq H(U_{ij}) && \text{for } (i, j) \in E, && (15.135) \end{aligned}$$

where $\varrho_t(n, \epsilon) \to 0$ as $\epsilon \to 0$ and $n \to \infty$. Moreover, upon specializing (15.95) for a single source, we also have

$$H(X_1) \geq n(\omega_1 - \epsilon). \quad (15.136)$$

Now fix a sink node $t \in \Xi$ and consider any cut W between node $Q(1)$ and node t, i.e., $Q(1) \in W$ and $t \notin W$, and let

$$E_W = \{(i, j) \in E : i \in W \text{ and } j \notin W\} \quad (15.137)$$

be the set of edges across the cut W. Then

$$\begin{aligned} &\sum_{(i,j)\in E_W} n(R_{ij} + 2\epsilon)\\ &\stackrel{a)}{\geq} \sum_{(i,j)\in E_W} H(U_{ij}) && (15.138)\\ &\geq H(U_{ij} : (i, j) \in E_W) && (15.139)\\ &\stackrel{b)}{=} H(U_{ij} : (i, j) \in E_W \text{ or both } i, j \notin W) && (15.140)\\ &\geq H(U_{it} : (i, t) \in E) && (15.141)\\ &= H(U_{it} : (i, t) \in E|X_1) + I(U_{it} : (i, t) \in E; X_1) && (15.142)\\ &= I(U_{it} : (i, t) \in E; X_1) && (15.143)\\ &= H(X_1) - H(X_1|U_{it} : (i, t) \in E) && (15.144)\\ &\stackrel{c)}{\geq} H(X_1) - n\varrho_t(n, \epsilon) && (15.145)\\ &\stackrel{d)}{\geq} n(\omega_1 - \epsilon) - n\varrho_t(n, \epsilon) && (15.146)\\ &= n(\omega_1 - \epsilon - \varrho_t(n, \epsilon)), && (15.147) \end{aligned}$$

where

a) follows from (15.135);

b) follows because U_{ij}, where both $i, j \notin W$, are functions of $U_{i,j}, (i, j) \in E_W$ by virtue of (15.133) and the acyclicity of the graph G;

c) follows from (15.134);

d) follows from (15.136).

Dividing by n and letting $\epsilon \to 0$ and $n \to \infty$ in (15.147), we obtain

$$\sum_{(i,j)\in E_W} R_{ij} \geq \omega_1. \tag{15.148}$$

In other words, if $\omega_1 \in \mathcal{R}_{out}$, then ω_1 satisfies the above inequality for every cut W between node $Q(1)$ and any sink node $t \in \Xi$, which is precisely the max-flow bound for single-source network coding. Since we know from Chapter 11 that this bound is both necessary and sufficient, we conclude that $\mathcal{R}_{out}$ is tight for single-source network coding. However, we note that the max-flow bound was discussed in Chapter 11 in the more general context that the graph G may be cyclic.

In fact, it has been proved in [236] that $\mathcal{R}_{LP}$ is tight for all other special cases of multi-source network coding for which the achievable information region is known. In addition to single-source network coding, these also include the models described in [98], [230], [175], [235], and [236]. Since $\mathcal{R}_{LP}$ encompasses all Shannon-type information inequalities and the converse proofs of the achievable information rate region for all these special cases do not involve non-Shannon-type inequalities, the tightness of $\mathcal{R}_{LP}$ for all these cases is expected.

15.7 ACHIEVABILITY OF $\mathcal{R}_{in}$

In this section, we prove the achievability of $\mathcal{R}_{in}$, namely Theorem 15.6. To facilitate our discussion, we first introduce a few functions regarding the auxiliary random variables in the definition of $\mathcal{R}'$ in Definition 15.4.

From (15.47), since U_{ij} is a function of $(Y_s : s \in J(i))$ and $(U_{i'i} : (i', i) \in E))$, we write

$$U_{ij} = u_{ij}((Y_s : s \in J(i)), (U_{i'i} : (i', i) \in E)). \tag{15.149}$$

Since the graph G is acyclic, we see inductively that all the auxiliary random variables $U_{ij}, (i, j) \in E$ are functions of the auxiliary random variables $Y_s, s \in S$. Thus we also write

$$U_{ij} = \tilde{u}_{ij}(Y_s : s \in S). \tag{15.150}$$

Equating (15.149) and (15.150), we obtain

$$u_{ij}((Y_s : s \in J(i)), (U_{i'i} : (i', i) \in E)) = \tilde{u}_{ij}(Y_s : s \in S). \tag{15.151}$$

For $s \in F(i)$, since Y_s is a function of $(U_{i'i} : (i', i) \in E)$ from (15.48), we write

$$Y_s = y_s^{(i)}(U_{i'i} : (i', i) \in E). \tag{15.152}$$

Substituting (15.150) into (15.152), we obtain

$$Y_s = y_s^{(i)}(\tilde{u}_{i'i}(Y_{s'} : s' \in S) : (i', i) \in E). \tag{15.153}$$

These relations will be useful in the proof of Theorem 15.6.

Before we prove Theorem 15.6, we first prove in the following lemma that strong typicality is preserved when a function is applied to a vector.

LEMMA 15.10 *Let $Y = f(X)$. If*

$$\mathbf{x} = (x_1, x_2, \cdots, x_n) \in T^n_{[X]\delta}, \tag{15.154}$$

then

$$f(\mathbf{x}) = (y_1, y_2, \cdots, y_n) \in T^n_{[Y]\delta}, \tag{15.155}$$

where $y_i = f(x_i)$ for $1 \leq i \leq n$.

Proof Consider $\mathbf{x} \in T^n_{[X]\delta}$, i.e.,

$$\sum_x \left| \frac{1}{n} N(x; \mathbf{x}) - p(x) \right| < \delta. \tag{15.156}$$

Since $Y = f(X)$,

$$p(y) = \sum_{x \in f^{-1}(y)} p(x) \tag{15.157}$$

for all $y \in \mathcal{Y}$. On the other hand,

$$N(y; f(\mathbf{x})) = \sum_{x \in f^{-1}(y)} N(x; \mathbf{x}) \tag{15.158}$$

for all $y \in \mathcal{Y}$. Then

$$\begin{aligned}
&\sum_y \left| \frac{1}{n} N(y; f(\mathbf{x})) - p(y) \right| \\
&= \sum_y \left| \sum_{x \in f^{-1}(y)} \left(\frac{1}{n} N(x; \mathbf{x}) - p(x) \right) \right| && (15.159) \\
&\leq \sum_y \sum_{x \in f^{-1}(y)} \left| \frac{1}{n} N(x; \mathbf{x}) - p(x) \right| && (15.160) \\
&= \sum_x \left| \frac{1}{n} N(x; \mathbf{x}) - p(x) \right| && (15.161) \\
&< \delta. && (15.162)
\end{aligned}$$

Therefore, $f(\mathbf{x}) \in T^n_{[Y]\delta}$, proving the lemma. □

Proof of Theorem 15.6 Consider an information rate tuple

$$\omega = (\omega_s : s \in S) \tag{15.163}$$

in $\mathcal{R}'$, i.e., there exist auxiliary random variables $Y_s, s \in S$ and $U_{ij}, (i,j) \in E$ which satisfy (15.46) to (15.50). We first impose the constraint that all the auxiliary random variables have finite alphabets. The purpose of imposing this additional constraint is to enable the use of strong typicality which applies only to random variables with finite alphabets. This constraint will be removed later.

Consider any fixed $\epsilon > 0$. We now construct an

$$(n, (\eta_{ij} : (i,j) \in E), (\omega_s - \epsilon : s \in S)) \tag{15.164}$$

random code on graph G for sufficiently large n such that

$$n^{-1} \log \eta_{ij} \leq R_{ij} + \epsilon \tag{15.165}$$

for all $(i,j) \in E$, and

$$\Delta_i \leq \epsilon \tag{15.166}$$

for all $i \in \Xi$. Let δ and δ' be small positive quantities such that

$$\delta' < \delta. \tag{15.167}$$

The actual values of δ and δ' will be specified later. The coding scheme is described in the following steps, where n is assumed to be a sufficiently large integer.

1. For $s \in S$, by the strong AEP (Theorem 5.2),

$$\zeta_s \stackrel{\text{def}}{=} |T^n_{[Y_s]\delta'}| \geq (1 - \delta')2^{n(H(Y_s) - \gamma_s)}, \tag{15.168}$$

where $\gamma_s \to 0$ as $\delta' \to 0$. By letting δ' be sufficiently small, by virtue of (15.50), we have

$$H(Y_s) - \gamma_s > \omega_s, \tag{15.169}$$

and

$$(1 - \delta')2^{n(H(Y_s) - \gamma_s)} > \theta_s, \tag{15.170}$$

where

$$\theta_s = \lceil 2^{n(\omega_s - \epsilon)} \rceil = |\mathcal{X}_s|. \tag{15.171}$$

Then it follows from (15.168) and (15.170) that

$$\zeta_s > \theta_s. \tag{15.172}$$

Denote the vectors in $T^n_{[Y_s]\delta'}$ by $\underline{\nu}_s(k), 1 \leq k \leq \zeta_s$.

2. Arbitrarily partition $T^n_{[Y_s]\delta'}$ into subsets $A_1(s), A_2(s), \cdots, A_{\theta_s}(s)$ almost uniformly, such that the size of each subset is either equal to $\lfloor \theta_s^{-1}\zeta_s \rfloor$ or $\lceil \theta_s^{-1}\zeta_s \rceil$. Define a random mapping

$$\rho_s : \{1, 2, \cdots, \theta_s\} \to \{1, 2, \cdots, \zeta_s\} \tag{15.173}$$

by letting $\rho_s(l)$ be an index randomly chosen in $A_{l_s}(s)$ according to the uniform distribution for $1 \le l_s \le \theta_s$.

3. For $(i, j) \in E$, denote the vectors in $T^n_{[U_{ij}]\delta}$ by $\underline{\mu}_{ij}(k)$, $1 \le k \le \zeta_{ij}$, where

$$\zeta_{ij} = |T^n_{[U_{ij}]\delta}|. \tag{15.174}$$

By the strong AEP,

$$\zeta_{ij} = |T^n_{[U_{ij}]\delta}| \le 2^{n(H(U_{ij})+\gamma_{ij})}, \tag{15.175}$$

where $\gamma_{ij} \to 0$ as $\delta \to 0$. Then by letting δ be sufficiently small, by virtue of (15.49), we have

$$H(U_{ij}) + \gamma_{ij} \le R_{ij} + \epsilon \tag{15.176}$$

and hence

$$2^{n(H(U_{ij})+\gamma_{ij})} \le 2^{n(R_{ij}+\epsilon)}. \tag{15.177}$$

Then we can choose an integer η_{ij} such that

$$2^{n(H(U_{ij})+\gamma_{ij})} \le \eta_{ij} \le 2^{n(R_{ij}+\epsilon)}. \tag{15.178}$$

The upper bound above ensures that η_{ij} satisfies (15.165).

4. Define the encoding function

$$f_{ij} : \prod_{s \in J(i)} \mathcal{X}_s \times \prod_{i':(i',i) \in E} \{0, 1, \cdots, \eta_{i'i}\} \to \{0, 1, \cdots, \eta_{ij}\} \tag{15.179}$$

as follows. Let x_s be the outcome of X_s. If

$$\Big((\underline{\nu}_s(\rho_s(x_s))) : s \in J(i)), (\underline{\mu}_{i'i}(\tilde{f}_{i'i}(x_s : s \in S) : (i', i) \in E)) \Big)$$
$$\in T_{[(Y_s : s \in J(i)), (U_{i'i} : (i',i) \in E)]\delta}, \tag{15.180}$$

where $\tilde{f}_{i'i}(x_s : s \in S)$ denotes the value of $f_{i'i}$ as a function of $(x_s : s \in S)$, then by Lemma 15.10,

$$u_{ij} \Big((\underline{\nu}_s(\rho_s(x_s))) : s \in J(i)), (\underline{\mu}_{i'i}(\tilde{f}_{i'i}(x_s : s \in S) : (i', i) \in E)) \Big) \tag{15.181}$$

is in $T^n_{[U_{ij}]\delta}$ (cf. (15.149) for the definition of u_{ij}). Then let

$$f_{ij}\left((\rho_s(x_s) : s \in J(i)), (\tilde{f}_{i'i}(x_s : s \in S) : (i', i) \in E)\right) \quad (15.182)$$

be the index of

$$u_{ij}\left((\underline{\nu}_s(\rho_s(x_s)) : s \in J(i)), (\underline{\mu}_{ij}(\tilde{f}_{i'i}(x_s : s \in S) : (i', i) \in E))\right) \quad (15.183)$$

in $T^n_{[U_{ij}]\delta}$ (which is greater than or equal to 1), i.e.,

$$\underline{\mu}_{ij}(\tilde{f}_{ij}(x_s : s \in S)) =$$

$$u_{ij}\left((\underline{\nu}_s(\rho_s(x_s)) : s \in J(i)), (\underline{\mu}_{ij}(\tilde{f}_{i'i}(x_s : s \in S) : (i', i) \in E))\right). \quad (15.184)$$

If (15.180) is not satisfied, then let

$$f_{ij}\left((\rho_s(x_s) : s \in J(i)), (\tilde{f}_{i'i}(x_s : s \in S) : (i', i) \in E)\right) = 0. \quad (15.185)$$

Note that the lower bound on η_{ij} in (15.178) ensures that f_{ij} is properly defined because from (15.175), we have

$$\eta_{ij} \geq 2^{n(H(U_{ij})+\gamma_{ij})} \geq |T^n_{[U_{ij}]\delta}| = \zeta_{ij}. \quad (15.186)$$

5. For $i \in \Xi$, define the decoding function

$$g_i : \prod_{i':(i',i)\in E} \{0, 1 \cdots, \eta_{i'i}\} \to \prod_{s \in F(i)} \mathcal{X}_s \quad (15.187)$$

as follows. If

$$\tilde{f}_{i'i}(x_s : s \in S) \neq 0 \quad (15.188)$$

for all $(i', i) \in E$, and if for all $s \in F(i)$ there exists $1 \leq l_s \leq \theta_s$ such that

$$y_s^{(i)}\left(\underline{\mu}_{i'i}(\tilde{f}_{i'i}(x_s : s \in S)) : (i', i) \in E\right) = \underline{\nu}_s(\rho_s(l_s)) \quad (15.189)$$

(cf. the definition of $y_s^{(i)}$ in (15.152) and note that $y_s^{(i)}$ is applied to a vector as in Lemma 15.10), then let

$$g_i\left(\tilde{f}_{i'i}(x_s : s \in S) : (i', i) \in E\right) = (l_s : s \in F(i)). \quad (15.190)$$

Note that if an l_s as specified above exists, then it must be unique because

$$\underline{\nu}_s(\rho_s(l)) \neq \underline{\nu}_s(\rho_s(l')) \quad (15.191)$$

if $l \neq l'$. For all other cases, let

$$g_i\left(\tilde{f}_{i'i}(x_s : s \in S) : (i', i) \in E\right) = (\underbrace{1, 1, \cdots, 1}_{|F(i)|}), \tag{15.192}$$

a constant vector in $\prod_{s \in F(i)} \mathcal{X}_s$.

We now analyze the performance of this random code. Specifically, we will show that (15.166) is satisfied for all $i \in \Xi$. We claim that for any $\delta > 0$, if

$$(\underline{\nu}_s(\rho_s(x_s)) : s \in S) \in T^n_{[Y_s : s \in S]\delta}, \tag{15.193}$$

then the following are true:

i) (15.180) holds for all $i \in V$;

ii) $\tilde{f}_{ij}(x_s : s \in S) \neq 0$ for all $(i, j) \in E$;

iii) $\underline{\mu}_{ij}(\tilde{f}_{ij}(x_s : s \in S)) = \tilde{u}_{ij}(\underline{\nu}_s(\rho_s(x_s)) : s \in S)$ for all $(i, j) \in E$.

We will prove the claim by induction on i, the index of a node. We first consider node 1, which is the first node to encode. It is evident that

$$\{i' \in V : (i', 1) \in E\} \tag{15.194}$$

is empty. Since $(Y_s : s \in J(1))$ is a function of $(Y_s : s \in S)$, (15.180) follows from (15.193) via an application of Lemma 15.10. This proves i), and ii) follows by the definition of the encoding function f_{ij}. For all $(1, j) \in E$, consider

$$\begin{aligned}
\underline{\mu}_{1j}(\tilde{f}_{1j}(x_s : s \in S)) &= u_{1j}(\underline{\nu}_s(\rho_s(x_s)) : s \in J(1)) && (15.195)\\
&= \tilde{u}_{1j}(\underline{\nu}_s(\rho_s(x_s)) : s \in S), && (15.196)
\end{aligned}$$

where the first and the second equality follow from (15.184) and (15.151), respectively. The latter can be seen by replacing the random variable Y_s by the vector $\underline{\nu}_s(\rho_s(x_s))$ in (15.151). This proves iii). Now assume that the claim is true for all node $j \leq i - 1$ for some $2 \leq i \leq |V|$, and we will prove that the claim is true for node i. For all $(i', i) \in E$,

$$\underline{\mu}_{i'i}(\tilde{f}_{i'i}(x_s : s \in S)) = \tilde{u}_{i'i}(\underline{\nu}_s(\rho_s(x_s)) : s \in S) \tag{15.197}$$

by iii) of the induction hypothesis. Since $((Y_s : s \in J(i)), (U_{i'i} : (i', i) \in E))$ is a function of $(Y_s : s \in S)$ (cf. (15.150)), (15.180) follows from (15.193) via an application of Lemma 15.10, proving i). Again, ii) follows immediately from i). For all $(i,j) \in E$, it follows from (15.184) and (15.151) that

$$\begin{aligned}
&\underline{\mu}_{ij}(\tilde{f}_{ij}(x_s : s \in S)) \\
&= u_{ij}\left((\underline{\nu}_s(\rho_s(x_s)) : s \in J(i)), (\underline{\mu}_{i'i}(\tilde{f}_{i'i}(x_s : s \in S) : (i',i) \in E))\right) \qquad (15.198)\\
&= \tilde{u}_{ij}(\underline{\nu}_s(\rho_s(x_s)) : s \in S), \qquad (15.199)
\end{aligned}$$

where the last equality is obtained by replacing the random variable Y_s by the vector $\underline{\nu}_s(\rho_s(x_s))$ and the random variable $U_{i'i}$ by the vector $\underline{\mu}_{i'i}(\tilde{f}_{i'i}(x_s : s \in S)$ in (15.151). This proves iii). Thus the claim is proved.

If (15.193) holds, for $s \in F(i)$, it follows from iii) of the above claim and (15.153) that

$$\begin{aligned}
&y_s^{(i)}(\underline{\mu}_{i'i}(\tilde{f}_{i'i}(x_{s'} : s' \in S)) : (i',i) \in E) \\
&= y_s^{(i)}(\tilde{u}_{i'i}(\underline{\nu}_{s'}(\rho_{s'}(x_{s'})) : s' \in S) : (i',i) \in E) \qquad (15.200)\\
&= \underline{\nu}_s(\rho_s(x_s)), \qquad (15.201)
\end{aligned}$$

where the last equality is obtained by replacing $Y_{s'}$ by $\underline{\nu}_{s'}(\rho_{s'}(x_{s'}))$ in (15.153). Then by the definition of the decoding function g_i, we have

$$g_i\left(\tilde{f}_{i'i}(x_s : s \in S) : (i',i) \in E\right) = (x_s : s \in F(i)). \qquad (15.202)$$

Hence, x_s, $s \in F(i)$ can be decoded correctly.

Therefore, it suffices to consider the probability that (15.193) does not hold, since this is the only case for a decoding error to occur. For $s \in S$, consider any $1 \le k_s \le \zeta_s$, where $k_s \in A_{l_s}(s)$ for some $1 \le l_s \le \theta_s$. Then

$$\begin{aligned}
&\Pr\{\rho_s(X_s) = k_s\} \\
&= \Pr\{\rho_s(X_s) = k_s, X_s = l_s\} \qquad (15.203)\\
&= \Pr\{\rho_s(X_s) = k_s | X_s = l_s\}\Pr\{X_s = l_s\} \qquad (15.204)\\
&= |A_{l_s}(s)|^{-1}\theta_s^{-1}. \qquad (15.205)
\end{aligned}$$

Since $|A_{l_s}(s)|$ is equal to either $\lfloor \theta_s^{-1}\zeta_s \rfloor$ or $\lceil \theta_s^{-1}\zeta_s \rceil$, we have

$$|A_{l_s}(s)| \ge \lfloor \theta_s^{-1}\zeta_s \rfloor \ge \theta_s^{-1}\zeta_s - 1. \qquad (15.206)$$

Therefore,

$$\Pr\{\rho_s(X_s) = k_s\} \le \frac{1}{(\theta_s^{-1}\zeta_s - 1)\theta_s} = \frac{1}{\zeta_s - \theta_s}. \qquad (15.207)$$

Since $\theta_s \approx 2^{n\omega_s}$ and $\zeta_s \approx 2^{nH(Y_s)}$, where

$$H(Y_s) > \omega_s \qquad (15.208)$$

from (15.50), θ_s is negligible compared with ζ_s when n is large. Then by the strong AEP, there exists $\gamma_s'(\delta') > 0$ such that

$$(1-\delta')2^{n(H(Y_s)-\gamma_s'(\delta'))} \le \zeta_s - \theta_s < \zeta_s \le 2^{n(H(Y_s)+\gamma_s'(\delta'))}, \qquad (15.209)$$

where $\gamma_s'(\delta') \to 0$ as $\delta' \to 0$. Therefore,

$$\Pr\{\rho_s(X_s) = k_s\} \le (1-\delta')^{-1}2^{-n(H(Y_s)-\gamma_s'(\delta'))}. \qquad (15.210)$$

Note that this lower bound does not depend on the value of k_s. It then follows that

$$\begin{aligned}
&\Pr\{\rho_s(X_s) = k_s : s \in S\} \\
&\le (1-\delta')^{-|S|}\prod_s 2^{-n(H(Y_s)-\gamma_s'(\delta'))} && (15.211)\\
&= (1-\delta')^{-|S|}2^{-n\sum_s(H(Y_s)-\gamma_s'(\delta'))} && (15.212)\\
&= (1-\delta')^{-|S|}2^{-n(\sum_s H(Y_s)-\sum_s \gamma_s'(\delta'))} && (15.213)\\
&= (1-\delta')^{-|S|}2^{-n(H(Y_s:s\in S)-\gamma'(\delta'))}, && (15.214)
\end{aligned}$$

where the last equality follows from (15.46) with

$$\gamma'(\delta') = \sum_s \gamma_s'(\delta') \qquad (15.215)$$

which tends to 0 as $\delta' \to 0$. In other words, for every $(\mathbf{y}_s : s \in S) \in \prod_s T^n_{[Y_s]\delta'}$,

$$\Pr\{\underline{\nu}_s(\rho_s(X_s)) = \mathbf{y}_s : s \in S\} \le (1-\delta')^{-|S|}2^{-n(H(Y_s:s\in S)-\gamma'(\delta'))}. \qquad (15.216)$$

Let $\mathbf{Y}_s$ denote n i.i.d. copies of the random variable Y_s. For every $\mathbf{y}_s \in T^n_{[Y_s]\delta'}$, by the strong AEP,

$$\Pr\{\mathbf{Y}_s = \mathbf{y}_s\} \ge 2^{-n[H(Y_s)+v_s(\delta')]}, \qquad (15.217)$$

where $v_s(\delta') \to 0$ as $\delta' \to 0$. Then for every $(\mathbf{y}_s : s \in S) \in \prod_s T^n_{[Y_s]\delta'}$,

$$\begin{aligned}
\Pr\{\mathbf{Y}_s = \mathbf{y}_s : s \in S\} &\ge \prod_s 2^{-n[H(Y_s)+v_s(\delta')]} && (15.218)\\
&= 2^{-n(\sum_s H(Y_s)+\sum_s v_s(\delta'))} && (15.219)\\
&= 2^{-n(H(Y_s:s\in S)+v(\delta'))}, && (15.220)
\end{aligned}$$

where

$$v(\delta') = \sum_s v_s(\delta') \to 0 \qquad (15.221)$$

as $\delta' \to 0$.

For every $(\mathbf{y}_s : s \in S) \in \prod_s T^n_{[Y_s]\delta'}$, from (15.216) and (15.220), we have

$$\begin{aligned} &\Pr\{\underline{\nu}_s(\rho_s(X_s)) = \mathbf{y}_s : s \in S\} \\ &\leq (1-\delta')^{-|S|} 2^{n(\gamma'(\delta')+\upsilon(\delta'))} 2^{-n(H(Y_s:s\in S)+\upsilon(\delta'))} && (15.222) \\ &\leq (1-\delta')^{-|S|} 2^{n(\gamma'(\delta')+\upsilon(\delta'))} \Pr\{\mathbf{Y}_s = \mathbf{y}_s : s \in S\} && (15.223) \\ &= (1-\delta')^{-|S|} 2^{n\varpi(\delta')} \Pr\{\mathbf{Y}_s = \mathbf{y}_s : s \in S\}, && (15.224) \end{aligned}$$

where

$$\varpi(\delta') = \gamma'(\delta') + \upsilon(\delta'). \tag{15.225}$$

Since

$$\Pr\{\underline{\nu}_s(\rho_s(X_s)) = \mathbf{y}_s : s \in S\} = 0 \tag{15.226}$$

for all $(\mathbf{y}_s : s \in S) \not\in \prod_s T^n_{[Y_s]\delta'}$, (15.224) is in fact true for all $(\mathbf{y}_s : s \in S) \in (\mathcal{Y}^n_s : s \in S)$. By Theorem 5.3,

$$\Pr\{(\mathbf{Y}_s : s \in S) \not\in T^n_{[Y_s:s\in S]\delta}\} \leq 2^{-n\varphi(\delta)}, \tag{15.227}$$

where $\varphi(\delta) > 0$ and $\varphi(\delta) \to 0$ as $\delta \to 0$. Then by summing over all $(\mathbf{y}_s : s \in S) \not\in T^n_{[Y_s:s\in S]\delta}$ in (15.224), we have

$$\begin{aligned} &\Pr\{(\underline{\nu}_s(\rho_s(X_s)) : s \in S) \not\in T^n_{[Y_s:s\in S]\delta}\} \\ &\leq (1-\delta')^{-|S|} 2^{n\varpi(\delta')} \Pr\{(\mathbf{Y}_s : s \in S) \not\in T^n_{[Y_s:s\in S]\delta}\} && (15.228) \\ &\leq (1-\delta')^{-|S|} 2^{-n(\varphi(\delta)-\varpi(\delta'))}. && (15.229) \end{aligned}$$

Since $\delta' < \delta$ (cf. (15.167)), we can let δ' be sufficiently small so that

$$\varphi(\delta) - \varpi(\delta') > 0 \tag{15.230}$$

and the upper bound in (15.229) tends to 0 as $n \to \infty$. Thus, when n is sufficiently large, for all $i \in \Xi$,

$$\Delta_i \leq \Pr\left\{(\underline{\nu}_s(\rho_s(X_s)) : s \in S) \not\in T^n_{[Y_s:s\in S]\delta}\right\} \leq \epsilon. \tag{15.231}$$

Hence, we have constructed a desired random code with the extraneous constraint that the all the auxiliary random variables have finite alphabets.

We now show that when some of the auxiliary random variables do not have finite supports[1], they can be approximated by ones with finite supports while all the independence and functional dependence constraints required

[1] These random variables nevertheless are assumed to have finite entropies.

continue to hold. Suppose some of the auxiliary random variables $Y_s, s \in S$ and $U_{ij}, (i,j) \in E$ do not have finite supports. Without loss of generality, assume that Y_s takes values in the set of positive integers. For all $s \in S$, define a random variable $Y_s'(m)$ which takes values in

$$\mathcal{N}_m = \{1, 2, \cdots, m\} \tag{15.232}$$

such that

$$\Pr\{Y_s'(m) = k\} = \frac{\Pr\{Y_s = k\}}{\Pr\{Y_s \in \mathcal{N}_m\}} \tag{15.233}$$

for all $k \in \mathcal{N}_m$, i.e., $Y_s'(m)$ is a truncation of Y_s up to m. It is intuitively clear that $H(Y_s'(m)) \to H(Y_s)$ as $m \to \infty$. The proof is given in Appendix 15.A.

Let $Y_s'(m), s \in S$ be mutually independent so that they satisfy (15.46) with all the random variables in (15.46) primed. Now construct random variables $U_{ij}'(m)$ inductively by letting

$$U_{ij}'(m) = u_{ij}((Y_s'(m) : s \in J(i)), (U_{i'i}'(m) : (i', i) \in E)) \tag{15.234}$$

for all $(i,j) \in E$ (cf. (15.149)). Then $Y_s'(m), s \in S$ and $U_{ij}'(m), (i,j) \in E$ satisfy (15.47) and (15.48) with all the random variables in (15.47) and (15.48) primed. Using the exactly the same argument for $Y_s'(m)$, we see that $H(U_{ij}'(m)) \to H(U_{ij})$ as $m \to \infty$.

If $Y_s, s \in S$ and $U_{ij}, (i,j) \in E$ satisfy (15.49) and (15.50), then there exists $\psi > 0$ such that

$$R_{ij} > H(U_{ij}) + \psi \tag{15.235}$$

and

$$H(Y_s) - \psi > \omega_s. \tag{15.236}$$

Then for sufficiently large m, we have

$$R_{ij} > H(U_{ij}) + \psi > H(U_{ij}'(m)) \tag{15.237}$$

and

$$H(Y_s'(m)) > H(Y_s) - \psi > \omega_s. \tag{15.238}$$

In other words, $Y_s'(m), s \in S$ and $U_{ij}'(m), (i,j) \in E$, whose supports are finite, satisfy (15.49) and (15.50) with all the random variables in (15.49) and (15.50) primed.

Therefore, we have proved that all information rate tuples $\boldsymbol{\omega}$ in $\mathcal{R}'$ are achievable. By means of a time-sharing argument similar to the one we used in the proof of Theorem 9.12, we see that if $\boldsymbol{\omega}^{(1)}$ and $\boldsymbol{\omega}^{(2)}$ are achievable, then for any rational number λ between 0 and 1,

$$\boldsymbol{\omega} = \lambda\boldsymbol{\omega}^{(1)} + (1 - \lambda)\boldsymbol{\omega}^{(2)} \tag{15.239}$$

is also achievable. As the remark immediately after Definition 15.3 asserts that if $\omega^{(k)}$, $k \geq 1$ are achievable, then

$$\omega = \lim_{k\to\infty} \omega^{(k)} \tag{15.240}$$

is also achievable, we see that every ω in $\overline{\text{con}}(\mathcal{R}')$ is achievable. In other words,

$$\mathcal{R}_{in} = \overline{\text{con}}(\mathcal{R}') \subset \mathcal{R}, \tag{15.241}$$

and the theorem is proved. □

APPENDIX 15.A: APPROXIMATION OF RANDOM VARIABLES WITH INFINITE ALPHABETS

In this appendix, we prove that $H(Y_s'(m)) \to H(Y_s)$ as $m \to \infty$, where we assume that $H(Y_s) < \infty$. For every $m \geq 1$, define the binary random variable

$$B(m) = \begin{cases} 1 & \text{if } Y_s \leq m \\ 0 & \text{if } Y_s > m. \end{cases} \tag{15.A.1}$$

Consider

$$\begin{aligned} H(Y_s) &= -\sum_{k=1}^{m} \Pr\{Y_s = k\} \log \Pr\{Y_s = k\} \\ &\quad - \sum_{k=m+1}^{\infty} \Pr\{Y_s = k\} \log \Pr\{Y_s = k\}. \end{aligned} \tag{15.A.2}$$

As $m \to \infty$,

$$-\sum_{k=1}^{m} \Pr\{Y_s = k\} \log \Pr\{Y_s = k\} \to H(Y_s). \tag{15.A.3}$$

Since $H(Y_s) < \infty$,

$$-\sum_{k=m+1}^{\infty} \Pr\{Y_s = k\} \log \Pr\{Y_s = k\} \to 0 \tag{15.A.4}$$

as $k \to \infty$. Now consider

$$\begin{aligned} &H(Y_s) \\ &= H(Y_s|B(m)) + I(Y_s; B(m)) && (15.A.5) \\ &= H(Y_s|B(m) = 1)\Pr\{B(m) = 1\} + H(Y_s|B(m) = 0) \\ &\quad \times\Pr\{B(m) = 0\} + I(Y_s; B(m)) && (15.A.6) \\ &= H(Y_s'(m))\Pr\{B(m) = 1\} + H(Y_s|B(m) = 0) \\ &\quad \times\Pr\{B(m) = 0\} + I(Y_s; B(m)). && (15.A.7) \end{aligned}$$

As $m \to \infty$, $H(B(m)) \to 0$ since $\Pr\{B(m) = 1\} \to 1$. This implies $I(Y_s; B(m)) \to 0$ because

$$I(Y_s; B(m)) \leq H(B(m)). \tag{15.A.8}$$

In (15.A.7), we further consider

$$
\begin{aligned}
& H(Y_s|B(m)=0)\Pr\{B(m)=0\} \\
&= -\sum_{k=m+1}^{\infty} \Pr\{Y_s=k\}\log\frac{\Pr\{Y_s=k\}}{\Pr\{B(m)=0\}} && (15.A.9) \\
&= -\sum_{k=m+1}^{\infty} \Pr\{Y_s=k\}(\log\Pr\{Y_s=k\} \\
&\quad -\log\Pr\{B(m)=0\}) && (15.A.10) \\
&= -\sum_{k=m+1}^{\infty} (\Pr\{Y_s=k\}\log\Pr\{Y_s=k\}) \\
&\quad +\left(\sum_{k=m+1}^{\infty} \Pr\{Y_s=k\}\right)\log\Pr\{B(m)=0\} && (15.A.11) \\
&= -\sum_{k=m+1}^{\infty} (\Pr\{Y_s=k\}\log\Pr\{Y_s=k\}) \\
&\quad +\Pr\{B(m)=0\}\log\Pr\{B(m)=0\}. && (15.A.12)
\end{aligned}
$$

As $m \to \infty$, the summation above tends to 0 by (15.A.4). Since $\Pr\{B(m)=0\} \to 0$, $\Pr\{B(m)=0\}\log\Pr\{B(m)=0\} \to 0$. Therefore,

$$H(Y_s|B(m)=0)\Pr\{B(m)=0\} \to 0, \quad (15.A.13)$$

and we see from (15.A.7) that $H(Y_s'(m)) \to H(Y_s)$ as $m \to \infty$.

PROBLEMS

1. Consider the following network.

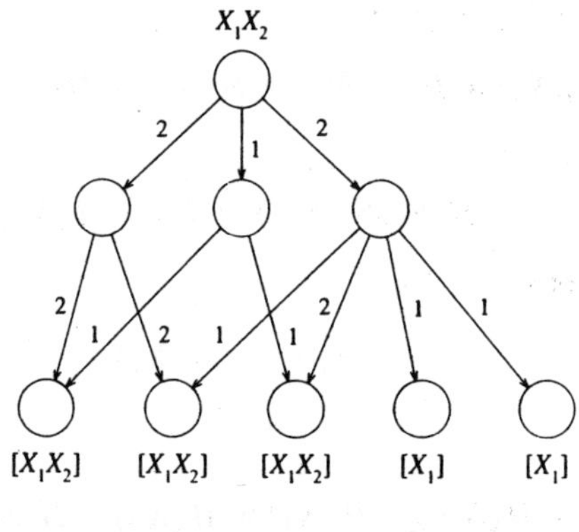

a) Let ω_i be the rate of information source X_i. Determine and illustrate the max-flow bounds.

b) Are the max-flow bounds achievable?

c) Is superposition coding always optimal?

2. Repeat Problem 1 for the following network in which the capacities of all the edges are equal to 1.

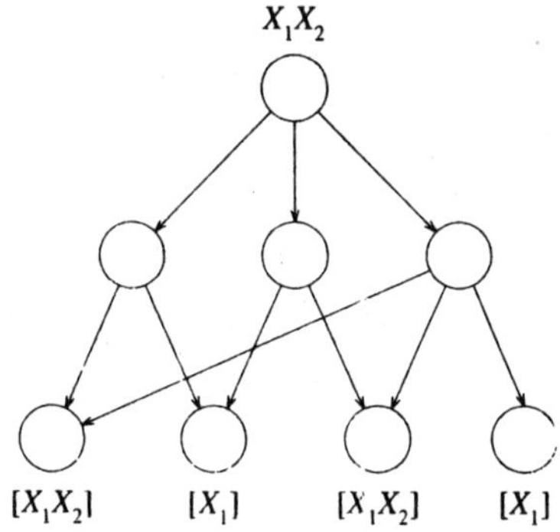

3. Consider a disk array with 3 disks. Let X_1, X_2, and X_3 be 3 mutually independent pieces of information to be retrieved from the disk array, and let S_1, S_2, and S_3 be the data to be stored separately in the 3 disks. It is required that X_1 can be retrieved from S_i, $i = 1, 2, 3$, X_2 can be retrieved from (S_i, S_j), $1 \leq i < j \leq 3$, and X_3 can be retrieved from (S_1, S_2, S_3).

   a) Prove that for $i = 1, 2, 3$,
   $$H(S_i) = H(X_1) + H(S_i|X_1).$$

   b) Prove that for $1 \leq i < j \leq 3$,
   $$H(S_i|X_1) + H(S_j|X_1) \geq H(X_2) + H(S_i, S_j|X_1, X_2).$$

   c) Prove that
   $$H(S_1, S_2, S_3|X_1, X_2) = H(X_3).$$

   d) Prove that for $i = 1, 2, 3$,
   $$H(S_i) \geq H(X_1).$$

   e) Prove that
   $$H(S_i) + H(S_j) \geq 2H(X_1) + H(X_2) + H(S_i, S_j|X_1, X_2).$$

   f) Prove that
   $$2S_i + S_{i\oplus1} + S_{i\oplus2} \geq 4H(X_1) + 2H(X_2) + H(X_3),$$

where $i = 1, 2, 3$ and

$$i \oplus j = \begin{cases} i + j & \text{if } i + j \leq 3 \\ i + j - 3 & \text{if } i + j > 3 \end{cases}$$

for $1 \leq i, j \leq 3$.

g) Prove that

$$H(S_1) + H(S_2) + H(S_3) \geq 3H(X_1) + \frac{3}{2}H(X_2) + H(X_3).$$

Parts d) to g) give constraints on $H(S_1)$, $H(S_2)$, and $H(S_3)$ in terms of $H(X_1)$, $H(X_2)$, and $H(X_3)$. It was shown in Roche *et al.* [175] that these constraints are the tightest possible.

4. Generalize the setup in Problem 3 to K disks and show that

$$\sum_{i=1}^{K} H(S_i) \geq K \sum_{\alpha=1}^{K} \frac{H(X_\alpha)}{\alpha}.$$

Hint: Use the inequalities in Problem 14 in Chapter 2 to prove that for $s = 0, 1, \cdots, K-1$,

$$\begin{aligned} \sum_{i=1}^{K} H(S_i) &\geq nK \sum_{\alpha=1}^{s} \frac{H(X_\alpha)}{\alpha} + \frac{K}{\binom{K}{s+1}} \\ &\quad \times \sum_{T:|T|=s+1} \frac{H(S_T|X_1, X_2, \cdots, X_s)}{s+1} \end{aligned}$$

by induction on s, where T is a subset of $\{1, 2, \cdots, K\}$.

5. Determine the regions $\mathcal{R}_{in}$, $\mathcal{R}_{out}$, and $\mathcal{R}_{LP}$ for the network in Figure 15.13.

6. Show that if there exists an

$$(n, (\eta_{ij} : (i, j) \in E), (\tau_s : s \in S))$$

code which satisfies (15.71) and (15.73), then there always exists an

$$(n, (\eta_{ij} : (i, j) \in E), (\tau'_s : s \in S))$$

code which satisfies (15.71) and (15.73), where $\tau'_s \leq \tau_s$ for all $s \in S$. Hint: use a random coding argument.

HISTORICAL NOTES

Multilevel diversity coding was introduced by Yeung [230], where it was shown that superposition coding is not always optimal. Roche *et al.* [175] showed that superposition coding is optimal for symmetrical three-level diversity coding. This result was extended to $K \geq 3$ levels by Yeung and Zhang [235] with a painstaking proof. Hau [98] studied all the one hundred configurations of a three-encoder diversity coding systems and found that superposition coding is optimal for eighty-six configurations.

Yeung and Zhang [236] introduced the distributed source coding model discussed in Section 15.2.2 which subsumes multilevel diversity coding. The regions Γ_n^* and Γ_n previously introduced by Yeung [231] for studying information inequalities enabled them to obtain inner and outer bounds on the coding rate region for a variety of networks.

Distributed source coding is equivalent to multi-source network coding in a two-tier acyclic network. Recently, the results in [236] have been generalized to an arbitrary acyclic network by Song *et al.* [198], on which the discussion in this chapter is based. Koetter and Médard [120] have developed an algebraic approach to network coding.

Chapter 16

ENTROPY AND GROUPS

The *group* is the first major mathematical structure in abstract algebra, while entropy is the most basic measure of information. Group theory and information theory are two seemingly unrelated subjects which turn out to be intimately related to each other. This chapter explains this intriguing relation between these two fundamental subjects. Those readers who have no knowledge in group theory may skip this introduction and go directly to the next section.

Let X_1 and X_2 be any two random variables. Then

$$H(X_1) + H(X_2) \geq H(X_1, X_2), \tag{16.1}$$

which is equivalent to the basic inequality

$$I(X_1; X_2) \geq 0. \tag{16.2}$$

Let G be any finite group and G_1 and G_2 be subgroups of G. We will show in Section 16.4 that

$$|G||G_1 \cap G_2| \geq |G_1||G_2|, \tag{16.3}$$

where $|G|$ denotes the *order* of G and $G_1 \cap G_2$ denotes the *intersection* of G_1 and G_2 ($G_1 \cap G_2$ is also a subgroup of G, see Proposition 16.13). By rearranging the terms, the above inequality can be written as

$$\log \frac{|G|}{|G_1|} + \log \frac{|G|}{|G_2|} \geq \log \frac{|G|}{|G_1 \cap G_2|}. \tag{16.4}$$

By comparing (16.1) and (16.4), one can easily identify the one-to-one correspondence between these two inequalities, namely that X_i corresponds to G_i, $i = 1, 2$, and (X_1, X_2) corresponds to $G_1 \cap G_2$. While (16.1) is true for any pair of random variables X_1 and X_2, (16.4) is true for any finite group G and subgroups G_1 and G_2.

Recall from Chapter 12 that the region Γ_n^* characterizes all information inequalities (involving n random variables). In particular, we have shown in Section 14.1 that the region $\overline{\Gamma}_n^*$ is sufficient for characterizing all unconstrained information inequalities, i.e., by knowing $\overline{\Gamma}_n^*$, one can determine whether any unconstrained information inequality always holds. The main purpose of this chapter is to obtain a characterization of $\overline{\Gamma}_n^*$ in terms of finite groups. An important consequence of this result is a one-to-one correspondence between unconstrained information inequalities and group inequalities. Specifically, for every unconstrained information inequality, there is a corresponding group inequality, and vice versa. A special case of this correspondence has been given in (16.1) and (16.4).

By means of this result, unconstrained information inequalities can be proved by techniques in group theory, and a certain form of inequalities in group theory can be proved by techniques in information theory. In particular, the unconstrained non-Shannon-type inequality in Theorem 14.7 corresponds to the group inequality

$$\begin{aligned} &|G_1 \cap G_3|^3 |G_1 \cap G_4|^3 |G_3 \cap G_4|^3 |G_2 \cap G_3||G_2 \cap G_4| \\ &\leq |G_1||G_1 \cap G_2||G_3|^2 |G_4|^2 |G_1 \cap G_3 \cap G_4|^4 |G_2 \cap G_3 \cap G_4|, \end{aligned} \tag{16.5}$$

where G_i are subgroups of a finite group G, $i = 1, 2, 3, 4$. The meaning of this inequality and its implications in group theory are yet to be understood.

16.1 GROUP PRELIMINARIES

In this section, we present the definition and some basic properties of a group which are essential for subsequent discussions.

DEFINITION 16.1 *A group is a set of objects G together with a binary operation on the elements of G, denoted by "$\circ$" unless otherwise specified, which satisfy the following four axioms:*

1. Closure *For every a, b in G, $a \circ b$ is also in G.*
2. Associativity *For every a, b, c in G, $a \circ (b \circ c) = (a \circ b) \circ c$.*
3. Existence of Identity *There exists an element e in G such that $a \circ e = e \circ a = a$ for every a in G.*
4. Existence of Inverse *For every a in G, there exists an element b in G such that $a \circ b = b \circ a = e$.*

PROPOSITION 16.2 *For any group G, the identity element is unique.*

Proof Let e and e' be both identity elements in a group G. Since e is an identity element,

$$e' \circ e = e, \tag{16.6}$$

and since e' is also an identity element,

$$e' \circ e = e'. \tag{16.7}$$

It follows by equating the right hand sides of (16.6) and (16.7) that $e = e'$, which implies the uniqueness of the identity element of a group. □

PROPOSITION 16.3 *For every element a in a group G, its inverse is unique.*

Proof Let b and b' be inverses of an element a, so that

$$a \circ b = b \circ a = e \tag{16.8}$$

and

$$a \circ b' = b' \circ a = e. \tag{16.9}$$

Then

$$\begin{aligned} b &= b \circ e && (16.10)\\ &= b \circ (a \circ b') && (16.11)\\ &= (b \circ a) \circ b' && (16.12)\\ &= e \circ b' && (16.13)\\ &= b', && (16.14) \end{aligned}$$

where (16.11) and (16.13) follow from (16.9) and (16.8), respectively, and (16.12) is by associativity. Therefore, the inverse of a is unique. □

Thus the inverse of a group element a is a function of a, and it will be denoted by a^{-1}.

DEFINITION 16.4 *The number of elements of a group G is called the order of G, denoted by $|G|$. If $|G| < \infty$, G is called a finite group, otherwise it is called an infinite group.*

There is an unlimited supply of examples of groups. Some familiar examples are: the integers under addition, the rationals excluding zero under multiplication, and the set of real-valued 2×2 matrices under addition, where addition and multiplication refer to the usual addition and multiplication for real numbers and matrices. In each of these examples, the operation (addition or multiplication) plays the role of the binary operation "$\circ$" in Definition 16.2.

All the above are examples of infinite groups. In this chapter, however, we are concerned with finite groups. In the following, we discuss two examples of finite groups in details.

EXAMPLE 16.5 (MODULO 2 ADDITION) *The trivial group consists of only the identity element. The simplest nontrivial group is the group of modulo 2 addition. The order of this group is 2, and the elements are* $\{0, 1\}$. *The binary operation, denoted by "+", is defined by following table:*

+	0	1
0	0	1
1	1	0

The four axioms of a group simply say that certain constraints must hold in the above table. We now check that all these axioms are satisfied. First, the closure axiom requires that all the entries in the table are elements in the group, which is easily seen to be the case. Second, it is required that associativity holds. To this end, it can be checked in the above table that for all a, b, and c,

$$a + (b + c) = (a + b) + c. \tag{16.15}$$

For example,

$$0 + (1 + 1) = 0 + 0 = 0, \tag{16.16}$$

while

$$(0 + 1) + 1 = 1 + 1 = 0, \tag{16.17}$$

which is the same as $0 + (1 + 1)$. *Third, the element 0 is readily identified as the unique identity. Fourth, it is readily seen that an inverse exists for each element in the group. For example, the inverse of 1 is 1, because*

$$1 + 1 = 0. \tag{16.18}$$

Thus the above table defines a group of order 2. It happens in this example that the inverse of each element is the element itself, which is not true for a group in general.

We remark that in the context of a group, the elements in the group should be regarded strictly as *symbols* only. In particular, one should not associate group elements with *magnitudes* as we do for real numbers. For instance, in the above example, one should not think of 0 as being less than 1. The element 0, however, is a special symbol which plays the role of the identity of the group.

We also notice that for the group in the above example, $a + b$ is equal to $b + a$ for all group elements a and b. A group with this property is called a *commutative* group or an *Abelian* group[1].

EXAMPLE 16.6 (PERMUTATION GROUP) *Consider a permutation of the components of a vector*

$$\mathbf{x} = (x_1, x_2, \cdots, x_r) \tag{16.19}$$

[1]The Abelian group is named after the Norwegian mathematician Niels Henrik Abel (1802-1829).

given by

$$\sigma[\mathbf{x}] = (x_{\sigma(1)}, x_{\sigma(2)}, \cdots, x_{\sigma(r)}), \tag{16.20}$$

where

$$\sigma : \{1, 2, \cdots, r\} \to \{1, 2, \cdots, r\} \tag{16.21}$$

is a one-to-one mapping. The one-to-one mapping σ is called a permutation on $\{1, 2, \cdots, r\}$, which is represented by

$$\sigma = (\sigma(1), \sigma(2), \cdots, \sigma(r)). \tag{16.22}$$

For two permutations σ_1 and σ_2, define $\sigma_1 \circ \sigma_2$ as the composite function of σ_1 and σ_2. For example, for $r = 4$, suppose

$$\sigma_1 = (2, 1, 4, 3) \tag{16.23}$$

and

$$\sigma_2 = (1, 4, 2, 3). \tag{16.24}$$

Then $\sigma_1 \circ \sigma_2$ is given by

$$\begin{array}{rcccccl} \sigma_1 \circ \sigma_2(1) & = & \sigma_1(\sigma_2(1)) & = & \sigma_1(1) & = & 2 \\ \sigma_1 \circ \sigma_2(2) & = & \sigma_1(\sigma_2(2)) & = & \sigma_1(4) & = & 3 \\ \sigma_1 \circ \sigma_2(3) & = & \sigma_1(\sigma_2(3)) & = & \sigma_1(2) & = & 1 \\ \sigma_1 \circ \sigma_2(4) & = & \sigma_1(\sigma_2(4)) & = & \sigma_1(3) & = & 4, \end{array} \tag{16.25}$$

or

$$\sigma_1 \circ \sigma_2 = (2, 3, 1, 4). \tag{16.26}$$

The reader can easily check that

$$\sigma_2 \circ \sigma_1 = (4, 1, 2, 3), \tag{16.27}$$

which is different from $\sigma_1 \circ \sigma_2$. Therefore, the operation "$\circ$" is not commutative.

We now show that the set of all permutations on $\{1, 2, \cdots, r\}$ and the operation "$\circ$" form a group, called the permutation group. First, for two permutations σ_1 and σ_2, since both σ_1 and σ_2 are one-to-one mappings, so is $\sigma_1 \circ \sigma_2$. Therefore, the closure axiom is satisfied. Second, for permutations σ_1, σ_2, and σ_3,

$$\begin{aligned} \sigma_1 \circ (\sigma_2 \circ \sigma_3)(i) &= \sigma_1(\sigma_2 \circ \sigma_3(i)) && (16.28) \\ &= \sigma_1(\sigma_2(\sigma_3(i))) && (16.29) \\ &= \sigma_1 \circ \sigma_2(\sigma_3(i)) && (16.30) \\ &= (\sigma_1 \circ \sigma_2) \circ \sigma_3(i) && (16.31) \end{aligned}$$

for $1 \leq i \leq r$. Therefore, associativity is satisfied. Third, it is clear that the identity map is the identity element. Fourth, for a permutation σ, it is clear *that its inverse is σ^{-1}, the inverse mapping of σ which is defined because σ is one-to-one. Therefore, the set of all permutations on $\{1, 2, \cdots, r\}$ and the operation "$\circ$" form a group. The order of this group is evidently equal to $(r!)$.*

DEFINITION 16.7 *Let G be a group with operation "$\circ$", and S be a subset of G. If S is a group with respect to the operation "$\circ$", then S is called a subgroup of G.*

DEFINITION 16.8 *Let S be a subgroup of a group G and a be an element of G. The left coset of S with respect to a is the set $a \circ S = \{a \circ s : s \in S\}$. Similarly, the right coset of S with respect to a is the set $S \circ a = \{s \circ a : s \in S\}$.*

In the sequel, only the left coset will be used. However, any result which applies to the left coset also applies to the right coset, and vice versa. For simplicity, $a \circ S$ will be denoted by aS.

PROPOSITION 16.9 *For a_1 and a_2 in G, a_1S and a_2S are either identical or disjoint. Further, a_1S and a_2S are identical if and only if a_1 and a_2 belong to the same left coset of S.*

Proof Suppose a_1S and a_2S are not disjoint. Then there exists an element b in $a_1S \cap a_2S$ such that

$$b = a_1 \circ s_1 = a_2 \circ s_2, \tag{16.32}$$

for some s_i in S, $i = 1, 2$. Then

$$a_1 = (a_2 \circ s_2) \circ s_1^{-1} = a_2 \circ (s_2 \circ s_1^{-1}) = a_2 \circ t, \tag{16.33}$$

where $t = s_2 \circ s_1^{-1}$ is in S. We now show that $a_1S \subset a_2S$. For an element $a_1 \circ s$ in a_1S, where $s \in S$,

$$a_1 \circ s = (a_2 \circ t) \circ s = a_2 \circ (t \circ s) = a_2 \circ u, \tag{16.34}$$

where $u = t \circ s$ is in S. This implies that $a_1 \circ s$ is in a_2S. Thus, $a_1S \subset a_2S$. By symmetry, $a_2S \subset a_1S$. Therefore, $a_1S = a_2S$. Hence, if a_1S and a_2S are not disjoint, then they are identical. Equivalently, a_1S and a_2S are either identical or disjoint. This proves the first part of the proposition.

We now prove the second part of the proposition. Since S is a group, it contains e, the identity element. Then for any group element a, $a = a \circ e$ is in aS because e is in S. If a_1S and a_2S are identical, then $a_1 \in a_1S$ and $a_2 \in a_2S = a_1S$. Therefore, a_1 and a_2 belong to the same left coset of S.

To prove the converse, assume a_1 and a_2 belong to the same left coset of S. From the first part of the proposition, we see that a group element belongs to one and only one left coset of S. Since a_1 is in a_1S and a_2 is in a_2S, and a_1 and a_2 belong to the same left coset of S, we see that a_1S and a_2S are identical. The proposition is proved. □

PROPOSITION 16.10 *Let S be a subgroup of a group G and a be an element of G. Then $|aS| = |S|$, i.e., the numbers of elements in all the left cosets of S are the same, and they are equal to the order of S.*

Proof Consider two elements $a \circ s_1$ and $a \circ s_2$ in $a \circ S$, where s_1 and s_2 are in S such that

$$a \circ s_1 = a \circ s_2. \tag{16.35}$$

Then

$$\begin{aligned} a^{-1} \circ (a \circ s_1) &= a^{-1} \circ (a \circ s_2) && (16.36) \\ (a^{-1} \circ a) \circ s_1 &= (a^{-1} \circ a) \circ s_2 && (16.37) \\ e \circ s_1 &= e \circ s_2 && (16.38) \\ s_1 &= s_2. && (16.39) \end{aligned}$$

Thus each element in S corresponds to a unique element in aS. Therefore, $|aS| = |S|$ for all $a \in G$. □

We are just one step away from obtaining the celebrated *Lagrange's Theorem* stated below.

THEOREM 16.11 (LAGRANGE'S THEOREM) *If S is a subgroup of G, then $|S|$ divides $|G|$.*

Proof Since $a \in aS$ for every $a \in G$, every element of G belongs to a left coset of S. Then from Proposition 16.9, we see that the distinct left cosets of S partition G. Therefore $|G|$, the total number of elements in G, is equal to the number of distinct cosets of S multiplied by the number of elements in each left coset, which is equal to $|S|$ by Proposition 16.10. This implies that $|S|$ divides $|G|$, proving the theorem. □

The following corollary is immediate from the proof of Lagrange's Theorem.

COROLLARY 16.12 *Let S be a subgroup of a group G. The number of distinct left cosets of S is equal to $\frac{|G|}{|S|}$.*

16.2 GROUP-CHARACTERIZABLE ENTROPY FUNCTIONS

Recall from Chapter 12 that the region Γ_n^* consists of all the entropy functions in the entropy space $\mathcal{H}_n$ for n random variables. As a first step toward establishing the relation between entropy and groups, we discuss in this section entropy functions in Γ_n^* which can be described by a finite group G and subgroups $G_1, G_2, \cdots, G_n$. Such entropy functions are said to be *group-characterizable*. The significance of this class of entropy functions will become clear in the next section.

In the sequel, we will make use of the intersections of subgroups extensively. We first prove that the intersection of two subgroups is also a subgroup.

PROPOSITION 16.13 *Let G_1 and G_2 be subgroups of a group G. Then $G_1 \cap G_2$ is also a subgroup of G.*

Proof It suffices to show that $G_1 \cap G_2$ together with the operation "$\circ$" satisfy all the axioms of a group. First, consider two elements a and b of G in $G_1 \cap G_2$. Since both a and b are in G_1, $(a \circ b)$ is in G_1. Likewise, $(a \circ b)$ is in G_2. Therefore, $a \circ b$ is in $G_1 \cap G_2$. Thus the closure axiom holds for $G_1 \cap G_2$. Second, associativity for $G_1 \cap G_2$ inherits from G. Third, G_1 and G_2 both contain the identity element because they are groups. Therefore, the identity element is in $G_1 \cap G_2$. Fourth, for an element $a \in G_i$, since G_i is a group, a^{-1} is in G_i, $i = 1, 2$. Thus for an element $a \in G_1 \cap G_2$, a^{-1} is also in $G_1 \cap G_2$. Therefore, $G_1 \cap G_2$ is a group and hence a subgroup of G. □

COROLLARY 16.14 *Let $G_1, G_2, \cdots, G_n$ be subgroups of a group G. Then $\cap_{i=1}^n G_i$ is also a subgroup of G.*

In the rest of the chapter, we let $\mathcal{N}_n = \{1, 2, \cdots, n\}$ and denote $\cap_{i \in \alpha} G_i$ by G_α, where α is a nonempty subset of $\mathcal{N}_n$.

LEMMA 16.15 *Let G_i be subgroups of a group G and a_i be elements of G, $i \in \alpha$. Then*

$$|\cap_{i \in \alpha} a_i G_i| = \begin{cases} |G_\alpha| & \text{if } \bigcap_{i \in \alpha} a_i G_i \neq \emptyset \\ 0 & \text{otherwise.} \end{cases} \tag{16.40}$$

Proof For the special case when α is a singleton, i.e., $\alpha = \{i\}$ for some $i \in \mathcal{N}_n$, (16.40) reduces to

$$|a_i G_i| = |G_i|, \tag{16.41}$$

which has already been proved in Proposition 16.10.

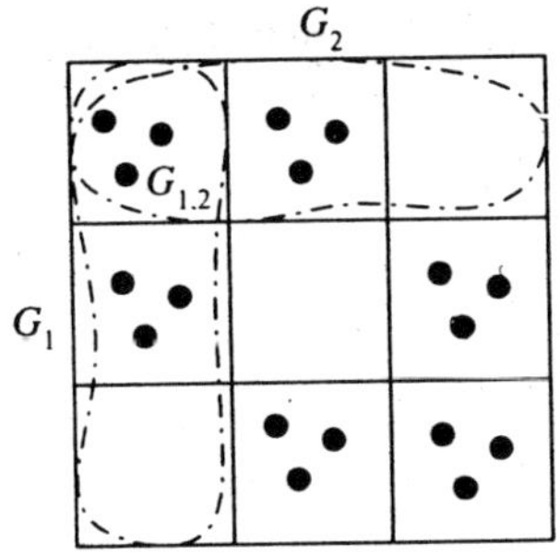

Figure 16.1. The membership table for a finite group G and subgroups G_1 and G_2.

Let α be any nonempty subset of $\mathcal{N}_n$. If $\bigcap_{i\in\alpha} a_iG_i = \emptyset$, then (16.40) is obviously true. If $\bigcap_{i\in\alpha} a_iG_i \neq \emptyset$, then there exists $x \in \bigcap_{i\in\alpha} a_iG_i$ such that for all $i \in \alpha$,

$$x = a_i \circ s_i, \tag{16.42}$$

where $s_i \in G_i$. For any $i \in \alpha$ and for any $y \in G_\alpha$, consider

$$x \circ y = (a_i \circ s_i) \circ y = a_i \circ (s_i \circ y). \tag{16.43}$$

Since both s_i and y are in G_i, $s_i \circ y$ is in G_i. Thus $x \circ y$ is in a_iG_i for all $i \in \alpha$, or $x \circ y$ is in $\bigcap_{i\in\alpha} a_iG_i$. Moreover, for $y, y' \in G_\alpha$, if $x \circ y = x \circ y'$, then $y = y'$. Therefore, each element in G_α corresponds to a unique element in $\bigcap_{i\in\alpha} a_iG_i$. Hence,

$$|\cap_{i\in\alpha} a_iG_i| = |G_\alpha|, \tag{16.44}$$

proving the lemma. □

The relation between a finite group G and subgroups G_1 and G_2 is illustrated by the *membership table* in Figure 16.1. In this table, an element of G is represented by a dot. The first column represents the subgroup G_1, with the dots in the first column being the elements in G_1. The other columns represent the left cosets of G_1. By Proposition 16.10, all the columns have the same number of dots. Similarly, the first row represents the subgroup G_2 and the other rows represent the left cosets of G_2. Again, all the rows have the same number of dots.

The upper left entry in the table represents the subgroup $G_1 \cap G_2$. There are $|G_1 \cap G_2|$ dots in this entry, with one of them representing the identity element. Any other entry represents the intersection between a left coset of G_1 and a left coset of G_2, and by Lemma 16.15, the number of dots in each of these entries is either equal to $|G_1 \cap G_2|$ or zero.

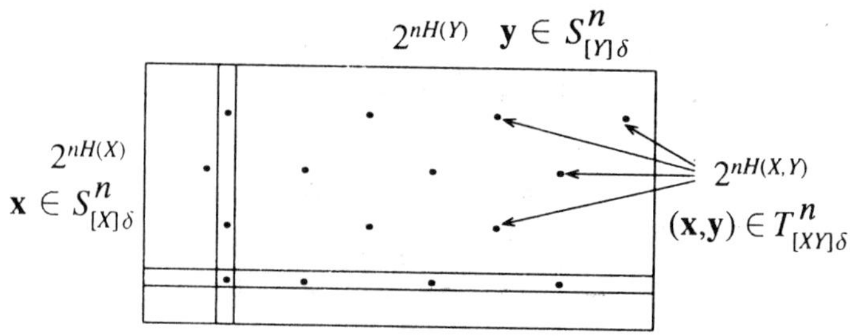

Figure 16.2. A two-dimensional strong typicality array.

Since all the column have the same numbers of dots and all the rows have the same number of dots, we say that the table in Figure 16.1 exhibits a *quasi-uniform* structure. We have already seen a similar structure in Figure 5.1 for the two-dimensional strong joint typicality array, which we reproduce in Figure 16.2. In this array, when n is large, all the columns have approximately the same number of dots and all the rows have approximately the same number of dots. For this reason, we say that the two-dimensional strong typicality array exhibits an *asymptotic* quasi-uniform structure. In a strong typicality array, however, each entry can contain only one dot, while in a membership table, each entry can contain multiple dots.

One can make a similar comparison between a strong joint typicality array for any $n \geq 2$ random variables and the membership table for a finite group with n subgroups. The details are omitted here.

THEOREM 16.16 *Let $G_i, i \in \mathcal{N}_n$ be subgroups of a group G. Then $\mathbf{h} \in \mathcal{H}_n$ defined by*

$$h_\alpha = \log \frac{|G|}{|G_\alpha|} \tag{16.45}$$

for all nonempty subset α of $\mathcal{N}_n$ is entropic, i.e., $\mathbf{h} \in \Gamma^*_n$.

Proof It suffices to show that there exists a collection of random variables $X_1, X_2, \cdots, X_n$ such that

$$H(X_\alpha) = \log \frac{|G|}{|G_\alpha|} \tag{16.46}$$

for all nonempty subset α of $\mathcal{N}_n$. We first introduce a uniform random variable Λ defined on the sample space G with probability mass function

$$\Pr\{\Lambda = a\} = \frac{1}{|G|} \tag{16.47}$$

for all $a \in G$. For any $i \in \mathcal{N}_n$, let random variable X_i be a function of Λ such that $X_i = aG_i$ if $\Lambda = a$.

Let α be a nonempty subset of $\mathcal{N}_n$. Since $X_i = a_i G_i$ for all $i \in \alpha$ if and only if Λ is equal to some $b \in \cap_{i\in\alpha} a_i G_i$,

$$\Pr\{X_i = a_i G_i : i \in \alpha\} = \frac{|\bigcap_{i\in\alpha} a_i G_i|}{|G|} \tag{16.48}$$

$$= \begin{cases} \frac{|G_\alpha|}{|G|} & \text{if } \bigcap_{i\in\alpha} a_i G_i \neq \emptyset \\ 0 & \text{otherwise} \end{cases} \tag{16.49}$$

by Lemma 16.15. In other words, $(X_i, i \in \alpha)$ distributes uniformly on its support whose cardinality is $\frac{|G|}{|G_\alpha|}$. Then (16.46) follows and the theorem is proved. □

DEFINITION 16.17 *Let G be a finite group and $G_1, G_2, \cdots, G_n$ be subgroups of G. Let $\mathbf{h}$ be a vector in $\mathcal{H}_n$. If $h_\alpha = \log \frac{|G|}{|G_\alpha|}$ for all nonempty subsets α of $\mathcal{N}_n$, then $(G, G_1, \cdots, G_n)$ is a group characterization of $\mathbf{h}$.*

Theorem 16.16 asserts that certain entropy functions in Γ_n^* have a group characterization. These are called group-characterizable entropy functions, which will be used in the next section to obtain a group characterization of the region $\overline{\Gamma}_n^*$. We end this section by giving a few examples of such entropy functions.

EXAMPLE 16.18 *Fix any subset β of $\mathcal{N}_3 = \{1, 2, 3\}$ and define a vector $\mathbf{h} \in \mathcal{H}_3$ by*

$$h_\alpha = \begin{cases} \log 2 & \text{if } \alpha \cap \beta \neq \emptyset \\ 0 & \text{otherwise.} \end{cases} \tag{16.50}$$

We now show that $\mathbf{h}$ has a group characterization. Let $G = \{0, 1\}$ be the group of modulo 2 addition in Example 16.5, and for $i = 1, 2, 3$, let

$$G_i = \begin{cases} \{0\} & \text{if } i \in \beta \\ G & \text{otherwise.} \end{cases} \tag{16.51}$$

Then for a nonempty subset α of $\mathcal{N}_3$, if $\alpha \cap \beta \neq \emptyset$, there exists an i in α such that i is also in β, and hence by definition $G_i = \{0\}$. Thus,

$$G_\alpha = \bigcap_{i\in\alpha} G_i = \{0\}. \tag{16.52}$$

Therefore,

$$\log \frac{|G|}{|G_\alpha|} = \log \frac{|G|}{|\{0\}|} = \log \frac{2}{1} = \log 2. \tag{16.53}$$

If $\alpha \cap \beta = \emptyset$, then $G_i = G$ for all $i \in \alpha$, and

$$G_\alpha = \bigcap_{i \in \alpha} G_i = G. \tag{16.54}$$

Therefore,

$$\log \frac{|G|}{|G_\alpha|} = \log \frac{|G|}{|G|} = \log 1 = 0. \tag{16.55}$$

Then we see from (16.50), (16.53), and (16.55) that

$$h_\alpha = \log \frac{|G|}{|G_\alpha|} \tag{16.56}$$

for all nonempty subset α of $\mathcal{N}_3$. Hence, (G, G_1, G_2, G_3) is a group characterization of **h**.

EXAMPLE 16.19 *This is a generalization of the last example. Fix any nonempty subset β of $\mathcal{N}_n$ and define a vector $\mathbf{h} \in \mathcal{H}_n$ by*

$$h_\alpha = \begin{cases} \log 2 & \text{if } \alpha \cap \beta \neq \emptyset \\ 0 & \text{otherwise.} \end{cases} \tag{16.57}$$

Then $(G, G_1, G_2, \cdots, G_n)$ is a group characterization of **h**, *where G is the group of modulo 2 addition, and*

$$G_i = \begin{cases} \{0\} & \text{if } i \in \beta \\ G & \text{otherwise.} \end{cases} \tag{16.58}$$

By letting $\beta = \emptyset$, $\mathbf{h} = 0$. Thus we see that $(G, G_1, G_2, \cdots, G_n)$ is a group characterization of the origin of $\mathcal{H}_n$, with $G = G_1 = G_2 = \cdots = G_n$.

EXAMPLE 16.20 *Define a vector $\mathbf{h} \in \mathcal{H}_3$ as follows:*

$$h_\alpha = \min(|\alpha|, 2). \tag{16.59}$$

Let F be the group of modulo 2 addition, $G = F \times F$, and

$$G_1 = \{(0,0), (1,0)\} \tag{16.60}$$
$$G_2 = \{(0,0), (0,1)\} \tag{16.61}$$
$$G_3 = \{(0,0), (1,1)\}. \tag{16.62}$$

Then (G, G_1, G_2, G_3) is a group characterization of **h**.

16.3 A GROUP CHARACTERIZATION OF $\overline{\Gamma}_n^*$

We have introduced in the last section the class of entropy functions in Γ_n^* which have a group characterization. However, an entropy function $\mathbf{h} \in \Gamma_n^*$ may not have a group characterization due to the following observation. Suppose $\mathbf{h} \in \Gamma_n^*$. Then there exists a collection of random variables $X_1, X_2, \cdots, X_n$ such that

$$h_\alpha = H(X_\alpha) \tag{16.63}$$

for all nonempty subset α of $\mathcal{N}_n$. If $(G, G_1, \cdots, G_n)$ is a group characterization of $\mathbf{h}$, then

$$H(X_\alpha) = \log \frac{|G|}{|G_\alpha|} \tag{16.64}$$

for all nonempty subset of $\mathcal{N}_n$. Since both $|G|$ and $|G_\alpha|$ are integers, $H(X_\alpha)$ must be the logarithm of a rational number. However, the joint entropy of a set of random variables in general is not necessarily the logarithm of a rational number (see Corollary 2.44). Therefore, it is possible to construct an entropy function $\mathbf{h} \in \Gamma_n^*$ which has no group characterization.

Although $\mathbf{h} \in \Gamma_n^*$ does not imply $\mathbf{h}$ has a group characterization, it turns out that the set of all $\mathbf{h} \in \Gamma_n^*$ which have a group characterization is almost good enough to characterize the region Γ_n^*, as we will see next.

DEFINITION 16.21 *Define the following region in $\mathcal{H}_n$:*

$$\Upsilon_n = \{\mathbf{h} \in \mathcal{H}_n : \mathbf{h} \text{ has a group characterization}\}. \tag{16.65}$$

By Theorem 16.16, if $\mathbf{h} \in \mathcal{H}_n$ has a group characterization, then $\mathbf{h} \in \Gamma_n^*$. Therefore, $\Upsilon_n \subset \Gamma_n^*$. We will prove as a corollary of the next theorem that $\overline{con}(\Upsilon_n)$, the convex closure of Υ_n, is in fact equal to $\overline{\Gamma}_n^*$, the closure of Γ_n^*.

THEOREM 16.22 *For any $\mathbf{h} \in \Gamma_n^*$, there exists a sequence $\{\mathbf{f}^{(r)}\}$ in Υ_n such that $\lim_{r\to\infty} \frac{1}{r}\mathbf{f}^{(r)} = \mathbf{h}$.*

We need the following lemma to prove this theorem. The proof of this lemma resembles the proof of Theorem 5.9. Nevertheless, we give a sketch of the proof for the sake of completeness.

LEMMA 16.23 *Let X be a random variable such that $|\mathcal{X}| < \infty$ and the distribution $\{p(x)\}$ is rational, i.e., $p(x)$ is a rational number for all $x \in \mathcal{X}$. Without loss of generality, assume $p(x)$ is a rational number with denominator q for all $x \in \mathcal{X}$. Then for $r = q, 2q, 3q, \cdots$,*

$$\lim_{r\to\infty} \frac{1}{r} \log \frac{r!}{\prod_x (rp(x))!} = H(X). \tag{16.66}$$

Proof Applying Lemma 5.10, we can obtain

$$\begin{aligned} & \frac{1}{r} \ln \frac{r!}{\prod_x (rp(x))!} \\ \leq & \ -\sum_x p(x) \ln p(x) + \frac{r+1}{r} \ln(r+1) - \ln r \quad (16.67) \\ = & \ H_e(X) + \frac{1}{r} \ln r + \left(1 + \frac{1}{r}\right) \ln \left(1 + \frac{1}{r}\right). \quad (16.68) \end{aligned}$$

This upper bound tends to $H_e(X)$ as $r \to \infty$. On the other hand, we can obtain

$$\begin{aligned} & \frac{1}{r} \ln \frac{r!}{\prod_x (rp(x))!} \\ \geq & \ -\sum_x \left(p(x) + \frac{1}{r}\right) \ln \left(p(x) + \frac{1}{r}\right) - \frac{\ln r}{r}. \quad (16.69) \end{aligned}$$

This lower bound also tends to $H_e(X)$ as $r \to \infty$. Then the proof is completed by changing the base of the logarithm if necessary. □

Proof of Theorem 16.22 For any $\mathbf{h} \in \Gamma_n^*$, there exists a collection of random variables $X_1, X_2, \cdots, X_n$ such that

$$h_\alpha = H(X_\alpha) \quad (16.70)$$

for all nonempty subset α of $\mathcal{N}_n$. We first consider the special case that $|\mathcal{X}_i| < \infty$ for all $i \in \mathcal{N}_n$ and the joint distribution of $X_1, X_2, \cdots, X_n$ is rational. We want to show that there exists a sequence $\{\mathbf{f}^{(r)}\}$ in Υ_n such that $\lim_{r\to\infty} \frac{1}{r}\mathbf{f}^{(r)} = \mathbf{h}$.

Denote $\prod_{i\in\alpha} \mathcal{X}_i$ by $\mathcal{X}_\alpha$. For any nonempty subset α of $\mathcal{N}_n$, let Q_α be the marginal distribution of X_α. Assume without loss of generality that for any nonempty subset α of $\mathcal{N}_n$ and for all $a \in \mathcal{X}_\alpha$, $Q_\alpha(a)$ is a rational number with denominator q.

For each $r = q, 2q, 3q, \cdots$, fix a sequence

$$\mathbf{x}_{\mathcal{N}_n} = (x_{\mathcal{N}_n,1}, x_{\mathcal{N}_n,2}, \cdots x_{\mathcal{N}_n,r})$$

where for all $j = 1, 2, \cdots, r$, $x_{\mathcal{N}_n,j} = (x_{i,j} : i \in \mathcal{N}_n) \in \mathcal{X}_{\mathcal{N}_n}$, such that $N(a|\mathbf{x}_{\mathcal{N}_n})$, the number of occurrences of a in sequence $\mathbf{x}_{\mathcal{N}_n}$, is equal to $rQ_{\mathcal{N}_n}(a)$ for all $a \in \mathcal{X}_{\mathcal{N}_n}$. The existence of such a sequence is guaranteed by that all the values of the joint distribution of $X_{\mathcal{N}_n}$ are rational numbers with denominator q. Also, we denote the sequence of r elements of $\mathcal{X}_\alpha$, $(x_{\alpha,1}, x_{\alpha,2}, \cdots x_{\alpha,r})$, where $x_{\alpha,j} = (x_{i,j} : i \in \alpha)$, by $\mathbf{x}_\alpha$. Let $a \in \mathcal{X}_\alpha$. It is easy to check that $N(a|\mathbf{x}_\alpha)$, the number of occurrences of a in the sequence $\mathbf{x}_\alpha$, is equal to $rQ_\alpha(a)$ for all $a \in \mathcal{X}_\alpha$.

Let G be the group of permutations on $\{1, 2, \cdots, r\}$. The group G depends on r, but for simplicity, we do not state this dependency explicitly. For any $i \in \mathcal{N}_n$, define

$$G_i = \{\sigma \in G : \sigma[\mathbf{x}_i] = \mathbf{x}_i\},$$

where

$$\sigma[\mathbf{x}_i] = (x_{i,\sigma(1)}, x_{i,\sigma(2)}, \cdots, x_{i,\sigma(r)}). \tag{16.71}$$

It is easy to check that G_i is a subgroup of G.

Let α be a nonempty subset of $\mathcal{N}_n$. Then

$$\begin{aligned} G_\alpha &= \bigcap_{i \in \alpha} G_i & (16.72)\\ &= \bigcap_{i \in \alpha} \{\sigma \in G : \sigma[\mathbf{x}_i] = \mathbf{x}_i\} & (16.73)\\ &= \{\sigma \in G : \sigma[\mathbf{x}_i] = \mathbf{x}_i \text{ for all } i \in \alpha\} & (16.74)\\ &= \{\sigma \in G : \sigma[\mathbf{x}_\alpha] = \mathbf{x}_\alpha\}, & (16.75) \end{aligned}$$

where

$$\sigma[\mathbf{x}_\alpha] = (x_{\alpha,\sigma(1)}, x_{\alpha,\sigma(2)}, \cdots, x_{\alpha,\sigma(r)}). \tag{16.76}$$

For any $a \in \mathcal{X}_\alpha$, define the set

$$L_{\mathbf{x}_\alpha}(a) = \{j \in \{1, 2, \cdots, r\} : x_{\alpha,j} = a\}. \tag{16.77}$$

$L_{\mathbf{x}_\alpha}(a)$ contains the "locations" of a in $\mathbf{x}_\alpha$. Then $\sigma[\mathbf{x}_\alpha] = \mathbf{x}_\alpha$ if and only if for all $a \in \mathcal{X}_\alpha$, $j \in L_{\mathbf{x}_\alpha}(a)$ implies $\sigma(j) \in L_{\mathbf{x}_\alpha}(a)$. Since

$$|L_{\mathbf{x}_\alpha}(a)| = N(a|\mathbf{x}_\alpha) = rQ_\alpha(a), \tag{16.78}$$

$$|G_\alpha| = \prod_{a \in \mathcal{X}_\alpha} (rQ_\alpha(a))! \tag{16.79}$$

and therefore

$$\frac{|G|}{|G_\alpha|} = \frac{r!}{\prod_{a \in \mathcal{X}_\alpha} (rQ_\alpha(a))!}. \tag{16.80}$$

By Lemma 16.23,

$$\lim_{r \to \infty} \frac{1}{r} \log \frac{|G|}{|G_\alpha|} = H(X_\alpha) = h_\alpha. \tag{16.81}$$

Recall that G and hence all its subgroups depend on r. Define $\mathbf{f}^{(r)}$ by

$$f_\alpha^{(r)} = \log \frac{|G|}{|G_\alpha|} \tag{16.82}$$

for all nonempty subset α of $\mathcal{N}_n$. Then $\mathbf{f}^{(r)} \in \Upsilon_n$ and

$$\lim_{r \to \infty} \frac{1}{r} \mathbf{f}^{(r)} = \mathbf{h}. \tag{16.83}$$

We have already proved the theorem for the special case that $\mathbf{h}$ is the entropy function of a collection of random variables $X_1, X_2, \cdots, X_n$ with finite alphabets and a rational joint distribution. To complete the proof, we only have to note that for any $\mathbf{h} \in \Gamma_n^*$, it is always possible to construct a sequence $\{\mathbf{h}^{(k)}\}$ in Γ_n^* such that $\lim_{k\to\infty} \mathbf{h}^{(k)} = \mathbf{h}$, where $\mathbf{h}^{(k)}$ is the entropy function of a collection of random variables $X_1^{(k)}, X_2^{(k)}, \cdots, X_n^{(k)}$ with finite alphabets and a rational joint distribution. This can be proved by techniques similar to those used in Appendix 15.A together with the continuity of the entropy function. The details are omitted here. □

COROLLARY 16.24 $\overline{con}(\Upsilon_n) = \overline{\Gamma}_n^*$.

Proof First of all, $\Upsilon_n \subset \Gamma_n^*$. By taking convex closure, we have $\overline{con}(\Upsilon_n) \subset \overline{con}(\Gamma_n^*)$. By Theorem 14.5, $\overline{\Gamma}_n^*$ is convex. Therefore, $\overline{con}(\Gamma_n^*) = \overline{\Gamma}_n^*$, and we have $\overline{con}(\Upsilon_n) \subset \overline{\Gamma}_n^*$. On the other hand, we have shown in Example 16.19 that the origin of $\mathcal{H}_n$ has a group characterization and therefore is in Υ_n. It then follows from Theorem 16.22 that $\overline{\Gamma}_n^* \subset \overline{con}(\Upsilon_n)$. Hence, we conclude that $\overline{\Gamma}_n^* = \overline{con}(\Upsilon_n)$, completing the proof. □

16.4 INFORMATION INEQUALITIES AND GROUP INEQUALITIES

We have proved in Section 14.1 that an unconstrained information inequality

$$\mathbf{b}^\top \mathbf{h} \geq 0 \tag{16.84}$$

always holds if and only if

$$\overline{\Gamma}_n^* \subset \{\mathbf{h} \in \mathcal{H}_n : \mathbf{b}^\top \mathbf{h} \geq 0\}. \tag{16.85}$$

In other words, all unconstrained information inequalities are fully characterized by $\overline{\Gamma}_n^*$. We also have proved at the end of the last section that $\overline{con}(\Upsilon_n) = \overline{\Gamma}_n^*$. Since $\Upsilon_n \subset \Gamma_n^* \subset \overline{\Gamma}_n^*$, if (16.85) holds, then

$$\Upsilon_n \subset \{\mathbf{h} \in \mathcal{H}_n : \mathbf{b}^\top \mathbf{h} \geq 0\}. \tag{16.86}$$

On the other hand, if (16.86) holds, since $\{\mathbf{h} \in \mathcal{H}_n : \mathbf{b}^\top \mathbf{h} \geq 0\}$ is closed and convex, by taking convex closure in (16.86), we obtain

$$\overline{\Gamma}_n^* = \overline{con}(\Upsilon_n) \subset \{\mathbf{h} \in \mathcal{H}_n : \mathbf{b}^\top \mathbf{h} \geq 0\}. \tag{16.87}$$

Therefore, (16.85) and (16.86) are equivalent.

Now (16.86) is equivalent to

$$\mathbf{b}^\top \mathbf{h} \geq 0 \text{ for all } \mathbf{h} \in \Upsilon_n. \tag{16.88}$$

Since $\mathbf{h} \in \Upsilon_n$ if and only if

$$h_\alpha = \log \frac{|G|}{|G_\alpha|} \tag{16.89}$$

for all nonempty subset α of $\mathcal{N}_n$ for some finite group G and subgroups G_1, $G_2, \cdots, G_n$, we see that the inequality (16.84) holds for all random variables $X_1, X_2, \cdots, X_n$ if and only if the inequality obtained from (16.84) by replacing h_α by $\log \frac{|G|}{|G_\alpha|}$ for all nonempty subset α of $\mathcal{N}_n$ holds for all finite group G and subgroups $G_1, G_2, \cdots, G_n$. In other words, for every unconstrained information inequality, there is a corresponding group inequality, and vice versa. Therefore, inequalities in information theory can be proved by methods in group theory, and inequalities in group theory can be proved by methods in information theory.

In the rest of the section, we explore this one-to-one correspondence between information theory and group theory. We first give a group-theoretic proof of the basic inequalities in information theory. At the end of the section, we will give an information-theoretic proof for the group inequality in (16.5).

DEFINITION 16.25 *Let G_1 and G_2 be subgroups of a finite group G. Define*

$$G_1 \circ G_2 = \{a \circ b : a \in G_1 \text{ and } b \in G_2\}. \tag{16.90}$$

$G_1 \circ G_2$ is in general not a subgroup of G. However, it can be shown that $G_1 \circ G_2$ is a subgroup of G if G is Abelian (see Problem 1).

PROPOSITION 16.26 *Let G_1 and G_2 be subgroups of a finite group G. Then*

$$|G_1 \circ G_2| = \frac{|G_1||G_2|}{|G_1 \cap G_2|}. \tag{16.91}$$

Proof Fix $(a_1, a_2) \in G_1 \times G_2$, Then $a_1 \circ a_2$ is in $G_1 \circ G_2$. Consider any $(b_1, b_2) \in G_1 \times G_2$ such that

$$b_1 \circ b_2 = a_1 \circ a_2. \tag{16.92}$$

We will determine the number of (b_1, b_2) in $G_1 \times G_2$ which satisfies this relation. From (16.92), we have

$$\begin{aligned} b_1^{-1} \circ (b_1 \circ b_2) &= b_1^{-1} \circ (a_1 \circ a_2) &(16.93)\\ (b_1^{-1} \circ b_1) \circ b_2 &= b_1^{-1} \circ a_1 \circ a_2 &(16.94)\\ b_2 &= b_1^{-1} \circ a_1 \circ a_2. &(16.95) \end{aligned}$$

Then

$$b_2 \circ a_2^{-1} = b_1^{-1} \circ a_1 \circ (a_2 \circ a_2^{-1}) = b_1^{-1} \circ a_1. \tag{16.96}$$

Let k be this common element in G, i.e.,

$$k = b_2 \circ a_2^{-1} = b_1^{-1} \circ a_1. \tag{16.97}$$

Since $b_1^{-1} \circ a_1 \in G_1$ and $b_2 \circ a_2^{-1} \in G_2$, k is in $G_1 \cap G_2$. In other words, for given $(a_1, a_2) \in G_1 \times G_2$, if $(b_1, b_2) \in G_1 \times G_2$ satisfies (16.92), then (b_1, b_2) satisfies (16.97) for some $k \in G_1 \cap G_2$. On the other hand, if $(b_1, b_2) \in G_1 \times G_2$ satisfies (16.97) for some $k \in G_1 \cap G_2$, then (16.96) is satisfied, which implies (16.92). Therefore, for given $(a_1, a_2) \in G_1 \times G_2$, $(b_1, b_2) \in G_1 \times G_2$ satisfies (16.92) if and only if (b_1, b_2) satisfies (16.97) for some $k \in G_1 \cap G_2$.

Now from (16.97), we obtain

$$b_1(k) = (k \circ a_1^{-1})^{-1} \tag{16.98}$$

and

$$b_2(k) = k \circ a_2, \tag{16.99}$$

where we have written b_1 and b_2 as $b_1(k)$ and $b_2(k)$ to emphasize their dependence on k. Now consider $k, k' \in G_1 \cap G_2$ such that

$$(b_1(k), b_2(k)) = (b_1(k'), b_2(k')). \tag{16.100}$$

Since $b_1(k) = b_1(k')$, from (16.98), we have

$$(k \circ a_1^{-1})^{-1} = (k' \circ a_1^{-1})^{-1}, \tag{16.101}$$

which implies

$$k = k'. \tag{16.102}$$

Therefore, each $k \in G_1 \cap G_2$ corresponds to a unique pair $(b_1, b_2) \in G_1 \times G_2$ which satisfies (16.92). Therefore, we see that the number of distinct elements in $G_1 \circ G_2$ is given by

$$|G_1 \circ G_2| = \frac{|G_1 \times G_2|}{|G_1 \cap G_2|} = \frac{|G_1||G_2|}{|G_1 \cap G_2|}, \tag{16.103}$$

completing the proof. □

THEOREM 16.27 *Let G_1, G_2, and G_3 be subgroups of a finite group G. Then*

$$|G_3||G_{123}| \geq |G_{13}||G_{23}|. \tag{16.104}$$

Proof First of all,

$$G_{13} \cap G_{23} = (G_1 \cap G_3) \cap (G_2 \cap G_3) = G_1 \cap G_2 \cap G_3 = G_{123}. \quad (16.105)$$

By Proposition 16.26, we have

$$|G_{13} \circ G_{23}| = \frac{|G_{13}||G_{23}|}{|G_{123}|}. \quad (16.106)$$

It is readily seen that $G_{13} \circ G_{23}$ is a subset of G_3, Therefore,

$$|G_{13} \circ G_{23}| = \frac{|G_{13}||G_{23}|}{|G_{123}|} \leq |G_3|. \quad (16.107)$$

The theorem is proved. □

COROLLARY 16.28 *For random variables X_1, X_2, and X_3,*

$$I(X_1; X_2|X_3) \geq 0. \quad (16.108)$$

Proof Let G_1, G_2, and G_3 be subgroups of a finite group G. Then

$$|G_3||G_{123}| \geq |G_{13}||G_{23}| \quad (16.109)$$

by Theorem 16.27, or

$$\frac{|G|^2}{|G_{13}||G_{23}|} \geq \frac{|G|^2}{|G_3||G_{123}|}. \quad (16.110)$$

This is equivalent to

$$\log \frac{|G|}{|G_{13}|} + \log \frac{|G|}{|G_{23}|} \geq \log \frac{|G|}{|G_3|} + \log \frac{|G|}{|G_{123}|}. \quad (16.111)$$

This group inequality corresponds to the information inequality

$$H(X_1, X_3) + H(X_2, X_3) \geq H(X_3) + H(X_1, X_2, X_3), \quad (16.112)$$

which is equivalent to

$$I(X_1; X_2|X_3) \geq 0. \quad (16.113)$$

□

The above corollary shows that all the basic inequalities in information theory has a group-theoretic proof. Of course, Theorem 16.27 is also implied by the basic inequalities. As a remark, the inequality in (16.3) is seen to be a special case of Theorem 16.27 by letting $G_3 = G$.

We are now ready to prove the group inequality in (16.5). The non-Shannon-type inequality we have proved in Theorem 14.7 can be expressed in canonical form as

$$\begin{aligned} &H(X_1) + H(X_1, X_2) + 2H(X_3) + 2H(X_4) \\ &+4H(X_1, X_3, X_4) + H(X_2, X_3, X_4) \\ &\quad \leq \; 3H(X_1, X_3) + 3H(X_1, X_4) + 3H(X_3, X_4) \\ &\qquad +H(X_2, X_3) + H(X_2, X_4), \end{aligned} \tag{16.114}$$

which corresponds to the group inequality

$$\begin{aligned} &\log \frac{|G|}{|G_1|} + \log \frac{|G|}{|G_{12}|} + 2\log \frac{|G|}{|G_3|} + 2\log \frac{|G|}{|G_4|} \\ &+4\log \frac{|G|}{|G_{134}|} + \log \frac{|G|}{|G_{234}|} \\ &\quad \leq \; 3\log \frac{|G|}{|G_{13}|} + 3\log \frac{|G|}{|G_{14}|} + 3\log \frac{|G|}{|G_{34}|} + \log \frac{|G|}{|G_{23}|} \\ &\qquad + \log \frac{|G|}{|G_{24}|}. \end{aligned} \tag{16.115}$$

Upon rearranging the terms, we obtain

$$\begin{aligned} &|G_1 \cap G_3|^3 |G_1 \cap G_4|^3 |G_3 \cap G_4|^3 |G_2 \cap G_3||G_2 \cap G_4| \\ &\leq |G_1||G_1 \cap G_2||G_3|^2 |G_4|^2 |G_1 \cap G_3 \cap G_4|^4 |G_2 \cap G_3 \cap G_4|, \end{aligned} \tag{16.116}$$

which is the group inequality in (16.5). The meaning of this inequality and its implications in group theory are yet to be understood.

PROBLEMS

1. Let G_1 and G_2 be subgroups of a finite group G. Show that $G_1 \circ G_2$ is a subgroup if G is Abelian.

2. Let $\mathbf{g}_1$ and $\mathbf{g}_2$ be group characterizable entropy functions.

   a) Prove that $m_1\mathbf{g}_1 + m_2\mathbf{g}_2$ is group characterizable, where m_1 and m_2 are any positive integers.

   b) For any positive real numbers a_1 and a_2, construct a sequence of group characterizable entropy functions $\mathbf{f}^{(k)}$ for $k = 1, 2, \cdots$, such that

   $$\lim_{k\to\infty} \frac{\mathbf{f}^{(k)}}{||\mathbf{f}^{(k)}||} = \frac{\mathbf{h}}{||\mathbf{h}||},$$

   where $\mathbf{h} = a_1\mathbf{g}_1 + a_2\mathbf{g}_2$.

3. Let $(G, G_1, G_2, \cdots, G_n)$ be a group characterization of $\mathbf{g} \in \Gamma_n^*$, where $\mathbf{g}$ is the entropy function for random variables $X_1, X_2, \cdots, X_n$. Fix any nonempty subset α of $\mathcal{N}_n$, and define $\mathbf{h}$ by

$$h_\beta = g_{\alpha \cup \beta} - g_\alpha$$

for all nonempty subsets β of $\mathcal{N}_n$. It can easily be checked that $h_\beta = H(X_\beta | X_\alpha)$. Show that $(K, K_1, K_2, \cdots, K_n)$ is a group characterization of $\mathbf{h}$, where $K = G_\alpha$ and $K_i = G_i \cap G_\alpha$.

4. Let $(G, G_1, G_2, \cdots, G_n)$ be a group characterization of $\mathbf{g} \in \Gamma_n^*$, where $\mathbf{g}$ is the entropy function for random variables $X_1, X_2, \cdots, X_n$. Show that if X_i is a function of $(X_j : j \in \alpha)$, then G_α is a subgroup of G_i.

5. Let G_1, G_2, G_3 be subgroups of a finite group G. Prove that

$$|G||G_1 \cap G_2 \cap G_3|^2 \geq |G_1 \cap G_2||G_2 \cap G_3||G_1 \cap G_3|.$$

Hint: Use the information-theoretic approach.

6. Let $\mathbf{h} \in \Gamma_2^*$ be the entropy function for random variables X_1 and X_2 such that $h_1 + h_2 = h_{12}$, i.e. X_1 and X_2 are independent. Let (G, G_1, G_2) be a group characterization of $\mathbf{h}$, and define a mapping $L : G_1 \times G_2 \rightarrow G$ by

$$L(a, b) = a \circ b.$$

   a) Prove that the mapping L is onto, i.e., for any element $c \in G$, there exists $(a, b) \in G_1 \times G_2$ such that $a \circ b = c$.

   b) Prove that $G_1 \circ G_2$ is a group.

7. Denote an entropy function $\mathbf{h} \in \Gamma_2^*$ by (h_1, h_2, h_{12}). Construct a group characterization for each of the following entropy functions:

   a) $\mathbf{h}_1 = (\log 2, 0, \log 2)$

   b) $\mathbf{h}_2 = (0, \log 2, \log 2)$

   c) $\mathbf{h}_3 = (\log 2, \log 2, \log 2)$.

   Verify that Γ_2 is the minimal convex set containing the above three entropy functions.

8. Denote an entropy function $\mathbf{h} \in \Gamma_3^*$ by $(h_1, h_2, h_3, h_{12}, h_{23}, h_{13}, h_{123})$. Construct a group characterization for each of the following entropy functions:

   a) $\mathbf{h}_1 = (\log 2, 0, 0, \log 2, 0, \log 2, \log 2)$

   b) $\mathbf{h}_2 = (\log 2, \log 2, 0, \log 2, \log 2, \log 2, \log 2)$

c) $\mathbf{h}_3 = (\log 2, \log 2, \log 2, \log 2, \log 2, \log 2, \log 2)$

d) $\mathbf{h}_4 = (\log 2, \log 2, \log 2, \log 4, \log 4, \log 4, \log 4)$.

9. *Ingleton inequality* Let G be a finite Abelian group and G_1, G_2, G_3, and G_4 be subgroups of G. Let (G, G_1, G_2, G_3, G_4) be a group characterization of $\mathbf{g}$, where $\mathbf{g}$ is the entropy function for random variables X_1, X_2, X_3, and X_4. Prove the following statements:

a)
$$|(G_1 \cap G_3) \circ (G_1 \cap G_4)| \leq |G_1 \cap (G_3 \circ G_4)|$$
Hint: Show that $(G_1 \cap G_3) \circ (G_1 \cap G_4) \subset G_1 \cap (G_3 \circ G_4)$.

b)
$$|G_1 \circ G_3 \circ G_4| \leq \frac{|G_1||G_3 \circ G_4||G_1 \cap G_3 \cap G_4|}{|G_1 \cap G_3||G_1 \cap G_4|}.$$

c)
$$|G_1 \circ G_2 \circ G_3 \circ G_4| \leq \frac{|G_1 \circ G_3 \circ G_4||G_2 \circ G_3 \circ G_4|}{|G_3 \circ G_4|}.$$

d)
$$\begin{aligned} & |G_1 \circ G_2 \circ G_3 \circ G_4| \\ \leq\ & \frac{|G_1||G_2||G_3||G_4||G_1 \cap G_3 \cap G_4||G_2 \cap G_3 \cap G_4|}{|G_1 \cap G_3||G_1 \cap G_4||G_2 \cap G_3||G_2 \cap G_4||G_3 \cap G_4|}. \end{aligned}$$

e)
$$\begin{aligned} & |G_1 \cap G_3||G_1 \cap G_4||G_2 \cap G_3||G_2 \cap G_4||G_3 \cap G_4| \\ \leq\ & |G_3||G_4||G_1 \cap G_2||G_1 \cap G_3 \cap G_4||G_2 \cap G_3 \cap G_4|. \end{aligned}$$

f)
$$\begin{aligned} & H(X_{13}) + H(X_{14}) + H(X_{23}) + H(X_{24}) + H(X_{34}) \\ \geq\ & H(X_3) + H(X_4) + H(X_{12}) + H(X_{134}) + H(X_{234}), \end{aligned}$$
where $H(X_{134})$ denotes $H(X_1, X_3, X_4)$, etc.

g) Is the inequality in f) implied by the basic inequalities? And does it always hold? Explain.

The Ingleton inequality [105] (see also [158]) was originally obtained as a constraint on the rank functions of vector spaces. The inequality in e) was obtained in the same spirit by Chan [40] for subgroups of a finite group. The inequality in f) is referred to as the Ingleton inequality for entropy in the literature. (See also Problem 7 in Chapter 14.)

HISTORICAL NOTES

The results in this chapter are due to Chan and Yeung [42], whose work was inspired by a one-to-one correspondence between entropy and quasi-uniform arrays previously established by Chan [40] (also Chan [41]). Romashchenko *et al.* [178] have developed an interpretation of Kolmogorov complexity similar to the combinatorial interpretation of entropy in Chan [40].

Bibliography

[1] J. Abrahams, "Code and parse trees for lossless source encoding," *Comm. Inform. & Syst.*, 1: 113-146, 2001 (http://www.ims.cuhk.edu.hk/~cis/).

[2] N. Abramson, *Information Theory and Coding,* McGraw-Hill, New York, 1963.

[3] Y. S. Abu-Mostafa, Ed., *Complexity in Information Theory*, Springer-Verlag, New York, 1988.

[4] J. Aczél and Z. Daróczy, *On Measures of Information and Their Characterizations*, Academic Press, New York, 1975.

[5] R. Ahlswede, B. Balkenhol and L. Khachatrian, "Some properties of fix-free codes," preprint 97-039, Sonderforschungsbereich 343, Universität Bielefeld, 1997.

[6] R. Ahlswede, N. Cai, S.-Y. R. Li, and R. W. Yeung, "Network information flow," *IEEE Trans. Inform. Theory*, IT-46: 1204-1216, 2000.

[7] R. Ahlswede and J. Körner, "Source coding with side information and a converse for degraded broadcast channels," *IEEE Trans. Inform. Theory*, IT-21: 629-637, 1975.

[8] R. Ahlswede and I. Wegener, *Suchprobleme*, Teubner Studienbcher. B. G. Teubner, Stuttgart, 1979 (in German). English translation: *Search Problems*, Wiley, New York, 1987.

[9] R. Ahlswede and J. Wolfowitz, "The capacity of a channel with arbitrarily varying cpf's and binary output alphabet," *Zeitschrift für Wahrscheinlichkeitstheorie und verwandte Gebiete*, 15: 186-194, 1970.

[10] P. Algoet and T. M. Cover, "A sandwich proof of the Shannon-McMillan-Breiman theorem," *Ann. Prob.*, 16: 899-909, 1988.

[11] S. Amari, *Differential-Geometrical Methods in Statistics*, Springer-Verlag, New York, 1985.

[12] J. B. Anderson and S. Mohan, *Source and Channel Coding: An Algorithmic Approach*, Kluwer Academic Publishers, Boston, 1991.

[13] S. Arimoto, "Encoding and decoding of p-ary group codes and the correction system," *Information Processing in Japan*, 2: 321-325, 1961 (in Japanese).

[14] S. Arimoto, "An algorithm for calculating the capacity of arbitrary discrete memoryless channels," *IEEE Trans. Inform. Theory*, IT-18: 14-20, 1972.

[15] S. Arimoto, "On the converse to the coding theorem for discrete memoryless channels," *IEEE Trans. Inform. Theory*, IT-19: 357-359, 1973.

[16] R. B. Ash, *Information Theory*, Interscience, New York, 1965.

[17] E. Ayanoglu, R. D. Gitlin, C.-L. I, and J. Mazo, "Diversity coding for transparent self-healing and fault-tolerant communication networks," 1990 IEEE International Symposium on Information Theory, San Diego, CA, Jan. 1990.

[18] A. R. Barron, "The strong ergodic theorem for densities: Generalized Shannon-McMillan-Breiman theorem," *Ann. Prob.*, 13: 1292-1303, 1985.

[19] L. A. Bassalygo, R. L. Dobrushin, and M. S. Pinsker, "Kolmogorov remembered," *IEEE Trans. Inform. Theory*, IT-34: 174-175, 1988.

[20] T. Berger, *Rate Distortion Theory: A Mathematical Basis for Data Compression*, Prentice-Hall, Englewood Cliffs, New Jersey, 1971.

[21] T. Berger, "Multiterminal source coding," in *The Information Theory Approach to Communications*, G. Longo, Ed., CISM Courses and Lectures #229, Springer-Verlag, New York, 1978.

[22] T. Berger and R. W. Yeung, "Multiterminal source coding with encoder breakdown," *IEEE Trans. Inform. Theory*, IT-35: 237-244, 1989.

[23] E. R. Berlekamp, "Block coding for the binary symmetric channel with noiseless, delayless feedback," in H. B. Mann, *Error Correcting Codes*, Wiley, New York, 1968.

[24] E. R. Berlekamp, Ed., *Key Papers in the Development of Coding Theory*, IEEE Press, New York, 1974.

[25] C. Berrou, A. Glavieux, and P. Thitimajshima, "Near Shannon limit error-correcting coding and decoding: Turbo codes," Proceedings of the 1993 International Conferences on Communications, 1064-1070, 1993.

[26] J. Berstel and D. Perrin, *Theory of Codes*, Academic Press, Orlando, 1985.

[27] D. Blackwell, L. Breiman, and A. J. Thomasian, "The capacities of certain channel classes under random coding," *Ann. Math. Stat.*, 31: 558-567, 1960.

[28] R. E. Blahut, "Computation of channel capacity and rate distortion functions," *IEEE Trans. Inform. Theory*, IT-18: 460-473, 1972.

[29] R. E. Blahut, "Information bounds of the Fano-Kullback type," *IEEE Trans. Inform. Theory*, IT-22: 410-421, 1976.

[30] R. E. Blahut, *Theory and Practice of Error Control Codes*, Addison-Wesley, Reading, Massachusetts, 1983.

[31] R. E. Blahut, *Principles and Practice of Information Theory*, Addison-Wesley, Reading, Massachusetts, 1987.

[32] R. E. Blahut, D. J. Costello, Jr., U. Maurer, and T. Mittelholzer, Ed., *Communications and Cryptography: Two Sides of One Tapestry*, Kluwer Academic Publishers, Boston, 1994.

[33] C. Blundo, A. De Santis, R. De Simone, and U. Vaccaro, "Tight bounds on the information rate of secret sharing schemes," *Designs, Codes and Cryptography*, 11: 107-110, 1997.

[34] J. A. Bondy and U. S. R. Murty, *Graph Theory with Applications*, North Holland, New York, 1976.

[35] R. C. Bose and D. K. Ray-Chaudhuri, "On a class of error correcting binary group codes," *Inform. Contr.*, 3: 68-79, Mar. 1960.

[36] L. Breiman, "The individual ergodic theorems of information theory," *Ann. Math. Stat.*, 28: 809-811, 1957.

[37] M. Burrows and D. J. Wheeler, "A block-sorting lossless data compression algorithm," Technical Report 124, Digital Equipment Corporation, 1994.

[38] R. Calderbank and N. J. A. Sloane, "Obituary: Claude Shannon (1916-2001)," *Nature*, 410: 768, April 12, 2001.

[39] R. M. Capocelli, A. De Santis, L. Gargano, and U. Vaccaro, "On the size of shares for secret sharing schemes," *J. Cryptology*, 6: 157-168, 1993.

[40] H. L. Chan (T. H. Chan), "Aspects of information inequalities and its applications," M.Phil. thesis, The Chinese University of Hong Kong, Jun. 1998.

[41] T. H. Chan, "A combinatorial approach to information inequalities," *Comm. Inform. & Syst.*, 1: 241 - 253, 2001 (http://www.ims.cuhk.edu.hk/~cis/).

[42] T. H. Chan and R. W. Yeung, "On a relation between information inequalities and group theory," to appear in *IEEE Trans. Inform. Theory*.

[43] T. H. Chan and R. W. Yeung, "Factorization of positive functions," in preparation.

[44] G. J. Chaitin, *Algorithmic Information Theory*, Cambridge Univ. Press, Cambridge, 1987.

[45] H. Chernoff, "A measure of the asymptotic efficiency of test of a hypothesis based on a sum of observations," *Ann. Math. Stat.*, 23: 493-507, 1952.

[46] K. L. Chung, "A note on the ergodic theorem of information theory," *Ann. Math. Stat.*, 32: 612-614, 1961.

[47] T. M. Cover, "A proof of the data compression theorem of Slepian and Wolf for ergodic sources," *IEEE Trans. Inform. Theory*, IT-21: 226-228, 1975.

[48] T. M. Cover, "An algorithm for maximizing expected log investment return," *IEEE Trans. Inform. Theory*, IT-30: 369-373, 1984.

[49] T. M. Cover, P. Gács, and R. M. Gray, "Kolmogorov's contribution to information theory and algorithmic complexity," *Ann. Prob.*, 17: 840-865, 1989.

[50] T. M. Cover and R. King, "A convergent gambling estimate of the entropy of English," *IEEE Trans. Inform. Theory*, IT-24: 413-421, 1978.

[51] T. M. Cover and S. K. Leung, "Some equivalences between Shannon entropy and Kolmogorov complexity," *IEEE Trans. Inform. Theory*, IT-24: 331-338, 1978.

[52] T. M. Cover and J. A. Thomas, *Elements of Information Theory*, Wiley, New York, 1991.

[53] I. Csiszár, "Information type measures of difference of probability distributions and indirect observations," *Studia Sci. Math. Hungar.*, 2: 229-318, 1967.

[54] I. Csiszár, "On the computation of rate-distortion functions," *IEEE Trans. Inform. Theory*, IT-20: 122-124, 1974.

[55] I. Csiszár and J. Körner, *Information Theory: Coding Theorems for Discrete Memoryless Systems*, Academic Press, New York, 1981.

[56] I. Csiszár and P. Narayan, "Arbitrarily varying channels with constrained inputs and states," *IEEE Trans. Inform. Theory*, IT-34: 27-34, 1988.

[57] I. Csiszár and P. Narayan, "The capacity of the arbitrarily varying channel revisited: Positivity, constraints," *IEEE Trans. Inform. Theory*, IT-34: 181-193, 1988.

[58] I. Csiszár and G. Tusnády, "Information geometry and alternating minimization procedures," *Statistics and Decisions*, Supplement Issue 1: 205-237, 1984.

[59] G. B. Dantzig, *Linear Programming and Extensions*, Princeton Univ. Press, Princeton, New Jersey, 1962.

[60] L. D. Davisson, "Universal noiseless coding," *IEEE Trans. Inform. Theory*, IT-19: 783-795, 1973.

[61] A. P. Dawid, "Conditional independence in statistical theory (with discussion)," *J. Roy. Statist. Soc., Series B*, 41: 1-31, 1979.

[62] A. P. Dempster, N. M. Laird, and D. B. Rubin, "Maximum likelihood form incomplete data via the EM algorithm," *Journal Royal Stat. Soc., Series B*, 39: 1-38, 1977.

[63] G. Dueck and J. Körner, "Reliability function of a discrete memoryless channel at rates above capacity," *IEEE Trans. Inform. Theory*, IT-25: 82-85, 1979.

[64] P. Elias, "Universal codeword sets and representations of the integers," *IEEE Trans. Inform. Theory*, IT-21: 194-203, 1975.

[65] *Encyclopedia Britannica*, http://www.britannica.com/.

[66] R. M. Fano, Class notes for Transmission of Information, Course 6.574, MIT, Cambridge, Massachusetts, 1952.

[67] R. M. Fano, *Transmission of Information: A Statistical Theory of Communication*, Wiley, New York, 1961.

[68] A. Feinstein, "A new basic theorem of information theory," *IRE Trans. Inform. Theory*, IT-4: 2-22, 1954.

[69] A. Feinstein, *Foundations of Information Theory*, McGraw-Hill, New York, 1958.

[70] W. Feller, *An Introduction to Probability Theory and Its Applications*, Vol. 1, Wiley, New York, 1950.

[71] B. M. Fitingof, "Coding in the case of unknown and changing message statistics," *PPI* 2: 3-11, 1966 (in Russian).

[72] L. K. Ford, Jr. and D. K. Fulkerson, *Flows in Networks*, Princeton Univ. Press, Princeton, New Jersey, 1962.

[73] G. D. Forney, Jr., "Convolutional codes I: Algebraic structure," *IEEE Trans. Inform. Theory*, IT-16: 720 - 738, 1970.

[74] G. D. Forney, Jr., Information Theory, unpublished course notes, Stanford University, 1972.

[75] G. D. Forney, Jr., "The Viterbi algorithm," *Proc. IEEE*, 61: 268-278, 1973.

[76] F. Fu, R. W. Yeung, and R. Zamir, "On the rate-distortion region for multiple descriptions," submitted to *IEEE Trans. Inform. Theory*.

[77] S. Fujishige, "Polymatroidal dependence structure of a set of random variables," *Inform. Contr.*, 39: 55-72, 1978.

[78] R. G. Gallager, "Low-density parity-check codes," *IEEE Trans. Inform. Theory*, IT-8: 21-28, Jan. 1962.

[79] R. G. Gallager, "A simple derivation of the coding theorem and some applications," *IEEE Trans. Inform. Theory*, IT-11: 3-18, 1965.

[80] R. G. Gallager, *Information Theory and Reliable Communication*, Wiley, New York, 1968.

[81] R. G. Gallager, "Variations on a theme by Huffman," *IEEE Trans. Inform. Theory*, IT-24: 668-674, 1978.

[82] Y. Ge and Z. Ye, "Information-theoretic characterizations of lattice conditional independence models," submitted.

[83] A. Gersho and R. M. Gray, *Vector Quantization and Signal Compression*, Kluwer Academic Publishers, Boston, 1992.

[84] S. Goldman, *Information Theory*, Prentice-Hall, Englewood Cliffs, New Jersey, 1953.

[85] J. Dj. Golić, "Noiseless coding for multiple channels," 1994 International Symposium on Information Theory and Its Applications, Sydney, 1994.

[86] S. W. Golomb, R. E. Peile, and R. A. Scholtz, *Basic Concepts in Information Theory and Coding : The Adventures of Secret Agent 00111*, Plenum Press, New York, 1994.

[87] R. M. Gray, *Entropy and Information Theory*, Springer-Verlag, New York, 1990.

[88] S. Guiasu, *Information Theory with Applications*, McGraw-Hill, New York, 1976.

[89] B. E. Hajek and T. Berger, "A decomposition theorem for binary Markov random fields," *Ann. Prob.*, 15: 1112-1125, 1987.

[90] D. Hammer, A. Romashchenko, A. Shen, and N. K. Vereshchagin, "Inequalities for Shannon Entropy and Kolmogorov Complexity," *J. Comp. & Syst. Sci.*, 60: 442-464, 2000.

[91] R. V. Hamming, "Error detecting and error correcting codes," *Bell Sys. Tech. Journal*, 29: 147-160, 1950.

[92] T. S. Han, "Linear dependence structure of the entropy space," *Inform. Contr.*, 29: 337-368, 1975.

[93] T. S. Han, "Nonnegative entropy measures of multivariate symmetric correlations," *Inform. Contr.*, 36: 133-156, 1978.

[94] T. S. Han, "A uniqueness of Shannon's information distance and related non-negativity problems," *J. Comb., Inform., & Syst. Sci.*, 6: 320-321, 1981.

[95] T. S. Han, "An information-spectrum approach to source coding theorems with a fidelity criterion," *IEEE Trans. Inform. Theory*, IT-43: 1145-1164, 1997.

[96] T. S. Han and K. Kobayashi, "A unified achievable rate region for a general class of multiterminal source coding systems," *IEEE Trans. Inform. Theory*, IT-26: 277-288, 1980.

[97] G. H. Hardy, J. E. Littlewood, and G. Polya, *Inequalities*, 2nd ed., Cambridge Univ. Press, London, 1952.

[98] K. P. Hau, "Multilevel diversity coding with independent data streams," M.Phil. thesis, The Chinese University of Hong Kong, Jun. 1995.

[99] C. Heegard and S. B. Wicker, *Turbo Coding*, Kluwer Academic Publishers, Boston, 1999.

[100] A. Hocquenghem, "Codes correcteurs d'erreurs," *Chiffres*, 2: 147-156, 1959.

[101] Y. Horibe, "An improved bound for weight-balanced tree," *Inform. Contr.*, 34: 148-151, 1977.

[102] Hu Guo Ding, "On the amount of Information," *Teor. Veroyatnost. i Primenen.*, 4: 447-455, 1962 (in Russian).

[103] D. A. Huffman, "A method for the construction of minimum redundancy codes," *Proc. IRE*, 40: 1098-1101, 1952.

[104] L. P. Hyvarinen, *Information Theory for Systems Engineers*, Springer-Verlag, Berlin, 1968.

[105] A. W. Ingleton, "Representation of matroids," in *Combinatorial Mathematics and Its Applications*, D. J. A. Welsh, Ed., 149-167, Academic Press, London, 1971

[106] P. Jacquet and W. Szpankowski, "Entropy computations via analytic depoissonization," *IEEE Trans. Inform. Theory*, IT-45: 1072-1081, 1999.

[107] E. T. Jaynes, "On the rationale of maximum entropy methods," *Proc. IEEE*, 70: 939-052, 1982.

[108] F. Jelinek, *Probabilistic Information Theory*, McGraw-Hill, New York, 1968.

[109] V. D. Jerohin, "ϵ-entropy of discrete random objects," *Teor. Veroyatnost. i Primenen*, 3: 103-107, 1958.

[110] O. Johnsen, "On the redundancy of binary Huffman codes," *IEEE Trans. Inform. Theory*, IT-26: 220-222, 1980.

[111] G. A. Jones and J. M. Jones, *Information and Coding Theory*, Springer, London, 2000.

[112] Y. Kakihara, *Abstract Methods in Information Theory*, World-Scientific, Singapore, 1999.

[113] J. Karush, "A simple proof of an inequality of McMillan," *IRE Trans. Inform. Theory*, 7: 118, 1961.

[114] T. Kawabata, "Gaussian multiterminal source coding," Master thesis, Math. Eng., Univ. of Tokyo, Japan, Feb. 1980.

[115] T. Kawabata and R. W. Yeung, "The structure of the I-Measure of a Markov chain," *IEEE Trans. Inform. Theory*, IT-38: 1146-1149, 1992.

[116] A. I. Khinchin, *Mathematical Foundations of Information Theory*, Dover, New York, 1957.

[117] J. C. Kieffer, "A survey of the theory of source coding with a fidelity criterion," *IEEE Trans. Inform. Theory*, IT-39: 1473-1490, 1993.

[118] J. C. Kieffer and E.-h. Yang, "Grammar-based codes: A new class of universal lossless source codes," *IEEE Trans. Inform. Theory*, IT-46: 737-754, 2000.

[119] R. Kindermann and J. Snell, *Markov Random Fields and Their Applications*, American Math. Soc., Providence, Rhode Island, 1980.

[120] R. Koetter and M. Médard, "An algebraic approach to network coding," 2001 IEEE International Symposium on Information Theory, Washington, D.C., Jun. 2001.

[121] A. N. Kolmogorov, "On the Shannon theory of information transmission in the case of continuous signals," *IEEE Trans. Inform. Theory*, IT-2: 102-108, 1956.

[122] A. N. Kolmogorov, "Three approaches to the quantitative definition of information," *Problems of Information Transmission*, 1: 4-7, 1965.

[123] A. N. Kolmogorov, "Logical basis for information theory and probability theory," *IEEE Trans. Inform. Theory*, IT-14: 662-664, 1968.

[124] L. G. Kraft, "A device for quantizing, grouping and coding amplitude modulated pulses," M.S. thesis, Dept. of E.E., MIT, 1949.

[125] F. R. Kschischang, B. J. Frey, and H.-A. Loeliger, "Factor graphs and the sum-product algorithm," *IEEE Trans. Inform. Theory*, IT-47: 498-519, 2001.

[126] S. Kullback, *Information Theory and Statistics*, Wiley, New York, 1959.

[127] S. Kullback, *Topics in Statistical Information Theory*, Springer-Verlag, Berlin, 1987.

[128] S. Kullback and R. A. Leibler, "On information and sufficiency," *Ann. Math. Stat.*, 22: 79-86, 1951.

[129] G. G. Langdon, "An introduction to arithmetic coding," *IBM J. Res. Devel.*, 28: 135-149, 1984.

[130] S. L. Lauritzen, *Graphical Models*, Oxford Science Publications, Oxford, 1996.

[131] M. Li and P. Vitányi, *An Introduction to Kolmogorov Complexity and Its Applications*, 2nd ed., Springer, New York, 1997.

[132] S.-Y. R. Li, R. W. Yeung and N. Cai, "Linear network coding," to appear in *IEEE Trans. Inform. Theory*.

[133] S. Lin and D. J. Costello, Jr., *Error Control Coding: Fundamentals and Applications*, Prentice-Hall, Englewood Cliffs, New Jersey, 1983.

[134] T. Linder, V. Tarokh, and K. Zeger, "Existence of optimal codes for infinite source alphabets," *IEEE Trans. Inform. Theory*, IT-43: 2026-2028, 1997.

[135] L. Lovasz, "On the Shannon capacity of a graph," *IEEE Trans. Inform. Theory*, IT-25: 1-7, 1979.

[136] D. J. C. MacKay, "Good error-correcting codes based on very sparse matrices," *IEEE Trans. Inform. Theory*, IT-45: 399-431, Mar. 1999.

[137] F. M. Malvestuto, "A unique formal system for binary decompositions of database relations, probability distributions, and graphs," *Inform. Sci.*, 59: 21-52, 1992; with Comment by F. M. Malvestuto and M. Studený, *Inform. Sci.*, 63: 1-2, 1992.

[138] M. Mansuripur, *Introduction to Information Theory*, Prentice-Hall, Englewood Cliffs, New Jersey, 1987.

[139] K. Marton, "Error exponent for source coding with a fidelity criterion," *IEEE Trans. Inform. Theory*, IT-20: 197 - 199, 1974.

[140] J. L. Massey, "Shift-register synthesis and BCH decoding," *IEEE Trans. Inform. Theory*, IT-15: 122-127, 1969.

[141] J. L. Massey, "Causality, feedback and directed information," in *Proc. 1990 Int. Symp. on Inform. Theory and Its Applications*, 303-305, 1990.

[142] J. L. Massey, "Contemporary cryptology: An introduction," in *Contemporary Cryptology: The Science of Information Integrity*, G. J. Simmons, Ed., IEEE Press, Piscataway, New Jersey, 1992.

[143] A. M. Mathai and P. N. Rathie, *Basic Concepts in Information Theory and Statistics: Axiomatic Foundations and Applications*, Wiley, New York, 1975.

[144] F. Matúš, "Probabilistic conditional independence structures and matroid theory: Background," *Int. J. of General Syst.*, 22: 185-196, 1994.

[145] F. Matúš, "Conditional independences among four random variables II," *Combinatorics, Probability & Computing*, 4: 407-417, 1995.

[146] F. Matúš, "Conditional independences among four random variables III: Final conclusion," *Combinatorics, Probability & Computing*, 8: 269-276, 1999.

[147] F. Matúš and M. Studený, "Conditional independences among four random variables I," *Combinatorics, Probability & Computing*, 4: 269-278, 1995.

[148] R. J. McEliece, *The Theory of Information and Coding*, Addison-Wesley, Reading, Massachusetts, 1977.

[149] W. J. McGill, "Multivariate information transmission," *Transactions PGIT, 1954 Symposium on Information Theory*, PGIT-4: pp. 93-111, 1954.

[150] B. McMillan, "The basic theorems of information theory," *Ann. Math. Stat.*, 24: 196-219, 1953.

[151] B. McMillan, "Two inequalities implied by unique decipherability," *IRE Trans. Inform. Theory*, 2: 115-116, 1956.

[152] S. C. Moy, "Generalization of the Shannon-McMillan theorem," *Pacific J. Math.*, 11: 705-714, 1961.

[153] J. K. Omura, "A coding theorem for discrete-time sources," *IEEE Trans. Inform. Theory*, IT-19: 490-498, 1973.

[154] J. M. Ooi, *Coding for Channels with Feedback*, Kluwer Academic Publishers, Boston, 1998.

[155] A. Orlitsky, "Worst-case interactive communication I: Two messages are almost optimal," *IEEE Trans. Inform. Theory*, IT-36: 1111-1126, 1990.

[156] A. Orlitsky, "Worst-case interactive communication—II: Two messages are not optimal," *IEEE Trans. Inform. Theory*, IT-37: 995-1005, 1991.

[157] D. S. Ornstein, "Bernoulli shifts with the same entropy are isomorphic," *Advances in Math.*, 4: 337-352, 1970.

[158] J. G. Oxley, *Matroid Theory*, Oxford Univ. Press, Oxford, 1992.

[159] A. Papoulis, *Probability, Random Variables and Stochastic Processes*, 2nd ed., McGraw-Hill, New York, 1984.

[160] J. Pearl, *Probabilistic Reasoning in Intelligent Systems*, Morgan Kaufman, San Meteo, California, 1988.

[161] A. Perez, "Extensions of Shannon-McMillan's limit theorem to more general stochastic processes," in Trans. Third Prague Conference on Information Theory, Statistical Decision Functions and Random Processes, 545-574, Prague, 1964.

[162] J. R. Pierce, *An Introduction to Information Theory : Symbols, Signals and Noise*, 2nd rev. ed., Dover, New York, 1980.

[163] J. T. Pinkston, "An application of rate-distortion theory to a converse to the coding theorem," *IEEE Trans. Inform. Theory*, IT-15: 66-71, 1969.

[164] M. S. Pinsker, *Information and Information Stability of Random Variables and Processes*, Vol. 7 of the series *Problemy Peredači Informacii*, AN SSSR, Moscow, 1960 (in Russian). English translation: Holden-Day, San Francisco, 1964.

[165] N. Pippenger, "What are the laws of information theory?" 1986 Special Problems on Communication and Computation Conference, Palo Alto, California, Sept. 3-5, 1986.

[166] C. Preston, *Random Fields*, Springer-Verlag, New York, 1974.

[167] M. O. Rabin, "Efficient dispersal of information for security, load balancing, and fault-tolerance," *J. ACM*, 36: 335-348, 1989.

[168] I. S. Reed and G. Solomon, "Polynomial codes over certain finite fields," *SIAM Journal Appl. Math.*, 8: 300-304, 1960.

[169] A. Rényi, *Foundations of Probability*, Holden-Day, San Francisco, 1970.

[170] F. M. Reza, *An Introduction to Information Theory*, McGraw-Hill, New York, 1961.

[171] J. Rissanen, "Generalized Kraft inequality and arithmetic coding," *IBM J. Res. Devel.*, 20: 198, 1976.

[172] J. Rissanen, "Universal coding, information, prediction, and estimation," *IEEE Trans. Inform. Theory*, IT-30: 629-636, 1984.

[173] J. R. Roche, "Distributed information storage," Ph.D. thesis, Stanford University, Mar. 1992.

[174] J. R. Roche, A. Dembo, and A. Nobel, "Distributed information storage," 1988 IEEE International Symposium on Information Theory, Kobe, Japan, Jun. 1988.

[175] J. R. Roche, R. W. Yeung, and K. P. Hau, "Symmetrical multilevel diversity coding," *IEEE Trans. Inform. Theory*, IT-43: 1059-1064, 1997.

[176] R. T. Rockafellar, *Convex Analysis*, Princeton Univ. Press, Princeton, New Jersey, 1970.

[177] S. Roman, *Coding and Information Theory*, Springer-Verlag, New York, 1992.

[178] A. Romashchenko, A. Shen, and N. K. Vereshchagin, "Combinatorial interpretation of Kolmogorov complexity," *Electronic Colloquium on Computational Complexity*, vol. 7, 2000.

[179] K. Rose, "A mapping approach to rate-distortion computation and analysis," *IEEE Trans. Inform. Theory*, IT-40: 1939-1952, 1994.

[180] F. Ruskey, "A survey of Venn diagrams," http://www.combinatorics.org/Surveys/ds5/VennEJC.html.

[181] S. Shamai, S. Verdú, "The empirical distribution of good codes," *IEEE Trans. Inform. Theory*, IT-43: 836-846, 1997.

[182] S. Shamai, S. Verdú, and R. Zamir, "Systematic lossy source/channel coding," *IEEE Trans. Inform. Theory*, IT-44: 564-579, 1998.

[183] A. Shamir, "How to share a secret," *Comm. ACM*, 22: 612-613, 1979.

[184] C. E. Shannon, "A Mathematical Theory of Communication," *Bell Sys. Tech. Journal*, 27: 379-423, 623-656, 1948.

[185] C. E. Shannon, "Communication theory of secrecy systems," *Bell Sys. Tech. Journal*, 28: 656-715, 1949.

[186] C. E. Shannon, "Prediction and entropy of printed English," *Bell Sys. Tech. Journal*, 30: 50-64, 1951.

[187] C. E. Shannon, "The zero-error capacity of a noisy channel," *IRE Trans. Inform. Theory*, IT-2: 8-19, 1956.

[188] C. E. Shannon, "Coding theorems for a discrete source with a fidelity criterion," *IRE National Convention Record, Part 4*, 142-163, 1959.

[189] C. E. Shannon, R. G. Gallager, and E. R. Berlekamp, "Lower bounds to error probability for coding in discrete memoryless channels," *Inform. Contr.*, 10: 65-103 (Part I), 522-552 (Part II), 1967.

[190] C. E. Shannon and W. W. Weaver, *The Mathematical Theory of Communication*, Univ. of Illinois Press, Urbana, Illinois, 1949.

[191] P. C. Shields, *The Ergodic Theory of Discrete Sample Paths*, American Math. Soc., Providence, Rhode Island, 1996.

[192] J. E. Shore and R. W. Johnson, "Axiomatic derivation of the principle of maximum entropy and the principle of minimum cross-entropy," *IEEE Trans. Inform. Theory*, IT-26: 26-37, 1980.

[193] I. Shunsuke, *Information theory for continuous systems*, World Scientific, Singapore, 1993.

[194] M. Simonnard, *Linear Programming*, translated by William S. Jewell, Prentice-Hall, Englewood Cliffs, New Jersey, 1966.

[195] D. S. Slepian, Ed., *Key Papers in the Development of Information Theory*, IEEE Press, New York, 1974.

[196] D. S. Slepian and J. K. Wolf, "Noiseless coding of correlated information sources," *IEEE Trans. Inform. Theory*, IT-19: 471-480, 1973.

[197] N. J. A. Sloane and A. D. Wyner, Ed., *Claude Elwood Shannon Collected Papers*, IEEE Press, New York, 1993.

[198] L. Song, R. W. Yeung and N. Cai, "Zero-error network coding for acyclic networks," submitted to *IEEE Trans. Inform. Theory*.

[199] F. Spitzer, "Random fields and interacting particle systems," M. A. A. Summer Seminar Notes, 1971.

[200] D. R. Stinson, "An explication of secret sharing schemes," *Designs, Codes and Cryptography*, 2: 357-390, 1992.

[201] D. R. Stinson, "New general lower bounds on the information rate of secret sharing schemes," in *Adv. in Cryptology – CRYPTO '92, Lecture Notes in Comput. Sci.*, vol. 740, 168-182, 1993.

[202] M. Studený, "Multiinformation and the problem of characterization of conditional-independence relations," *Problems Control Inform. Theory*, 18: 1, 3-16, 1989.

[203] W. Szpankowski, "Asymptotic average redundancy of Huffman (and other) block codes," *IEEE Trans. Inform. Theory*, IT-46: 2434-2443, 2000.

[204] J. C. A. van der Lubbe, *Information Theory*, Cambridge Univ. Press, Cambridge, 1997 (English translation).

[205] E. C. van der Meulen, "A survey of multi-way channels in information theory: 1961-1976," *IEEE Trans. Inform. Theory*, IT-23: 1-37, 1977.

[206] E. C. van der Meulen, "Some reflections on the interference channel," in *Communications and Cryptography: Two Side of One Tapestry*, R. E. Blahut, D. J. Costello, Jr., U. Maurer, and T. Mittelholzer, Ed., Kluwer Academic Publishers, Boston, 1994.

[207] M. van Dijk, "On the information rate of perfect secret sharing schemes," *Designs, Codes and Cryptography*, 6: 143-169, 1995.

[208] M. van Dijk, "Secret key sharing and secret key generation," Ph.D. thesis, Eindhoven University of Technology, Dec. 1997.

[209] S. Vembu, S. Verdú, and Y. Steinberg, "The source-channel separation theorem revisited," *IEEE Trans. Inform. Theory*, IT-41: 44-54, 1995.

[210] S. Verdú and T. S. Han, "A general formula for channel capacity," *IEEE Trans. Inform. Theory*, IT-40: 1147-1157, 1994.

[211] S. Verdú and T. S. Han, "The role of the asymptotic equipartition property in noiseless source coding," *IEEE Trans. Inform. Theory*, IT-43: 847-857, 1997.

[212] S. Verdú and S. W. McLaughlin, Ed., *Information Theory : 50 Years of Discovery*, IEEE Press, New York, 2000.

[213] A. J. Viterbi, "Error bounds for convolutional codes and an asymptotically optimum decoding algorithm," *IEEE Trans. Inform. Theory*, IT-13: 260-269, 1967.

[214] A. J. Viterbi and J. K. Omura, *Principles of Digital Communications and Coding*, McGraw-Hill, New York, 1979.

[215] T. A. Welch, "A technique for high-performance data compression," *Computer*, 17: 8-19, 1984.

[216] P. M. Woodard, *Probability and Information Theory with Applications to Radar*, McGraw-Hill, New York, 1953.

[217] S. B. Wicker, *Error Control Systems for Digital Communication and Storage*, Prentice-Hall, Englewood Cliffs, New Jersey, 1995.

[218] S. B. Wicker and V. K. Bhargava, Ed., *Reed-Solomon Codes and Their Applications*, IEEE Press, Piscataway, New Jersey, 1994.

[219] F. M. J. Willems, Y. M. Shtarkov, and T. J. Tjalkens, "The context-tree weighting method: basic properties," *IEEE Trans. Inform. Theory*, IT 41: 653-664, 1995

[220] J. Wolfowitz, "The coding of messages subject to chance errors," *Illinois Journal of Mathematics*, 1: 591-606, 1957.

[221] J. Wolfowitz, *Coding Theorems of Information Theory*, Springer, Berlin-Heidelberg, 2nd ed., 1964, 3rd ed., 1978.

[222] A. D. Wyner, "On source coding with side information at the decoder," *IEEE Trans. Inform. Theory*, IT-21: 294-300, 1975.

[223] A. D. Wyner and J. Ziv, "The rate-distortion function for source coding with side information at the decoder," *IEEE Trans. Inform. Theory*, IT-22: 1-10, 1976.

[224] E.-h. Yang and J. C. Kieffer, "Efficient universal lossless data compression algorithms based on a greedy sequential grammar transform – Part one: Without context models," *IEEE Trans. Inform. Theory*, IT-46: 755-777, 2000.

[225] C. Ye and R. W. Yeung, "Some basic properties of fix-free codes," *IEEE Trans. Inform. Theory*, IT-47: 72-87, 2001.

[226] C. Ye and R. W. Yeung, "A simple upper bound on the redundancy of Huffman codes," to appear in *IEEE Trans. Inform. Theory*.

[227] Z. Ye and T. Berger, *Information Measures for Discrete Random Fields*, Science Press, Beijing/New York, 1998.

[228] R. W. Yeung, "A new outlook on Shannon's information measures," *IEEE Trans. Inform. Theory*, IT-37: 466-474, 1991.

[229] R. W. Yeung, "Local redundancy and progressive bounds on the redundancy of a Huffman code," *IEEE Trans. Inform. Theory*, IT-37: 687-691, 1991.

[230] R. W. Yeung, "Multilevel diversity coding with distortion," *IEEE Trans. Inform. Theory*, IT-41: 412-422, 1995.

[231] R. W. Yeung, "A framework for linear information inequalities," *IEEE Trans. Inform. Theory*, IT-43: 1924-1934, 1997.

[232] R. W. Yeung and T. Berger, "Multi-way alternating minimization," 1995 IEEE Internation Symposium on Information Theory, Whistler, British Columbia, Canada, Sept. 1995.

[233] R. W. Yeung, T. T. Lee and Z. Ye, "Information-theoretic characterization of conditional mutual independence and Markov random fields," to appear in *IEEE Trans. Inform. Theory*.

[234] R. W. Yeung and Y.-O. Yan, Information-Theoretic Inequality Prover (ITIP), http://user-www.ie.cuhk.edu.hk/~ITIP/.

[235] R. W. Yeung and Z. Zhang, "On symmetrical multilevel diversity coding," *IEEE Trans. Inform. Theory*, IT-45: 609-621, 1999.

[236] R. W. Yeung and Z. Zhang, "Distributed source coding for satellite communications," *IEEE Trans. Inform. Theory*, IT-45: 1111-1120, 1999.

[237] R. W. Yeung and Z. Zhang, "A class of non-Shannon-type information inequalities and their applications," *Comm. Inform. & Syst.*, 1: 87-100, 2001 (http://www.ims.cuhk.edu.hk/~cis/).

[238] Z. Zhang and R. W. Yeung, "A non-Shannon-type conditional inequality of information quantities," *IEEE Trans. Inform. Theory*, IT-43: 1982-1986, 1997.

[239] Z. Zhang and R. W. Yeung, "On characterization of entropy function via information inequalities," *IEEE Trans. Inform. Theory*, IT-44: 1440-1452, 1998.

[240] S. Zimmerman, "An optimal search procedure," *Am. Math. Monthly*, 66: 8, 690-693, 1959.

[241] K. Sh. Zigangirov, "Number of correctable errors for transmission over a binary symmetrical channel with feedback," *Problems Inform. Transmission*, 12: 85-97, 1976. Translated from *Problemi Peredachi Informatsii*, 12: 3-19 (in Russian).

[242] J. Ziv and A. Lempel, "A universal algorithm for sequential data compression," *IEEE Trans. Inform. Theory*, IT-23: 337-343, 1977.

[243] J. Ziv and A. Lempel, "Compression of individual sequences via variable-rate coding," *IEEE Trans. Inform. Theory*, IT-24: 530-536, 1978.

Index